The Scientific and Regulatory Basis for the Geological Disposal of Radioactive Waste

The Scientific and Regulatory Basis for the Geological Disposal of Radioactive Waste

Edited by David Savage
QuantiSci Ltd., Melton Mowbray, Leicestershire, UK

JOHN WILEY & SONS
Chichester · New York · Brisbane · Toronto · Singapore

Baffins Lane, Chichester,
West Sussex PO19 1UD, England

Telephone National (01243) 779777
International (+44) 1243 779777

Other Wiley Editorial Offices

John Wiley & Sons, Inc., 605 Third Avenue,
New York, NY 10158-0012, USA

Jacaranda Wiley Ltd, 33 Park Road, Milton,
Queensland 4064, Australia

John Wiley & Sons (Canada) Ltd, 22 Worcester Road,
Rexdale, Ontario M9W 1L1, Canada

John Wiley & Sons (SEA) Pte Ltd, 37 Jalan Pemimpin #05-04,
Block B, Union Industrial Building, Singapore 2057

British Library Cataloguing in Publication Data

A catalogue record for this book is available from the British Library

ISBN 0-471-96090-X

Produced from camera-ready copy supplied by the editor.
Printed and bound in Great Britain by Biddles Ltd, Guildford and King's Lynn.
This book is printed on acid-free paper responsibly manufactured from sustainable forestation,
for which at least two trees are planted for each one used for paper production.

Contents

Preface

Over the last 15-20 years, staff working for QuantiSci Ltd. have been working on a wide variety of scientific and regulatory problems to do with the geological disposal of radioactive wastes. This has involved them closely, and frequently at a high level, in the national waste management programmes of around twenty countries, as well as in working with several international agencies. In 1992 this experience was brought together in the form of a one-week technical course on *The Scientific and Regulatory Basis for the Geological Disposal of Radioactive Wastes*, aimed at a broad spectrum of scientists and engineers working in the environmental sector and in the nuclear industry and related organisations. This course has now become a regular event and is held in various countries around the world.

One of the items of support material prepared for the course was an integrated set of notes, explaining in detail the background concepts and issues underlying geological disposal. It was clear that the information contained in these notes would be of interest to a wider audience than the course participants, and it was decided to publish the 1994 version in book form.

Although the material presented in this book is thoroughly integrated, as it would be in an authored publication, there have, in fact, been many contributors involved. These were (in alphabetical order) as follows:

Chapter 1	Neil Chapman, Graham Smith;
Chapter 2	Mike Stenhouse;
Chapter 3	Mick Apted;
Chapter 4	Mick Apted, Peter Grindrod, David Savage;
Chapter 5	Peter Grindrod, Tim McEwen, Peter Robinson, David Savage;
Chapter 6	Graham Smith;
Chapter 7	Tim McEwen;
Chapter 8	Tim McEwen;
Chapter 9	Neil Chapman, David Hodgkinson, Philip Maul;
Chapter 10	Neil Chapman, Peter Grindrod, Richard Little, Peter Robinson;

Chapter 11 Neil Chapman.

We hope that the reader will find this a thorough, if not exhaustive, review of the whole field of geological disposal, and will be able to see parallels and contrasts with the way in which we deal with other types of technical and environmental problem.

List of acronyms and abbreviations

AECB	Atomic Energy Control Board of Canada
AECL	Atomic Energy of Canada Ltd.
AGR	advanced gas-cooled reactor
ALARA	as low as reasonably achievable
ANDRA	Agence nationale pour la gestion des déchets radioactifs (France)
AVH	Atelier de Vitrification de La Hague (La Hague vitrification facility)
AVM	Atelier de Vitrification de Marcoule (Marcoule vitrification facility)
BGR	Bundesanstalt für Geowissenschaften und Rohstoffe ([German] Federal Institute of Geosciences and Natural Resources)
BGS	British Geological Survey
BPEO	best practical environmental option
BUSC	basement underneath sedimentary cover
BVG	Borrowdale Volcanic Group
BWR	boiling water reactor
CAEX	computer aided exploration system
CANDU	Canada deuterium uranium (reactor)
CEC	Commission of the European Communities
CLAB	Swedish interim storage facility for spent nuclear fuel
CSP	cooperative siting policy
DAO	decisions, assumptions and omissions
EBS	engineered barrier system
EDZ	excavation damaged zone
EFEP	external feature, event, process
EM	electromagnetic (geophysical technique)
ENRESA	Empresa Nacional de Residuos Radioactivos S.A. (Spain)
EPA	Environmental Protection Agency (U.S.A.)
FEP	feature, event, process
FMI	formation micro imager
GSF	Gesellschaft für Strahlen und Umweltforschung (Germany)

GIS	geographical information system
HDPE	high density polyethylene
HEPA	high efficiency particulate air (filter)
HGW	heat generating waste
HLW	high-level waste
HMIP	Her Majesty's Inspectorate of Pollution (U.K.)
HRL	Hard Rock Laboratory
HSK	Hauptabteilung für die Sicherheit der Kernanlagen ([Swiss] Federal Nuclear Safety Inspectorate)
HTGR	high temperature gas-cooled reactor
IAEA	International Atomic Energy Agency
ICRP	International Commission on Radiological Protection
ILW	intermediate-level waste
INWAC	International Radioactive Waste Management Advisory Committee
KSA	Eidgenössische Kommission für die Sicherheit von Kernanlagen ([Swiss] Federal Commission for the Safety of Nuclear Installations)
LLW	low-level waste
LMFR	liquid metal fast reactor
LWR	light water reactor
Mox	mixed oxide (fuel)
MUA	multi utility attribute analysis
Nagra	Nationale Genossenschaft für die Lagerung radioaktiver Abfälle ([Swiss] national cooperative for the disposal of radioactive waste)
NBS	natural barrier system
NEA	Nuclear Energy Agency (of the OECD)
NRC	Nuclear Regulatory Commission (U.S.A.)
OECD	Organisation for Economic Cooperation and Development
PA	performance assessment
PID	process influence diagram
PRA	probabilistic risk assessment
PRZ	potential repository zone
PSA	probabilistic safety assessment
P-T	partitioning and transmutation; pressure-temperature
PWR	pressurised water reactor
QA	quality assurance
RCF	rock characterisation facility
REE	rare earth element
REV	representative elementary volume
RFT	repeat formation tester
RWMAC	Radioactive Waste Management Advisory Committee (U.K.)
SCV	site characterisation and validation
SFR	Swedish final repository for reactor waste
SKB	Svensk Kärnbränslehantering AB

	(Swedish Nuclear Fuel and Waste Management Co.)
SKI	Statens Kärnkraftinspektion (Swedish Nuclear Power Inspectorate)
STP	standard temperature and pressure
TAMS	tandem accelerator mass spectrometry
TEM	transmission electron microscope
TRU	transuranic (element)
TVO	Tellisuuden Voima Oy (Finland)
UCVS	underground characterisation and validation stage
URL	underground research laboratory
USGS	United States Geological Survey
VLF	very low frequency (geophysical technique)
WHO	World Health Organisation
WIPP	Waste Isolation Pilot Plant

1

Introduction

There are very few substances on Earth which receive the same care and attention, or generate such strong feelings as do radioactive wastes. An exceptional, and increasing amount of resources has been directed towards finding a permanent solution to their disposal. In purely intellectual terms, this effort dwarfs anything that has been applied to other wastes arising from human activities. There are a number of reasons for this:

- There is a strong link, in almost every person's mind, between nuclear wastes, the horror and mystique of nuclear weapons and the obvious connection between radiation exposure and cancer. Regardless of any comparisons that might be thought appropriate with the dangers and risks associated with other substances and other activities, for the forseeable future anything to do with radioactivity will tend to be treated with the utmost circumspection.
- It is generally assumed that the nuclear industry, which in many countries posesses a strong central link with government, can afford to invest considerable resources in solving its waste problems. The volumes of waste are small compared to those arising from many other industries; minute compared to those from most other serious forms of power generation. The investment might thus be considered worthwhile, since the size of the problem is apparently relatively tractable.
- A strong political imperative has been placed on the nuclear industry by the green movement, demanding the highest environmental standards. Some might say that the constant spotlight played on the nuclear waste issue by the environmental lobby over the last decade has, at the least, kept the nuclear industry honest. No other industry shouldering a burden of potential environmental impacts would willingly swap with them their centre stage position.

For whatever reason, a considerable amount of scientific effort has thus been applied to finding a solution to closing the nuclear fuel cycle. This effort has been very largely focussed on conditioning, packaging and burying the majority of the waste by-products in a suitable geological environment, at an appropriate depth. It has brought together scientists and engineers from many fields of endeavour, and generated investigational and interpretational techniques which are, in a number of cases, leading their particular fields. It is instructive to see how the various disciplines involved have interacted to influence design, experimentation, predictive modelling and regulatory philosophy, and how each area has reacted to the strictures of having to throw their data into the common pool to help demonstrate that a particular solution is credibly safe and acceptable.

In this book we explore the whole spectrum of geological disposal in what we hope the reader will find to be a logical sequence. We begin with an overview of regulatory criteria, the aims of waste management and the options that are available, and then concentrate on the geological solution, with much of the emphasis being on deep repositories, situated at depths of some hundreds of metres below the surface.

After looking at the wastes themselves and the designs of repositories, the various components of the repository system are examined, from the inside outwards. This is the core section of the book, describing the processes which lead inevitably to the degradation of the repository and the mobilisation of the wastes. We move on to discuss how repository sites are selected and characterised, and then address the issue which focusses all the preceeding information – that of quantitatively assessing the long-term behaviour of the whole system, with the objective of demonstrating that the chosen system and site are safe. This exercise is generally called *Performance Assessment*, or *PA* for short, and can be regarded both as the unifying theme behind the science of disposal, and the engine which should drive and direct much of the work.

At the end of this process, the overiding issue that must be faced by the presenter of a safety case is that of its validity and, consequently, its credibility. The final Chapters of the book thus cover issues such as dealing with uncertainty and building confidence in the results of a safety assessment. In concluding, we take a look at some of the residual concerns which are always going to create problems in presenting a disposal concept, draw some comparisons with other environmental matters, and look at issues, both technical and political, which are likely to arise over the next few years.

1.1 What are radioactive wastes?

Radioactive wastes arise in many forms from a wide range of activities, most notably from plant and processes associated with nuclear power research and production, and with military applications. There is also a large number of industrial, research and medical uses of radioactive materials giving rise to wastes, albeit in total on a relatively small scale and in most cases giving rise to only small volumes, containing small amounts and low concentrations of activity.

Apart from solids waste, some waste streams arise as liquid or gaseous effluents which, after treatment, are discharged to the environment. Their relevance to the discussion here derives from the alternative treatments employed prior to release which can modify the solid waste source term. For example, different types of filter may be applied to atmospheric discharges. In the overall management of waste, it may be desirable to take into account the effectiveness of alternative filters at reducing discharges and relate the differences to the various implications for solid waste management. The implications involve not only modifications to the radioactive waste source term but also the chemical and physical properties of the wastes. For example, filters may include organic material which, on degradation, could mobilise radionuclides, increasing the chance of significant release in groundwater. Similarly, some filters contain aluminium frames which could corrode, producing hydrogen gas. Such issues are considered further in later chapters.

Another important factor about on-going discharges is that, for better or worse, they provide information about how radionuclides move through the environment and hence give rise to radiological impact.

The range of different waste streams arising is large. The properties relevant to categorization schemes include, at least:

- the processes giving rise to the waste, *e.g.* reactor operation, fuel reprocessing, decommissioning, medical procedures,
- the activity levels,
- the physical and chemical forms,
- the particular radionuclides involved,
- the anticipated disposal route.

Table 1.1 sets out the categories from most to least active as used in the UK, according to the U.K. Radioactive Waste Management Advisory Committee (RWMAC, 1984). The definitions given here are not used everywhere. The different features of various national power programmes and other factors mean that no single categorisation is entirely satisfactory in all countries. It may be noted that:

- there is a vast range in activity levels, even within any one category,
- it is not precisely clear at what point the heat generated in wastes becomes significant,
- there is a need to be able to distinguish, from a regulatory control point of view, between radioactive and non-radioactive waste.

Exemption levels vary from country to country, mostly hovering around 1 Bq g^{-1}. The International Atomic Energy Agency (IAEA) have developed

Table 1.1 Categories of radioactive waste (RWMAC, 1984). Crown copyright is reproduced with the permission of the Controller of HMSO.

1	**High Level or heat-Generating Wastes (HLW or HGW)**
•	Generate significant decay heat.
•	Spent fuel or products from the immobilisation of highly active liquid wastes arising during fuel reprocessing.
•	Typically in activity range 10^{16}-10^{18} Bq t^{-1}.
2	**Low Level Wastes (LLW)**
•	Less than 4×10^9 Bq t^{-1} α activity and less than 12×10^9 Bq t^{-1} β-γ activity, but more active than 'very low-level' wastes.
•	Wide range of waste streams from fuel fabrication, reprocessing and reactor operation, as well as from outside the nuclear power industry.
•	Most bulk waste arising at considerably lower activity levels than the α and β-γ LLW limits given above.
3	**Intermediate Level Wastes (ILW)**
•	Between categories 1 and 2.
•	Wide range of activity levels in various waste streams, from just above LLW to approaching HLW.
4	**Very Low Level Wastes**
•	Single items of less than 4×10^4 Bq β-γ, or less than 0.12 m^3 containing less than 4×10^5 Bq β-γ.

general principles on exemption (IAEA, 1988), and an expert group under the auspices of the Commission of the European Communities (CEC) has suggested exemption levels for decommissioning waste (CEC, 1988a). Among the most difficult issues are defining reasonable values for the volume or mass over which to define the exempt level and then being able to demonstrate that the level is being met. Implementation of exemption principles is currently under wide international discussion.

The importance for waste disposal assessment is that exemption levels may have a significant influence on the volume of waste arising (Charles and Smith, 1993). This may be noted against the background that, in the U.K. for example, the LLW inventory is of the order of 10^6 t, the ILW is about 10^5 t, and the HLW is about 5×10^3 t.

The definitions given in Table 1.1, and the similar titles applied in other countries, only touch on the level of inventory detail necessary for waste disposal performance assessment. Key features which need to be specified for each waste stream include:

- waste volumes,
- gross activity levels,
- activity of particular radionuclides,
- heat output (for the more active wastes),
- materials composition,
- waste encapsulation form
- containment requirements,
- shielding requirements,
- other (non-radioactive) hazardous content.

These features are relevant to the assessment of the post disposal impact, but also have to be taken into account in the overall waste management programme, which has to include interim storage, waste treatment and transport. Transport regulations (IAEA, 1990) place strict requirements on packaging and containment which may affect disposal strategies. A relatively comprehensive breakdown of the U.K. waste inventory is provided in Electrowatt (1990). This includes a wide variety of information on various LLW and ILW streams. Typical HLW characteristics are described in CEC (1988b) and unusually comprehensive details of reactor decommissioning inventories relevant to waste management are provided in Miller and Steffes (1986).

This introduction to radioactive wastes is necessarily short. The references given can illustrate the complications involved and fuller explanation of waste sources, types and characteristics is given in the next chapter. For this chapter, a few particulars may be noted.

Sometimes the waste material contains radionuclides throughout its bulk (see Section 4.5.2), whereas sometimes the contamination is only on surfaces. Sometimes the waste has other potentially hazardous contents, not only in the context of safe handling and operations but also with respect to potential for long term pollution. Various alternatives for treatment and encapsulation may be suitable for different waste streams. However, interactions between different final waste forms may complicate long term predictions of waste behaviour in the repository. Lastly it is important to consider the basic characteristics of the radionuclides involved. For post-disposal safety assessment these characteristics include:

- mobility in the near-field of the repository,
- mobility in the wider geosphere,
- mobility and potential for accumulation in the components of the biosphere,
- radioactive half-life,
- the radiation dose arising following intake into the body,

- the potential for external irradiation by the radionuclide.

All these factors will come up again in the following chapters.

1.2 What are the disposal options?

A number of different solid waste disposal options have been investigated, and to some extent practiced in the past. These include:

- shallow land burial, in variously engineered facilities,
- deep geological disposal,
- disposal on the sea bed in continental shelf waters,
- disposal beneath the sea bed in continental shelf waters
- deep ocean bed disposal,
- disposal beneath the sea bed of the deep ocean,
- disposal into space,
- storage until decay to below exemption levels.

Concerning storage until decay, it should be noted that it is only suitable for relatively short-lived radionuclides. For further discussion, see IAEA (1992a). An important distinction between storage and disposal is that storage generally implies some further management action before control over the waste is given up, whereas disposal does not.

Sea bed disposal has been carried out in the past, most substantially in deep waters under international agreement. The practice has been under a moratorium since 1983, and under the London Dumping Convention, has been prohibited since February 1994 by international agreement (Sjoeblom and Linsley, 1994). It is interesting to note, however, that a major study by the Nuclear Energy Agency (NEA) concluded that, radiologically speaking, the northeast Atlantic dumpsite would be suitable for continued use for the range of less active wastes for which it had been used. The ban on seabed disposal was not entered into by those involved (both regulators and operators) because of any doubts they had about safety. The IAEA has recently been investigating sea bed disposals made outside the international agreement by the former Soviet Union (Linsley and Sjoeblom, 1994).

In most countries, the main focus today is upon various forms of land based disposal at various depths and including various engineered and other barriers to the migration of radionuclides back to the human environment. Such land based disposal includes disposal under the sea with access from the land. Not surprisingly, deeper disposal is the tendency for more highly active waste, especially HLW. For LLW, shallow burial (or near surface) disposal is planned in some countries, whereas in others it is planned to send LLW to deep facilities. A recent review of options is provided in IAEA (1992b). Interim storage may be a useful component of the management programme. Delay before completion of disposal operations can have advantages in reduction of activity levels, shielding requirements and heat output. Interim storage is necessary if no final disposal route is available at the time waste arises. The NEA nuclear waste bulletin (*e.g.* NEA, 1992) provides a very useful occasional update on national policies and programmes.

1.3 Post-disposal safety targets and criteria

Waste management planning has to take account of a whole range of factors, such as costs, safety during operations, safety of workers and members of the public, as well as the long term implications of possible release of contaminants from a closed repository (see DOE, 1986). The focus here relates to the latter, but, as noted above, the development of an assessment is important at each stage of the development of the corresponding waste management and disposal programme. In explaining the rationale behind the whole programme, it is important not to neglect any component of the environmental implications of various options. For example, targets for post disposal performance should not be set which result in an undue burden on workers involved in operations. The following discussion of post disposal criteria should be seen against this wider background of protection requirements.

1.3.1 International and national recommendations

Radioactive waste disposal criteria have been under development by the international community for over a decade, and the various national regulatory development programmes have been in existence for even longer. Details of the history of progress are tedious to say the least. However, it is appropriate to reflect a little on how things have developed.

A good summary of the situation in 1989 is provided in Hill (1990). This paper discusses the gradual introduction of the concept of a risk criterion as opposed to one based on dose, reflecting ideas developed in publications from the IAEA (1983), the World Health Organisation (WHO, 1982), the NEA (1984), and the International Commission on Radiological Protection (ICRP, 1985). The reason for introducing a risk criterion is to allow some credit for the fact that while some situations following disposal might give rise to high doses, these situations may be extremely unlikely to occur. These high doses could be unacceptable if they were certain to occur, but may not be if their likelihood of occurrence is sufficiently small. As will be seen, a large part of repository safety assessment is concerned with evaluation of unlikely circumstances and how to build a robust assessment against this background.

A NEA workshop considered radiation protection and safety criteria specifically for disposal of HLW (NEA, 1991). A summary of the quantitative criteria of the various organisations is shown in Tables 1.2 and 1.3. A measure of the pace of change is that new regulations have already been approved in France (Ministère de l'Industrie et du Commerce Extérieur, 1991) and in Switzerland (HSK, KSA, 1993). Many of the radiologically based criteria may need review (if not modification) following the revision of ICRP's basic radiation protection recommendations (ICRP, 1991). The U.K. National Radiological Protection Board's (NRPB) latest waste disposal guidance (NRPB, 1992) was issued recently in the light of these recommendations (ICRP, 1991). Further international developments may be expected from the European Community and the IAEA and the work of its International Radioactive Waste Management Advisory Committee (INWAC) sub-group on principles and criteria for radioactive waste disposal.

In developing a performance assessment strategy, account should be taken of the scope for changes in our basic understanding of radiation effects which might lead to variations in basic radiation protection standards. The papers presented in the NEA workshop (NEA, 1991) also present guidelines *etc.* on qualitative features for acceptability. However, performance assessments focus on demonstrating compliance with quantitative criteria.

Most of the references above discuss protection of man. Protection of other biota has also been examined, at the species or population level rather than with respect to individuals. Investigations carried out through the IAEA suggest that biota are adequately protected, provided that radionuclide levels are limited sufficiently to meet human protection requirements (IAEA, 1992c). However this conclusion is qualified by provisos concerning the possible need to look more closely at site specific situations.

Radioactive waste may contain non-radioactive constituents which could give rise to long term pollution problems. The question of whether the same principles can be applied to the management of radioactive and non-radioactive waste has been raised in an international forum (Bengtsson, 1990). At first, the different history and practices in different areas of waste management suggest this would be impractical. However it is argued (Bengtsson, 1990) that some wastes present radioactive and non-radioactive hazards, so-called mixed hazardous waste. It would seem equally impractical to apply two different sets of principles to the management of the same wastes. The issue of consistent methodology development for radioactive and other hazardous wastes is being investigated within the European Communities (Little *et al.*, 1993).

1.3.2 Components of a regulatory framework

It would be time consuming to examine each of the different sets of national criteria. It is clear that, while some general consensus emerges on protection

Table 1.2 Quantitative high-level waste disposal objectives/criteria at international level and in OECD countries (from NEA, 1991). Reproduced by permission of NEA/OECD.

Organisation/ Country	**Objective/ criteria**	**Main Feature(s)**	**Comments**
NEA (1984)	Max. individ. risk 10^{-5} /y (all sources)	Individual risk/dose = best criterion to judge long-term acceptability	No consensus on ALARA/ optimization
ICRP Publication 46 (1985)	1 mSv/y (normal evolution scenarios) 10^{-5}/y (probabilistic scenarios) for individuals (all sources)	Both prob. and doses should be taken into account in ALARA	ALARA useful, notably to compare alternatives, but may not be the most important siting factor
IAEA Safety Series 99 (1989)	*idem* ICRP Publication 46		Also includes qualitative technical criteria on disposal system features and role of safety analysis and QA
Canada (AECB regul. Document R.104 (1987)	Max. individ. risk obj. 10^{-6} /y	Period of time for demonstrating = 10^4 y. No sudden and and dramatic increase for times > 10^4 y	Additional non-prescriptive requirement and guidelines in regulatory documents. No explicit optimisation required
France	Under development; likely ref. to ICRP Publ. 46		Technical criteria for siting established 1987
Germany (Section 45, para 1 of Radiation Protection Ordinance (1989)	Individ. dose < 0.3 mSv/y for all reasonable scenarios	Calculation of individ. doses limited to ~ 10^4 y, but isolation potential beyond 10^4 y may be assessed	Additional technical criteria in guidelines and regulatory documents
NORDIC COUNTRIES Consultative document (1989)	Individ. dose < 0.1 mSv/y (normal scenarios) Individ. risk < 10^{-6} /y (disruptive events)	Additional criterion on 'total activity inflow' limiting releases to biosphere, based on inflow of natural α radionuclides	– Under revision following broad consultation – Includes other qualitative criteria
SPAIN (Statement by Nuclear Safety Council, 1987)	Individ. dose < 0.1 mSv/y Individ. risk < 10^{-6} /y in any situation		Further development under study

Table 1.3 Quantitative HLW disposal objectives and criteria in OECD countries (from NEA, 1991). Reproduced by permission of NEA/OECD.

Country	Objective/ criteria	Main Feature(s)	Comments
SWITZERLAND Regulatory Document R-21 (1980)	Individ. dose < 0.1 mSv/y at any time for reasonably probable scenarios	repository must be designed in such a way that it can be sealed within a few years without the need for institutional control	Revision of Regulatory Document R-21 envisaged
UNITED KINGDOM	No specific criteria for HLW – but likely application of principles similar to objectives for L/ILW; $< 10^{-6}$ y target for individual risk from a single facility	No time-frame for quantitative assessment specified	ALARA to be used to the extent practical and reasonable
USA EPA 40 CFR Part 191 (1985)	Limits on projected radionuclide releases to the accessible environment for 10^4 y, based on objective to limit serious health effects to < 10 in the first 10^4 y after disposal for each 1000 t of spent fuel or HLW disposed	– Individ. dose (over 1000 y) < 0.25 mSv/y – Other requirements on drinking water contamination	1985 EPA standards were remanded and are in the process of being revised
NRC 10 CFR Part 60	Minimum levels of performance: – waste package ('substantially complete' containment for 300-1000 y) –Engineered barriers (releases $< 10^{-5}$ /y of the inventory at at 1000 y after repository closure) –pre-waste emplacement groundwater travel time between 'disturbed zone' and 'accessible' environment > 1000 y		NRC sub-system requirements are intended to help achieve compliance with the EPA Standard and alternative criteria may be approved if appropriate

principles and objectives, the mixture of quantitative and qualitative criteria arrived at in different countries show significant differences. The options for quantitative criteria include:

- Individual dose, to man,
- Individual risk, to man,
- Collective dose, to man,
- Collective risk, to man,
- Doses and dose rates to non-human biota,
- Environmental quality standards (generally, concentration limits below which some feature of environmental quality is not degraded),
- Derived performance levels (*e.g.* constraints on performance of barriers which, if met, imply compliance with the basic protection principles),
- Releases or fluxes of repository derived radionuclides across various boundaries relative to natural background.

Any of these quantities adopted for criteria should preferably be very clearly defined before an assessment is started. However, such definition is not trivial. Notably, calculation and definition of risks present special difficulties concerning how to combine probabilities of events and probabilities of ensuing consequences.

In addition, a range of qualitative criteria can be considered, such as avoiding sites likely to be disturbed in geological exploration. Any of these criteria could be limited in time; for example, releases occurring after 10^4 y might be ignored (see Tables 1.2 and 1.3).

In developing a set of criteria it is appropriate to consider how compliance with those criteria can be demonstrated. Equally, in developing or designing a performance assessment, it is appropriate to consider whether all the issues and calculational end-points of interest are being properly addressed. The following items raise a series of questions about issues for criteria development and, hence, what might be required from the performance assessment:

1. The first, and very general question which has to be asked is *what were the objectives of setting the criteria?* Is the intention to provide a set of rules for use by regulators to judge the adequacy of proposals for disposal? Or, is the intention to provide criteria which are more widely based, so allowing for interaction with other groups with other interests and less technically based ideas on safety? If a broad approach is adopted, then special care is required to set out the criteria and the associated performance assessment in terms which are understandable to those for whom they are intended to be relevant.
2. The next question is *what is the scope of the criteria?* Do they relate to post-disposal radiological impact, or are other post-disposal impacts considered simultaneously? Should worker and public impacts arising

during operations also be taken into account? The danger of acting independently, or at least without reference to other potentially relevant criteria or hazards, is that inconsistent standards, assumptions and data may be adopted. Should the criteria apply only to future facilities or to existing facilities as well? Should the same criteria apply to deep and shallow facilities? If a distinction is to be made, is a rational explanation for the difference required? Should the criteria apply only to wastes from the nuclear power industry or to all radioactive wastes arising for disposal? If the latter, then comparisons with release from uranium, which would otherwise have been left in the ground, are less directly relevant.

3. *Are the recommendations of ICRP and similarly minded international authorities going to be taken fully on board?* In particular, is a limit on individual risk to be adopted for probabilistic scenarios, and is there to be a specific requirement to demonstrate compliance with the ALARA (as low as reasonably achievable) principle? If so, how will the performance assessment address that issue?
4. *Assuming that either a limit on individual dose or risk is adopted, what value has been chosen, and what value for application to a single facility (source upper bound) has been chosen?* Is a time limit set on the application of the limit, and if so what? What qualifications are added with respect to the very long term? How precisely has the risk been defined? How are the individuals defined?
5. *If there is to be a specific requirement to demonstrate that disposal has been optimised, what guidance, if any, has been provided as to how to demonstrate compliance?* Even if formalised optimisation is not required, is it appropriate to provide some kind of explanation for the choice of waste encapsulation, near-field design and facility location?
6. *Are there derived performance features which can be identified as providing useful guidance to regulators and others contributing to a widely understandable safety basis?* Examples might include presentation of the fraction of initial inventory which decays within the repository, or the fraction of the initial inventory which has escaped.
7. *Comparisons with natural radiation may well prove useful in a safety case. How exactly should the comparison be constructed?* For example, the exposure of the local population due to the facility could be limited to a small fraction of the natural exposure, or to a small fraction of the variation of natural background across a wider area. Very simplistic criteria can be understood widely, but to some extent can also be used to misrepresent hazards. Miller and Smith (1993) provide relevant estimates of natural fluxes for a particular application and discuss some of the problems of applying and justifying this type of criterion.
8. *To what extent do criteria reflect the wishes of those who want to retain an element of control and retrievability?* Possibilities include monitoring

requirements, and design for re-excavation. If such requirements are adopted, over what time period should they be required to be effective? Designing for retrievability, or even designing so as not to foreclose options on retrievability may reduce the long term integrity of the facility, *e.g.* by allowing a higher permeability flow path to the surface.

9. *How should the performance assessment results be presented?* The regulatory framework and the need for legal clarity are important. The layout must provide regulators with clear and unambiguous results relevant to that legal framework. To what extent should further effort be made to provide documentation more relevant to the public and non-technical groups? Overall, the safety objectives of industry and the public are not that different, but the expression of objectives in terms of criteria which have to be met often is. Even if it is decided to keep the formal presentation of the assessment short and legally exact, consideration could be given to providing supplementary information and explanation.

1.4 References

Bengtsson, G. (1990) Can the same principles be used for the management of radioactive and non-radioactive waste? In *Safety Assessment of Radioactive Waste Repositories*, Proceedings of the Paris Symposium, 1989. NEA/IAEA/CEC.

CEC (1988a) Radiological protection criteria for the recycling of materials from the dismantling of nuclear installations. *Radiation Protection Report* No 43. Commission of the European Communities, Luxembourg.

CEC (1988b) PAGIS. Performance assessment of geological isolation systems. Summary Report. *CEC Report* EUR 9220. Commission of the European Communities Joint Research Centre, Ispra.

Charles, D. and Smith, G.M. (1993) Waste management study for large volumes of very low level waste from decommissioning of nuclear installations. *CEC Report* EUR 14950. Commission of the European Communities, Luxembourg.

DOE (1986) Assessment of best practicable environmental options (BPEOs) for management of low-and intermediate-level solid radioactive waste. U.K. Department of the Environment, HMSO, London.

Electrowatt (1990) Electrowatt Engineering Services (U.K.) Ltd. The 1989 United Kingdom radioactive waste inventory. DoE/RW/89-110; U.K. Nirex Report No 87. UK Department of the Environment, London, U.K; U.K. Nirex Ltd, Harwell, U.K.

Hill, M.D. (1990) Safety objectives and criteria for waste disposal. In *Safety Assessment of Radioactive Waste Repositories*, Proceedings of the Paris Symposium, 1989. NEA/IAEA/CEC.

HSK, KSA (1993) Protection objectives for the disposal of radioactive waste. Swiss Federal Nuclear Safety Inspectorate (HSK) and Federal Commission for the Safety of Nuclear Installations (KSA) Villigen, Switzerland.

IAEA (1983) Criteria for underground disposal of radioactive wastes. *IAEA Safety Series Report* No 60. International Atomic Energy Agency, Vienna, Austria.

IAEA (1988) Principles for the exemption of radiation sources and practices from regulatory control. *IAEA Safety Series Report* 89. International Atomic Energy Agency, Vienna, Austria.

IAEA (1990) Regulations for the safe transport of radioactive material. *IAEA Safety Series Report* 6, 1985 edition, updated 1990. International Atomic Energy Agency, Vienna, Austria.

IAEA (1992a) Storage of radioactive waste. *IAEA Report* IAEA-TecDoc-653. International Atomic Energy Agency, Vienna, Austria.

IAEA (1992b) Review of options for low level radioactive waste disposal. *IAEA Report* IAEA-TecDoc-661. International Atomic Energy Agency, Vienna, Austria.

IAEA (1992c) Effects of ionizing radiation on plants and animals at levels implied by current radiation protection standards *IAEA Technical Report* No 332. International Atomic Energy Agency, Vienna, Austria.

ICRP (1985) International Commission on Radiological Protection. Radiation protection. Principles for the disposal of solid radioactive waste. *ICRP Publication* 46, Ann. ICRP 15(4).

ICRP (1991) International Commission on Radiological Protection. 1990 Recommendations of the ICRP. *ICRP Publication* 60, Ann. ICRP 21, no 1-3.

Linsley, G.S. and Sjoeblom, K-L. (1994) The international Arctic seas assessment project. *Radwaste Magazine* **1(3):** 64-68.

Little, R.H., Torres, C., Charles, D., Grogan H.A., Simón, I., Smith, G.M., Sumerling, T.J. and Watkins, B.M. (1993) Post-disposal safety assessment of toxic and radioactive waste: waste types, disposal practices, disposal criteria, assessment methods and post-disposal impacts. *CEC Report* EUR 14627. Commission of the European Communities, Luxembourg.

Miller, W.M. and Smith, G.M. (1993) Fluxes of elements and radionuclides from the geosphere. *Intera report prepared for the Swedish Radiation Protection Institute*, IE3438-1. Intera Information Technologies Ltd, Henley-on-Thames, UK.

Miller, R.L. and Steffes, J.M. (1986) Radionuclide inventory and source terms for the surplus production reactors at Hanford. UNC Nuclear Industries report for the US DOE, doc no UNI-3714.

Ministère de l'Industrie et du Commerce Extérieur (1991) Règle Fondamentale de Sûreté, no III.2.f, 1991.

NEA (1984) Long term radiation protection objectives for radioactive waste disposal. NEA/OECD, Paris.

NEA (1991) Radiation protection and safety criteria, disposal of high level radioactive waste. Note distributed with Proceedings of an NEA workshop, November 1990. NEA/OECD, Paris.

NEA (1992) Nuclear Waste Bulletin, No 7 July 1992. Update on waste management Policies and Programmes. NEA/OECD, Paris.

NRPB (1992) Statement on radiological protection objectives for the land-based disposal of solid radioactive wastes. *Documents of the NRPB*, Vol 3 No 3. National Radiological Protection Board, Chilton, UK.

RWMAC (1984) *Radioactive Waste Management Advisory Committee. Fifth Annual Report.* HMSO, London.

Sjoeblom, K-L. and Linsley, G.S. (1994) Sea disposal of radioactive wastes: The London Convention 1972. *IAEA Bulletin* 2/1994. International Atomic Energy

Agency, Vienna, Austria.
WHO (1982) World Health Organisation Regional Office for Europe. Management of High-Level Radoactive Waste. *WHO Regional European Series* no 13. World Health Organisation, Copenhagen, Denmark.

2

Waste characteristics

This chapter provides an overview of the major sources of radioactive waste, the types of waste derived from each source, and the important characteristics of different waste forms in the context of their disposal implications.

2.1 Sources of radioactive waste

Typical sources of radioactive waste are identified in Figure 2.1. The individual contributions, in terms of both volume and activity, will vary from country to country, particularly the contribution from defence activities. For example, the sources shown in Figure 2.1 are based on U.S. data but do *not* include spent fuel. Waste sources identified in the diagram contribute ~ 85 % by volume but only ~ 0.1 % in terms of activity to the *total* waste inventory in the U.S. (*i.e.* including high-level waste, defined in Section 2.2).

In terms of both radioactivity and volume, the major component of radioactive waste arises from activities associated with nuclear reactors, whether for commercial power or for defence purposes. Consequently, the first section of this chapter focuses on relevant aspects of the nuclear fuel cycle. Nuclear fuel cycle is the term used to encompass all activities associated with the production of nuclear power.

Currently, about 17 % of the world's electricity is derived from just over 400 nuclear power plants, although the percentage contribution varies considerably from one country to another. Figure 2.2 shows the individual nuclear energy contributions for a number of countries, with projections to the year 2000.

Waste is produced from all aspects of the nuclear fuel cycle (see Figure 2.3), although the major source, in terms of radioactivity, is derived from spent fuel – both the irradiated fuel itself and wastes generated from its reprocessing (during the first extraction cycle), as discussed below.

At the front end of the fuel cycle, uranium mining and milling operations generate massive quantities of natural material (displaced overburden) and discarded waste, mainly as mill tailings – the residual material from ore processing. On average, 86,000 tonnes of uranium-depleted tailings are

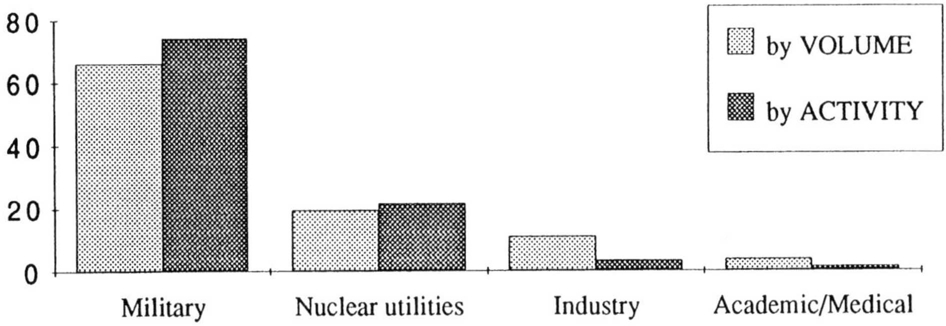

Figure 2.1 Major sources of radioactive waste (U.S.A.) (based on data from U.S. Department of Energy, 1988).

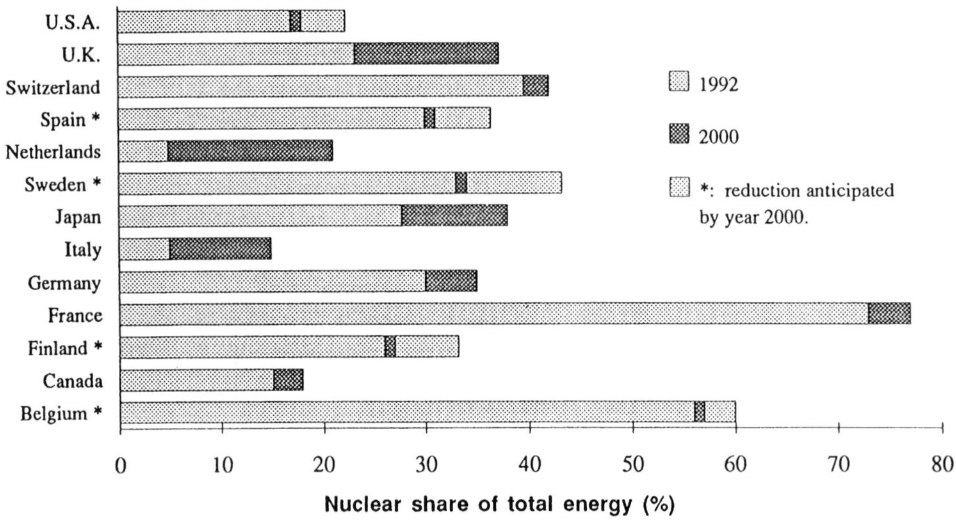

Figure 2.2 Nuclear contribution (%) to total energy generated by country.

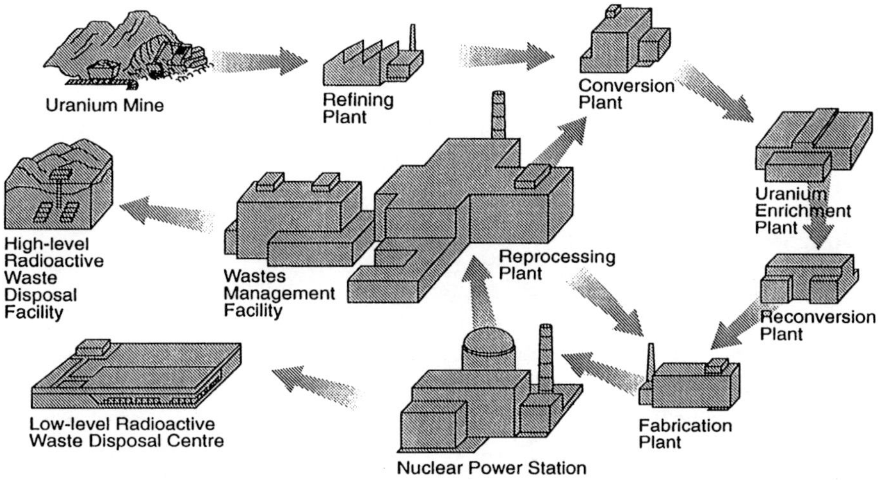

Figure 2.3 Example of nuclear fuel cycle: light water reactor cycle with reprocessing. Reproduced by permission of JNFL.

produced every year for each reactor (U.S. National Research Council, 1990). Such waste is not normally considered for disposal in a deep repository, however, but is treated independently, often being governed by separate legislation. Since the tailings contain Th-230 and Ra-226 plus daughter products, care is required to ensure safe emplacement of this processed material.

Besides the by-products of raw ore processing, waste is produced at each stage of uranium conversion and enrichment processes and of fuel fabrication. Aggressive inorganic and organic solvents are used during fuel fabrication, generating a small volume of low-activity alpha waste in the process. In particular, when plutonium is incorporated in mixed oxide (MOX) fuel (PuO_2-UO_2), plutonium-contaminated material contributes to overall waste from fuel fabrication.

2.1.1 Power generation

Waste generated during the production of nuclear power may be conveniently identified as:

- operational waste,

- spent fuel, or waste derived from spent fuel *i.e.* including waste from reprocessing, and
- decommissioning waste.

The major types of waste produced within each of these categories are discussed below.

Operational waste

Many different waste streams are generated during routine operations of a commercial nuclear power plant, liquid and solid wastes being the major contributors to the waste inventory ultimately assigned for geological disposal. Of these, ion exchange resins dominate operational waste, both in terms of volume and activity. In general, segregation of different categories of waste is best performed at the waste production stage.

Liquid wastes are generated directly as part of routine decontamination procedures, either from chemical processes (primary coolant clean-up, fuel storage pond clean-up, radiochemistry laboratories, ion exchange effluent and regenerant liquid), or from detergent waste, *e.g.* laundry and personnel decontamination. Additional radioactive liquid streams are produced during solid waste treatment and volume reduction processes. Since liquid wastes are not accepted at most disposal sites, they must be treated/processed in a variety of ways depending on the type of liquid (such conditioning methods are described in Section 2.4)

Within the nuclear plant, raw solid waste is segregated initially as 'wet' (sludges, ion exchange resins) or 'dry' (rubber gloves, paper tissue) material for the purposes of in-plant waste management. Treatment and final conditioning of solid waste is normally the responsibility of the waste producer (see Section 2.4), however, compressibility and combustibility are practical criteria for segregating waste streams.

Non-combustible, compressible wastes are normally compacted, reducing waste volume by factors of up to 10. Combustible wastes may be incinerated, although this is dependent on the level of alpha activity present in the waste. If incinerated, the resultant ash must be conditioned prior to disposal. Note that, depending on the type of gas purification system in operation, various secondary wastes are generated during incineration, *e.g.* scrubber chemicals or high efficiency particulate air (HEPA) filters.

Spent fuel

Once irradiated fuel assemblies are removed from the reactor, they are stored, either in dedicated convection-cooled, dry storage vaults or under water in special handling pools, for extended time periods (generally at least 6 months)

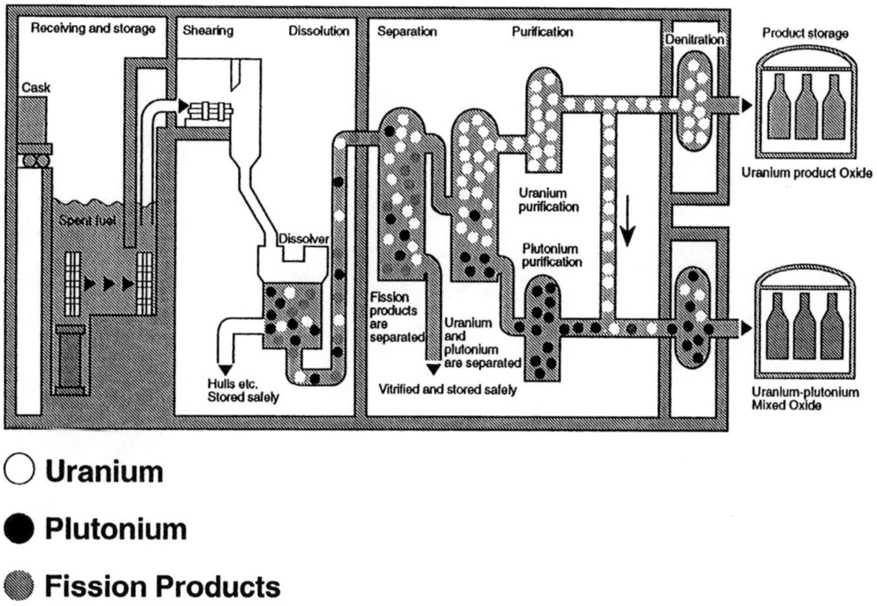

Figure 2.4 Schematic diagram for a uranium-plutonium reprocessing facility. Reproduced by permission of JNFL.

to allow the substantial decay heat to diminish to a level which allows safe handling and transportation. Thereafter, the spent fuel assemblies are either forwarded for reprocessing, or retained complete for interim storage (30-50 years) prior to final disposal.

At the reprocessing plant (see Figure 2.4), spent fuel is disassembled from its associated hardware. Fuel rods are normally sectioned into small pieces, subjected to aggressive inorganic solvents to dissolve the bulk of the fuel, and subsequently to a mixture of organic and aqueous solvents to separate uranium and plutonium from unwanted radioactive products. Such a facility may be regarded as one stage of the back end of a complete fuel cycle, the final stage being disposal. Whether reprocessed or not, waste derived from spent fuel is highly radioactive and heat producing due to the energy produced by decay of short-lived radionuclides.

The types and proportions of waste derived from the reprocessing of spent fuel may vary, depending on the particular fuel cycle which applies, although in general, the wastes from different fuel cycles are very similar in their nuclear, chemical and physical characteristics. For example, the only significant differences in waste from a high-temperature gas-cooled reactor (HTGR) compared with waste from a light-water reactor (LWR) are, in the former case, silicon carbide hull residues rather than metallic cladding, an increased inventory of C-14, and predominantly U-232 and U-233 plus daughter products rather than the actinide/transuranic elements (see Section 2.3 for definition of actinide) from the LWR cycle.

Alternative spent fuel processing techniques

A relatively advanced spent fuel treatment concept is again being given serious consideration having been introduced over two decades ago, following a report by Claiborne (1972). The method involves an actinide-burning cycle where the transuranic component of LWR spent fuel (mainly Np, Pu, Am and Cm) is separated for subsequent transmutation during power generation in either a special fission reactor system (*e.g.* a liquid-metal fast reactor, LMFR) or an accelerator-driven system. The proposed benefits of such a partitioning and transmutation (P-T) technology include a reduced health risk to future generations due to removal/depletion of long-lived alpha activity, reduced heat output in the repository, and an improved acceptance of geological disposal of radioactive waste by the public. The topic is reviewed in detail by Skålberg and Liljenzin (1992), and current proponents of P-T include France, Japan and the C.I.S.

A quantitative assessment of potential benefits has been performed by Lee and Choi (1992). These workers adopted an equal energy production comparison and examined waste characteristics and inventories. They also included in the comparison/assessment possible future pyrochemical reprocessing technology (see, for example, Lineberry and Phipps, 1989) as well as conventional aqueous reprocessing.

The significance of this work in the context of this Chapter is the *possible* future generation of significant quantities of wastes or waste forms such as silver iodide (AgI), zeolite, cadmium or copper metal and borosilicate glass, all of which must be assessed in terms of radionuclide release rates as well as chemical and physical characteristics, including barrier compatibility (the concept of barriers to radionuclide mobility is discussed in Chapter 3).

Decommissioning waste

At the end of a nuclear facility's operational life, the facility must be decommissioned, which involves

> 'removal from service in a manner which provides adequate protection for the health and safety of the decommissioning

workers and the general public, and for the environment' (IAEA, 1989).

Several decommissioning options are available under this description including dismantlement. By 1987, 565 research and test reactors had been placed in operation in the world, over 400 having been commissioned prior to 1970 (IAEA, 1988). Currently, over 300 units are still operating, but many of these are > 20 years old and, therefore, potential candidates for decommissioning.

Over 140 nuclear facilities including research, test and prototype reactors have been decommissioned (NEA/OECD, 1986), but to date, no large power reactor has been completely dismantled. However, ongoing large-scale decommissioning projects involving the dismantling of nuclear reactors (*e.g.* the 33 MWe advanced gas-cooled reactor at Sellafield, U.K.; the 72 MWe demonstration light water reactor at Shippingport, U.S.) demonstrate that the dismantlement option can be carried out safely.

Table 2.1 gives some examples of decommissioning projects being carried out under the NEA/OECD cooperative programme. Three decommissioning stages are identified in this table, after removal of fuel from the facility, *viz.*:

- Stage 1: leaving the facility intact, with continuing surveillance and maintenance,
- Stage 2: removal of all components except those associated with the primary containment (including biological shield), which are 'entombed', and
- Stage 3: full decommissioning and return of the site to unrestricted use.

Although the number of power reactors planned for decommissioning in the next 10 years is small, it has been estimated that by 2010, reactors equivalent to 200 GWe will be candidates for decommissioning or refurbishment in OECD countries (NEA/OECD, 1986). In addition to reactors, a wide variety of other nuclear fuel cycle facilities (*e.g.* hot cells) will have to be decommissioned.

Radioactivity in decommissioning wastes from nuclear power plants arises principally from neutron irradiation of the reactor vessel, its internals and a portion of the surrounding reinforced concrete (biological shield). Contamination arises from the spread of neutron-activated corrosion products which are carried in the coolant through the reactor core. Fuel element failures or leakage may also contribute to this contamination, releasing fission products and transuranics into the reactor coolant system and the fuel handling system.

With regard to residual radioactivity, what remains to be decommissioned is the shell of the reactor building, since operational liquid and solid wastes and the spent fuel will have been removed long before decommissioning starts. Thus, the bulk of residual radioactivity is found in the pressure vessel and primary coolant loop, and is due to activation of structural materials, mainly concrete and steel.

As an example, Figure 2.5 shows typical ranges of residual concentrations of contaminant radionuclides from pieces of primary coolant piping (in Bq/cm^2)

Table 2.1 Sample decommissioning projects with the OECD/NEA cooperative programme (abstracted from SKB, 1992). Reproduced by permission of SKB.

Facility	**Type**	**Planned final stage**
Eurochemic, Belgium	Reprocessing plant	Stage 3
BR-3, Belgium	PWR, 41 MWt	Stage 3
Gentilly-1, Canada	Heavy water reactor, 250 MWe	Stage 2
Rapsodie, France	Sodium-cooled fast reactor, 20 MWt	Stage 2
AT1, France	Reprocessing plant for fast reactor fuel	Stage 3
Niederaichbach, Germany	Gas-cooled heavy water moderated reactor, 106 MWe	Stage 3
Garigliano, Italy	BWR, 160 MWe	Stage 1
Japan Power Demonstration Reactor (JPDR), Japan	BWR, 13 MWe	Stage 3
Windscale Advanced Gas Cooled Reactor, United Kingdom	AGR, 33 MWe	Stage 3
West Valley Demonstration Project, USA	Reprocessing plant for LWR fuel	Stage 3

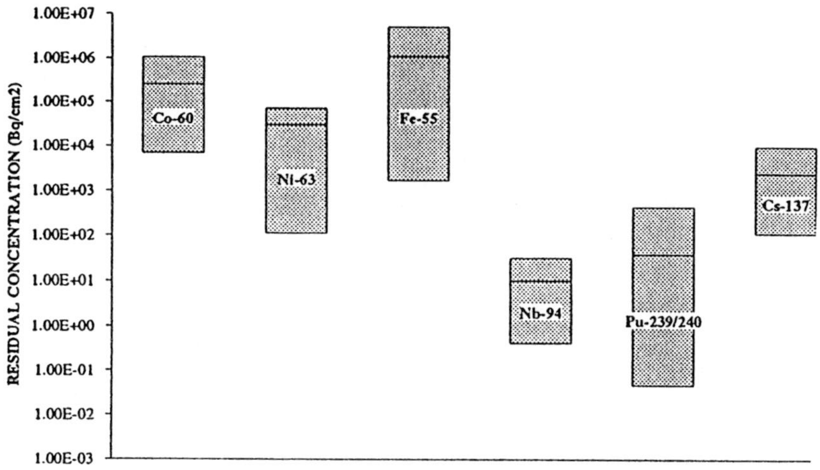

Figure 2.5 Typical residual radionuclide concentrations (Bq/cm^2) on primary coolant piping. Shown above are the ranges in concentration of the various radionuclides, from minimum to maximum; the horizontal lines represent the average residual concentration for each radionuclide (based on data of Robertson *et al.*, 1991).

sampled from several commercial power reactors. Much of this contamination, which consists predominantly of β- and γ-emitters, can be removed by specialised decontamination agents, with resultant benefits in terms of reduced occupational exposure during dismantling and reduced volume of waste for disposal. Note that the higher radionuclide activities of such material, though of small volume, correspond to 0.1–1 TBq/m^3, assuming a thickness of piping of 1 cm.

Finally, with respect to decommissioning waste, it is likely that the decommissioning of fission plants will give rise to waste containing some less commonly considered activation products. In this category, for example, is Be-10, a significant radionuclide in graphite (White and Smith, 1984).

2.1.2 Other sources of radioactive waste

Other contributions to the radioactive waste inventory for geological disposal come from military or defence applications of nuclear energy, and from industry, medicine and research, and these additional sources are summarised briefly below.

Military/defence applications

Nuclear weapons represent the most significant military application of nuclear energy. The facilities used to produce the nuclear component of nuclear weapons, in particular fissionable plutonium, are similar to civilian power plants in many respects, the major difference being the desired end product – nuclear-generated electricity in one case, and the fissionable product in the other. Elsewhere, nuclear-powered submarines generate radioactive waste, but mainly at the end of their operational lifetime, when the reactor is decommissioned, in the same way as a power plant reactor.

Thus, in terms of the type of waste generated, radioactive waste produced as a result of military applications is similar to that from commercial nuclear power plants.

Industry, medicine, and research

A wide range of materials and radioactive isotopes contribute to waste generated from these sources. As examples, the main applications of radioactivity and radioisotopes in industry, research and medicine are shown in Table 2.2, together with typical activity levels of the radioisotopes used.

In the medical field, the radioisotopes used, particularly for clinical and diagnostic purposes, tend to be short-lived with non-penetrating radiation and do not, therefore, normally pose a problem. On the other hand, sources of high activity are used in radiotherapy, including radium (radium needles), leading to a significant long-lived waste component. In addition, medical and biological research experiments (*e.g.* drug metabolism investigations) frequently make use of H-3 and C-14, and the latter provides a significant contribution to medical radioactive waste. Occasionally in medical applications, radioisotopes may be associated with pathogenic substances which create special waste handling and treatment concerns.

At nuclear research centres, radioisotopes are produced in research reactors, particle accelerators and cyclotron facilities. In general, such facilities generate relatively small amounts of waste containing long-lived radioisotopes, especially accelerator and cyclotron facilities which do not possess nuclear fuel. For the latter, the main source of waste is derived from liquids produced during chemical processing or etching of target materials.

2.1.3 Waste sources for special consideration

Several types of waste sources are identified here for particular comment, either because the sources themselves are unusual, or the specific radionuclides

Table 2.2 Examples of applications for radioactive sources.

Application	Radio-nuclide	Half-life (y)	Source strength	Comments
INDUSTRIAL				
radiography	Co-60 (Cs-137)	5.3	0.1–5 TBq	often portable units
gauging	Cs-137	30	0.1–40 GBq	fixed installations
(density, thickness)	Am-241	432	1–10 GBq	
	Sr-90	29.1	0.1–10 GBq	
sterilisation and	Co-60	5.3	0.1–400 PBq	fixed installation
food preservation	Cs-137	30	0.1–400 PBq	
well logging	Cs-137	30	–100 GBq	portable units
	Am-241/Be	432	1–100 GBq	
moisture detector	Am-241/Be	432	0.1–10 GBq	portable units
	Ra-226/Be	1600	0.1–10 GBq	
RESEARCH/TEACHING				
calibration	numerous	–	< 0.1 GBq	small, portable
electron capture	H-3	12.3	1–50 GBq	can be used in
	Ni-63	100	1–50 GBq	portable units
irradiator	Co-60	5.3	1–1000 TBq	fixed installations
target for neutron generators	H-3	12.3	1–10 TBq	fixed installation
MEDICAL				
teletherapy	Co-60 (Cs-137)	5.3	50–500 TBq	fixed installation
brachytherapy	Cs-137	30	50–500 MBq	small, portable
	Ra-226	1600	30–300 MBq	
bone densitometry	Am-241	432	1–10 GBq	mobile units

present in the wastes are less commonly considered.

For example, without predicting specific radionuclides, it is likely, given existing radioisotope production technologies combined with the already wide range of radioisotope applications, that some novel uses of various radionuclides will occur outside the nuclear industry – for example in the areas of medicine and biotechnology. Correspondingly, wastes generated from such work are likely to present new challenges to waste management.

In addition, considerable research effort has been devoted to fusion power development, and from these activities and possible future projects, many different additional radionuclides may arise for waste disposal compared with fission reactor wastes (Smith and Charles, 1991). One such example is Re-186m, which is both long-lived and potentially mobile in the geosphere.

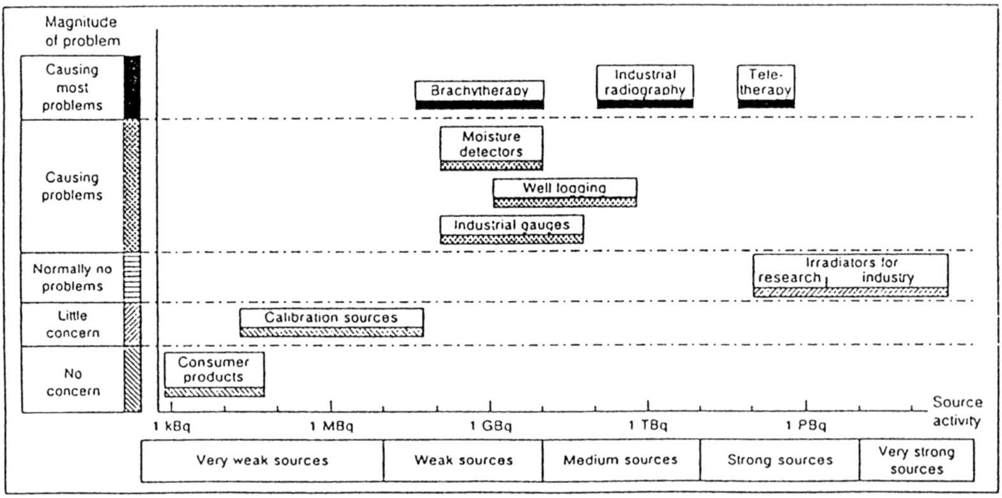

Figure 2.6 Assessment of problem associated with spent radiation sources (from Pettersson, 1992). Reproduced by permission of IAEA.

In terms of other future developments, the special wastes associated with actinide burning of light water reactor spent fuel have been discussed previously together with activation products in decommissioning waste (both in Section 2.1.1).

Small spent sources

Particular attention is given here to the relatively small (volume) component of radioactive waste – small, spent, radioactive sources. Control and waste management of such sources can be problematic (Pettersson, 1992; see Figure 2.6), particularly old radium sources which were manufactured to standards lower than those of today. Specific disposal problems may also exist, however, for the current generation of spent sources, for example, those used for food preservation and sterilisation. Key characteristics of such sources are:

- high specific activity,
- small size,
- high integrity of materials used.

The combination of these characteristics leads to certain future exposure scenarios, particularly those associated with human intrusion, which could result in doses to members of the population in excess of those predicted from uniformly contaminated radioactive waste.

For example, double encapsulation is often used in the fabrication of small sources, resulting in a product which is likely to resist degradation much more than other waste components. In this situation, the specific activity of such a source will remain high, possibly up to the time of its re-entry into the biosphere, if the half-life of the radioisotope is sufficiently long. Under these circumstances, the 'dilute and disperse' function of the geosphere may be circumvented, leading to higher doses than expected.

In most cases, provided that a shallow land repository is properly sited, constructed and operated, it may be used for the disposal of spent sources which have been conditioned properly. The main exceptions are Ra-226 and Am-241 sources and large sources used in teletherapy or irradiation facilities. Long-lived Am-241 and Ra-226 pose a potential future radiological threat, and deserve special attention when developing waste acceptance criteria. Ultimately, such sources may best be managed as high-level waste since they will contribute minimally to the overall volume.

2.2 Waste classification

At this stage, it is appropriate to address the ways in which radioactive waste may be classified, since the terms 'high', 'low', and 'intermediate' have already been introduced. In developing working definitions for waste, what is important is that the classification of different wastes is linked to the method of disposal, including waste conditioning to ensure adequate isolation of component radionuclides. Although the classification of waste tends to vary from one country to another, there is also a good deal of common ground.

Thus, high-level waste (HLW) refers to spent fuel, the liquid waste generated from its dissolution at a reprocessing facility, and the encapsulated/solidified forms of this liquid waste stream. In each case, there is significant heat output from the waste, even after interim storage. For a 1000 MWe reactor, about 30 tonnes of spent fuel are produced each year which, after conditioning, yields 4–11 m^3 HLW and up to 400 m^3 non-HLW. Some of the cladding material which remains after fuel dissolution will contain sufficient spent fuel residues to be heat generating, and will, therefore, be included as high-level waste.

The terms 'low' (LLW) and 'intermediate' or 'medium' (ILW) are somewhat arbitrary, the distinction between the two categories varying from country to country. In some countries, for example the U.S.A., commercial waste has been designated only as HLW or LLW. Irrespective of terminology, the principal objective is to ensure that the timescale over which any waste is isolated from the biosphere is compatible with the radionuclide content of that waste (see Section 2.3).

Thus, although the IAEA classification system shown in Figure 2.7 provides a general philosophy for radioactive waste management, the ultimate

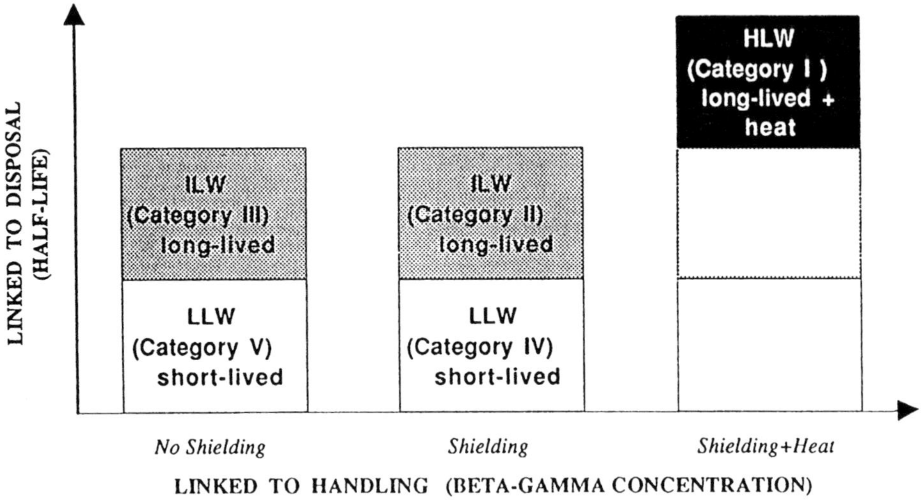

Figure 2.7 IAEA categories of radioactive waste (adapted from IAEA, 1990a). Reproduced by permission of IAEA.

disposal route for any waste type will depend on its specific performance characteristics.

2.3 Radionuclide inventory

The nature of the waste in terms of radionuclides present, their half-life and mode(s) of decay, is important, since the disposal route selected will often depend on the presence of several key radionuclides. The combination of level of activity (concentration) and half-life of each radionuclide determines the maximum amount of that radionuclide which is available at any time. On the other hand, mode of decay (alpha-, beta-, gamma-emission) is important for radiological considerations.

For radioactive waste management, radionuclides produced in a nuclear reactor are conveniently considered under the following categories:

- *fission products*, lighter elements created by the fission of the fissile component of the fuel (U-235, Pu-239, U-233). Fission products tend to be short-lived with some notable exceptions, (*e.g.* Cs-135, I-129), and β-γ-emitters,
- *activation products*, created when existing nuclides (stable or radioactive) absorb neutrons. The description 'activation product' is preferably used in

the context of light elements, such as those created in structural materials *e.g.* Fe-55 created from the (n, γ) reaction involving Fe-54. As such, these activation products are also predominantly short-lived and β-γ-emitters, with notable exceptions (*e.g.* C-14, Cl-36, Ni-59, Zr-93 and Nb-94),

- *actinides*, strictly only applicable to actinium and elements with atomic number $\geq$90, but the term generally includes the daughter products of these elements. Most actinides are alpha emitters and, as such, represent the significant radiological hazard from internal biological pathways.

Table 2.3 contains the radionuclide inventory for LLW and ILW assumed for disposal in the U.K. over 50 years. For spent fuel, activity levels depend primarily on the time since fuel discharge and burnup, the latter normally expressed in units of electric power$\times$time/metric ton of U.

The information provided in Tables 2.4 and 2.5 contains the computed radionuclide inventory for a fuel assembly (spent fuel plus structural materials) after a burnup of 33,000 MWe-days per MTU (Roddy *et al.*, 1986). Such inventories also depend to some extent on the type of reactor and the power density distribution, but Tables 2.4 and 2.5 may be used as the basis for safety assessment considerations, certainly with respect to key radionuclides and activity levels.

As a general rule, the largest contribution, in terms of volume, arises from low-activity waste (tissues, gloves, protective clothing, *etc.*). Indeed, one of the major concerns regarding this waste is the total bulk volume which must be controlled to some extent by volume reduction (compaction).

With respect to L/ILW, the influence of component radionuclides depends on the timescale of interest. For example, Co-60 is a major contributor to the gamma dose at short times after power has been shut down, and is of particular concern during the operational phase of the repository. Consequently, the emphasis in waste management of this radionuclide is on shielding. With a half-life of 5.27 years, however, the Co-60 contribution soon decays to negligible levels, almost within the construction phase of the repository. In contrast, radionuclides with long half-life, such as I-129, Se-79, pose a significant radiological hazard in the future.

The importance of timescales in determining what are key radionuclides is shown in Figure 2.8 (from Chapman and McEwen, 1991).

2.4 Waste conditioning

Prior to final conditioning, every effort should be made to minimise the volume of waste, either at source by sound management practices, or by some form of volume reduction depending on the nature of the waste.

Compaction is ideal for low-activity, compressible waste such as paper and

Table 2.3 Radionuclide inventory of U.K. L/ILW wastes. Volume of wastes assumed for disposal over 50 years is 2×10^6 m^3. Inventory figures are in TBq (adapted from Chapman and McEwen, 1991). Reproduced by permission of Lawrence Berkeley Laboratory, U.S.A.

Nuclide	Half-life (years)	Total Inventory		Operational Waste		Decommissioning	
		LLW	ILW	LLW	ILW	LLW	ILW
H-3	1.24×10^{1}	1.3×10^{2}	6.8×10^{3}	3.4×10^{0}	6.3×10^{3}	1.30×10^{2}	5.4×10^{2}
C-14	5.73×10^{3}	4.8×10^{1}	6.6×10^{3}	4.5×10^{1}	6.5×10^{3}	3.8×10^{0}	7.9×10^{1}
Cl-36	3.01×10^{5}	5.7×10^{-2}	2.6×10^{0}	3.0×10^{-3}	1.3×10^{0}	5.4×10^{-2}	1.3×10^{0}
Ca-41	1.40×10^{5}	5.8×10^{-1}	1.3×10^{0}	0	3.0×10^{-2}	5.8×10^{-1}	1.3×10^{0}
Co-60	5.27×10^{0}	3.7×10^{1}	2.9×10^{6}	3.3×10^{0}	2.4×10^{6}	3.3×10^{1}	4.3×10^{5}
Ni-59	7.50×10^{4}	6.0×10^{-2}	6.1×10^{4}	3.1×10^{-3}	6.1×10^{4}	5.7×10^{-2}	4.4×10^{2}
Ni-63	1.00×10^{2}	8.3×10^{0}	4.2×10^{6}	2.7×10^{0}	4.2×10^{6}	5.6×10^{0}	4.4×10^{4}
Se-79	6.50×10^{4}	3.2×10^{-2}	1.1×10^{1}	1.1×10^{-5}	1.1×10^{1}	3.2×10^{-2}	9.4×10^{-2}
Sr-90	2.91×10^{1}	1.0×10^{1}	1.4×10^{6}	8.4×10^{0}	1.3×10^{6}	1.9×10^{0}	8.7×10^{4}
Zr-93	1.53×10^{6}	8.6×10^{-2}	2.0×10^{3}	0	2.0×10^{3}	8.6×10^{-2}	7.0×10^{0}
Nb-93m	1.36×10^{1}	0	1.6×10^{3}	0	1.6×10^{3}		
Nb-94	2.03×10^{4}	5.4×10^{-2}	7.7×10^{3}	2.5×10^{-9}	7.7×10^{3}	5.4×10^{-2}	1.0×10^{0}
Tc-99	2.13×10^{5}	4.3×10^{-1}	5.7×10^{2}	1.1×10^{-1}	5.3×10^{2}	3.2×10^{-1}	4.7×10^{1}
Pd-107	6.50×10^{6}	4.3×10^{-3}	5.7×10^{0}	1.1×10^{-3}	5.3×10^{0}	3.2×10^{-3}	4.7×10^{-1}
Ag-108m	1.27×10^{2}	3.0×10^{-2}	4.2×10^{-1}	3.0×10^{-2}	4.2×10^{-1}		
Sn-126	1.00×10^{5}	3.2×10^{-2}	2.0×10^{1}	1.8×10^{-5}	2.0×10^{1}	3.2×10^{-2}	9.4×10^{-2}
I-129	1.57×10^{7}	3.6×10^{-5}	1.5×10^{0}	3.4×10^{-5}	1.5×10^{0}	1.6×10^{-6}	4.7×10^{-2}
Cs-135	2.30×10^{6}	3.2×10^{-1}	3.1×10^{1}	$13.\times10^{-5}$	2.9×10^{1}	3.2×10^{-1}	1.7×10^{0}
Cs-137	3.0×10^{1}	2.9×10^{1}	2.3×10^{6}	2.4×10^{1}	2.2×10^{6}	4.5×10^{0}	8.7×10^{4}
Sm-151	9.0×10^{1}	4.3×10^{-3}	1.4×10^{4}	4.3×10^{-3}	1.4×10^{4}		
Pb-210	2.23×10^{1}	1.6×10^{-2}	1.9×10^{0}	1.6×10^{-2}	1.9×10^{0}		
Ra-226	1.60×10^{3}	1.6×10^{-1}	2.0×10^{0}	5.1×10^{-2}	1.2×10^{0}	1.0×10^{-1}	8.7×10^{-1}
Ra-228	5.75×10^{0}	3.7×10^{-3}	1.9×10^{0}	3.7×10^{-3}	1.9×10^{0}		
Ac-227	2.18×10^{1}	5.4×10^{-6}	6.2×10^{-2}	5.4×10^{-6}	6.2×10^{-2}		
Th-228	1.91×10^{0}	3.7×10^{-3}	1.9×10^{0}	3.7×10^{-3}	1.9×10^{0}		
Th-229	7.34×10^{3}	0	3.6×10^{-1}	0	3.6×10^{-1}		
Th-230	7.70×10^{4}	4.8×10^{-5}	4.0×10^{-2}	4.8×10^{-5}	4.0×10^{-2}		
Th-232	1.40×10^{10}	3.7×10^{-3}	3.0×10^{0}	0	2.1×10^{0}	3.7×10^{-3}	8.7×10^{-1}
Pa-231	3.28×10^{4}	1.0×10^{-5}	9.9×10^{-2}	1.0×10^{-5}	9.9×10^{-2}		
U-233	1.59×10^{5}	0	9.5×10^{-2}	0	9.5×10^{-2}		
U-234	2.45×10^{5}	1.1×10^{1}	5.8×10^{1}	1.6×10^{0}	3.5×10^{1}	9.6×10^{0}	2.4×10^{1}
U-235	7.04×10^{8}	5.9×10^{-1}	8.1×10^{1}	2.7×10^{-1}	7.8×10^{1}	3.2×10^{-1}	2.5×10^{0}
U-236	2.34×10^{7}	8.8×10^{-5}	3.7×10^{0}	8.8×10^{-5}	3.7×10^{0}		
U-238	4.47×10^{9}	2.5×10^{1}	2.0×10^{2}	1.6×10^{1}	1.4×10^{2}	9.6×10^{0}	5.9×10^{1}
Np-237	2.14×10^{6}	3.2×10^{-2}	3.4×10^{1}	8.8×10^{-5}	3.4×10^{1}	3.2×10^{-2}	1.7×10^{-1}
Pu-238	8.77×10^{1}	1.9×10^{0}	1.7×10^{4}	8.5×10^{-1}	1.5×10^{4}	1.1×10^{0}	1.3×10^{3}
Pu-239	2.41×10^{4}	2.9×10^{0}	1.7×10^{4}	1.7×10^{0}	1.1×10^{4}	1.2×10^{0}	6.0×10^{3}
Pu-240	6.54×10^{3}	3.0×10^{0}	1.5×10^{4}	1.7×10^{0}	1.1×10^{4}	1.3×10^{0}	3.9×10^{3}
Pu-241	1.44×10^{1}	2.1×10^{1}	5.7×10^{5}	3.1×10^{0}	2.5×10^{5}	1.8×10^{1}	3.2×10^{5}
Pu-242	3.76×10^{5}	1.0×10^{-1}	1.1×10^{2}	3.5×10^{-4}	1.1×10^{2}	1.0×10^{-1}	3.8×10^{0}
Am-241	4.32×10^{2}	2.8×10^{0}	5.0×10^{4}	1.8×10^{0}	4.8×10^{4}	9.9×10^{-1}	1.5×10^{3}
Am-242m	1.52×10^{2}	2.3×10^{-3}	7.6×10^{1}	2.3×10^{-3}	7.6×10^{1}		
Am-243	7.38×10^{3}	8.8×10^{-4}	9.7×10^{1}	8.8×10^{-4}	9.7×10^{1}		
Cm-244	1.81×10^{1}	2.5×10^{-2}	1.9×10^{3}	2.5×10^{-2}	1.9×10^{3}		

Table 2.4 Radionuclide inventory for reference spent fuel (fission and activation products). Activity in TBq per metric ton U (33,000 MW–days per metric ton U). A.P. = activation product; F.P. = fission product. Based on data from Roddy *et al.* (1986). Reproduced by permission of Oak Ridge National Laboratory, U.S.A.

Radionuclide	Time since discharge (years) 1.0×10^1	1.0×10^3	1.0×10^5
Fission Products and Activation Products			
H-3	1.7×10^1	–	–
C-14	5.7×10^{-2}	5.1×10^{-2}	–
Fe-55	1.4×10^1	–	–
Co-60	7.8×10^1	–	–
Ni-59	1.9×10^{-1}	1.9×10^{-1}	8.0×10^{-2}
Ni-63	2.4×10^1	1.4×10^{-2}	–
Se-79	3.2×10^{-1}	3.2×10^{-1}	5.2×10^{-3}
Kr-85	1.8×10^2	–	–
Sr-90	2.1×10^3	–	–
Y-90	2.1×10^3	–	–
Zr-93	7.1×10^{-2}	7.1×10^{-2}	6.8×10^{-2}
Nb-93m	6.8×10^{-2}	6.8×10^{-2}	6.5×10^{-2}
Nb-94	4.8×10^{-2}	4.6×10^{-2}	1.6×10^{-3}
Tc-99	4.8×10^{-1}	4.8×10^{-1}	3.5×10^{-1}
Ru-106	2.0×10^1	–	–
Rh-106	2.0×10^1	–	–
Pd-107	4.0×10^{-3}	4.0×10^{-3}	4.1×10^{-3}
Sn-126	2.9×10^{-2}	2.9×10^{-2}	1.4×10^{-2}
Sb-125	4.8×10^1	–	–
Sb-126	4.0×10^{-3}	4.0×10^{-3}	2.0×10^{-3}
Sb-126m	2.9×10^{-2}	2.9×10^{-2}	1.4×10^{-2}
Te-125m	1.2×10^1	–	–
I-129	1.2×10^{-3}	1.2×10^{-3}	1.2×10^{-3}
Cs-134	1.9×10^2	–	–
Cs-137	3.0×10^3	–	–
Ba-137m	2.9×10^3	–	–
Ce-144	5.5×10^0	–	–
Pr-144	5.5×10^0	–	–
Pr-144m	6.6×10^{-2}	–	–
Pm-147	6.6×10^{-2}	–	–
Sm-151	1.2×10^1	6.0×10^{-3}	–
Eu-155	5.9×10^1	–	–
SUB–TOTALS			
A.P.	1.3×10^2	3.1×10^{-1}	9.1×10^{-2}
F.P.	1.1×10^4	7.1×10^{-1}	5.3×10^{-1}

Table 2.5 Radionuclide inventory for reference spent fuel (heavy nuclides). Activity in TBq per metric ton U (33,000 MW–days per metric ton U). Based on data from Roddy *et al.* (1986). Reproduced by permission of Oak Ridge National Laboratory, U.S.A.

Radionuclide	**Time since discharge (years)**		
	1.0×10^{1}	1.0×10^{3}	1.0×10^{5}
Heavy Nuclides			
Ra-226	9.8×10^{-7}	1.2×10^{-4}	4.0×10^{-2}
U-233	3.0×10^{-6}	1.5×10^{-4}	2.0×10^{-2}
U-234	7.5×10^{-2}	7.5×10^{-2}	6.0×10^{-2}
U-235	5.2×10^{-4}	5.2×10^{-4}	8.9×10^{-4}
U-236	1.0×10^{-2}	1.0×10^{-2}	1.3×10^{-2}
U-238	1.2×10^{1}	1.2×10^{1}	1.2×10^{1}
Np-237	–	3.7×10^{-2}	4.2×10^{-2}
Np-239	6.3×10^{-1}	5.8×10^{-1}	–
Pu-238	8.6×10^{1}	4.0×10^{-2}	–
Pu-239	1.2×10^{1}	1.1×10^{1}	6.7×10^{-1}
Pu-240	1.9×10^{1}	1.8×10^{1}	–
Pu-241	2.9×10^{3}	–	–
Pu-242	6.4×10^{-2}	6.4×10^{-2}	5.3×10^{-2}
Am-241	6.3×10^{1}	3.3×10^{1}	–
Am-243	6.3×10^{-1}	5.8×10^{-1}	–
Cm-242	2.1×10^{-1}	–	–
Cm-243	6.1×10^{-1}	–	–
Cm-244	4.9×10^{1}	–	–
SUB–TOTAL	3.1×10^{3}	6.4×10^{1}	1.4×10^{0}

cloth, although some sorting may be necessary to protect the compacting equipment from metal and other non-compressible items. In addition, incineration is used routinely, and on a large scale, to reduce the volume of combustible material to ash (*e.g.* incineration plants at Karlsruhe and CEN/Cadarache). The benefits of volume reduction must always be balanced against radiological safety aspects (ALARA principles), in particular, additional occupational exposure and potential airborne release.

As discussed in Section 2.1, a wide variety of solid waste types exists, but these raw wastes must be conditioned prior to acceptance for shipment to the disposal site. The type of conditioning necessary to demonstrate acceptability will be determined by considerations regarding:

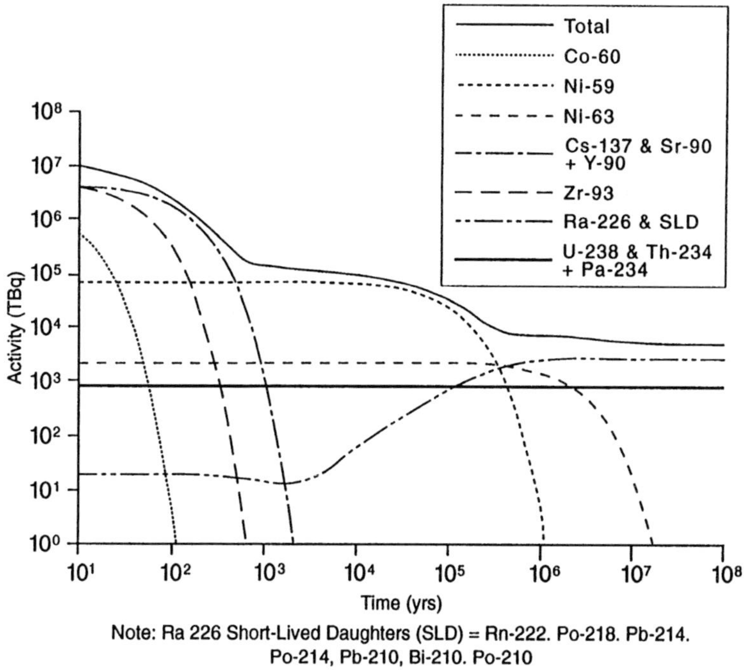

Figure 2.8 Activity of key radionuclides as a function of time. From Chapman and McEwen (1991). Reproduced by permission of Lawrence Berkeley Laboratory, U.S.A.

- ease and safety of handling and transport,
- interim storage requirements,
- geological disposal route,

and will depend, therefore, on the type of waste and its radionuclide content.

The most straightforward conditioning, for example, involves immobilisation, primarily to provide a monolithic waste form (*e.g.* cement) which is resistant to degradation. Incorporation of waste in glass also represents a monolithic waste form, but the conditioning process is more complex. Although it is the responsibility of the waste producer to ensure that wastes are properly conditioned according to pre-established guidelines, in the context of waste acceptance criteria, the waste disposer must be aware of what conditioning method has been used.

In the following sections, methods of conditioning HLW and L/ILW are discussed in detail.

2.4.1 High-level waste forms

Besides the heat content of this waste category, due mainly to short-lived beta-gamma emitters, the properties of the waste which demand a greater integrity of immobilisation or containment are the high concentrations of alpha emitters and the long half-lives of these radionuclides. Thus, the timescale associated with its isolation is at least 10^4 years (regulatory issues are discussed in Chapter 1). Research on suitable waste forms over the last 35 years has focused on incorporating radionuclides directly into readily fabricated solid phases, ideally each radioactive atom occupying a site in the structural framework of the solid. Results of this massive research effort indicate that certain glasses, in particular borosilicate glass, ceramics, such as SYNROC, and other crystalline materials are suitable matrices for incorporating HLW.

Borosilicate glass

Borosilicate glass was first proposed as a HLW matrix in the 1950's, and since then, has been investigated in depth. However, some experimental efforts have been misdirected, particularly with respect to emphasis on short-term dissolution properties.

At the present time, borosilicate glass is the waste form of choice for HLW streams produced in Europe, the U.S.A., and Japan. This is largely due to the fact that full-scale vitrification facilities incorporating HLW are operating successfully – at Marcoule (AVM, since 1978) and, more recently, two units at La Hague (AVH, R7 and T7, since 1989), both in France, and also recently at Sellafield (Windscale Vitrification Facility) in the U.K. Glassification facilities for HLW will soon be operational in the U.S.A. and Japan. Such facilities incorporate high-level liquid waste into the glass via calcining and vitrification. The composition and key properties of an example glass from the AVH facility are shown in Table 2.6, and successful production of this waste form is evident in the cumulative vitrification results for AVM and R7, shown in Table 2.7 (Moncouyoux and Sombret, 1992). With a view to reducing the volume of high-level waste associated with spent fuel assemblies, ongoing development work in France is aimed at constructing a full-scale active facility capable of melting zircaloy cladding hulls which are currently embedded in cement.

The advantages of a glass matrix are:

- good industrial-scale feasibility,
- good performance characteristics for both interim storage and adequate performance after final disposal.

With respect to long-term behaviour, borosilicate glass has a well understood dissolution mechanism which is substantiated by a large database under a wide range of repository-related conditions and for which extrapolated behaviour can be verified by comparison with natural analogues (Petit, 1991).

Table 2.6 Chemical composition of an example glass from the vitrification facility at La Hague. From Moncouyoux, J.P. and Sombret, C.G., 'French high-level waste management research and development programme', in *Proceedings of 3rd International Conference on High Level Radioactive Waste Management*, 1992, pages 2406-2409, reproduced by permission of American Society of Civil Engineers.

Composition	% (weight)
SiO_2	45.2
B_2O_3	13.9
Na_2O	9.8
Al_2O_3	4.9
CaO	4.0
Fe_2O_3	2.9
NiO	0.4
Cr_2O_3	0.5
P_2O_5	0.3
ZrO_2	1.0
LiO_2	2.0
ZnO	2.5
Fission product oxides	11.0
Actinide oxides	0.9
Metallic particles	0.7
TOTAL	100.0

The glass is least reactive at *(a)* temperatures below 100 °C and *(b)* under low groundwater flow conditions. The former condition may be achieved by lower waste loadings or longer periods of interim storage. The latter condition may be met by the low groundwater flow rates which are expected of 'typical repositories', by surrounding the waste glass by a low-permeability flow barrier, or both (see Chapters 3 and 4).

Alternative waste forms

Given the preponderance of relatively unreactive, insoluble minerals in rocks, the production of a synthetic material which behaves in a similar way to minerals has been investigated. The end result was the development of a multi-phase ceramic waste form – SYNROC (Ringwood *et al.*, 1981). SYNROC is the generic name for a number of titania-based wasteforms designed to immobilise high-level waste. These ceramics are synthetic mineral assemblages

Table 2.7 Cumulative vitrification results: AVM and R7 facilities, France. From Moncouyoux, J.P. and Sombret, C.G., 'French high-level waste management research and development programme', in *Proceedings of 3rd International Conference on High Level Radioactive Waste Management*, 1992, pages 2406-2409, reproduced by permission of American Society of Civil Engineers.

Facility	**AVM** (as of Nov. 1, 1991)	**R7** (as of Sept. 30, 1991)
Solution volume	$\sim$1502 m^3	$\sim$728 m^3
Glass weight	$\sim$672 tonnes	$\sim$334 tonnes
Total activity	$\sim 1.2\times10^{19}$ Bq	$\sim 11\times1.0^{19}$ Bq
Glass canisters	1916	855

which consist of several oxide structure types including fluorites, perovskites, hollandites, reduced rutile and magnetoplumbite (White, 1986).

Originally, it was envisaged that all the chemical species in HLW could be structurally incorporated (by isomorphic substitution) in three titanate phases – zirconolite ($BaAl_2Ti_6O_{16}$), perovskite ($CaTiO_3$), and hollandite ($CaZrTi_2O_7$). Although most of the components of HLW do enter these phases, it is now realised that the original concept is an oversimplification which is approached only at relatively low waste loadings. Furthermore, some radionuclides are reduced to the metallic state and form alloys during SYNROC fabrication.

The SYNROC phase assemblage is dependent upon waste loading, and a number of closely related, but distinct, phases appear in response to variations in radwaste composition and concentration. Approximate proportions (wt %) of different structure types are (White, 1986): fluorites (30 %), hollandites (30 %), perovskite (15 %), reduced rutile (15 %), magnetoplumbite ($\leq$ 5 %) and alloy ($\leq$ 5 %).

As a result, such structural modification should provide SYNROC considerable flexibility to respond to chemical changes in the waste-stream. SYNROC has improved mechanical properties over glass, and also has excellent thermodynamic and chemical durability. However, research involving 'active' (actinide-doped) material and under repository conditions has yet to be completed. SYNROC can accommodate waste loadings comparable to those of glass, requiring only 60 to 80 % of the volume. Its higher resistance to elevated temperatures allows higher waste loading, and its higher fracture toughness and thermal conductivity reduce the increase in surface area due to thermally-induced fractures.

Three mechanisms are expected to control the long-term release of radionuclides from SYNROC: (1) base-catalysed hydrolysis of the titanate structure; (2) diffusion-controlled ion-exchange; (3) precipitation and layer formation.

Sphene-based glass ceramics

This material has been studied in detail with a view to serving as host material for CANDU fuel (*CANada Deuterium Uranium*) recycling wastes. Glass-ceramics based on crystalline sphene ($CaTiSiO_5$) were selected for development because of the well-documented persistence of sphene as a naturally occurring mineral, with an ability to take a wide variety of ions into solid solution (Deer *et al.* 1982). The aluminosilicate glass matrix remaining after sphene crystallisation is a highly durable material for immobilising those waste ions which do not partition into the sphene phase (*e.g.* Cs-135, Cs-137, Se-79).

Spent fuel as waste form

Rather than reprocess the spent fuel and immobilise the heat producing liquid in glass or crystalline matrix, several countries including Canada, Sweden, Finland and the U.S.A., are considering the merits of direct disposal of spent fuel, *i.e.* without reprocessing. The attractive feature of this waste form is the lack of further processing of the waste (spent fuel) itself, *i.e.* there is no requirement for a dedicated reprocessing facility with associated highly radioactive liquid waste streams, resulting in reduced occupational exposure.

Because irradiated fuel is > 95 % by weight uranium oxide, the remainder comprising fission products, actinides and daughter products, performance of this waste form will be dominated by the solubility of UO_2. Dissolution of the waste will occur with time, but will be considerably slower under reducing than oxidising conditions (see Chapter 4).

2.4.2 L/ILW immobilisation matrices

A number of different materials have been tried and tested for their suitability as a matrix for immobilising L/ILW. Of these, the most routinely used materials are:

- cement/concrete,
- bitumen,
- polymer resins.

Table 2.8 Example of 'equilibrium leach test' (ELT) experimental matrix (from Heafield *et al.*, 1989). Reproduced by permission of Materials Research Society.

Waste	Matrix	Backfill	Cover Gas
Magnox swarf	Polymer-modified cement	PFA/OPC*	N_2/O_2
Advanced gas-cooled reactor hulls	Blast furnace slag/OPC*	Limestone flour/OPC*	Air
Zircaloy hulls	Polyester polymer	Bentonite	Ar/H_2
Ferric floc	Epoxide polymer		
Clinoptilolite	PFA/OPC*		
Plutonium-contaminated material (combustible)			
Plutonium-contaminated material (non-combustible)			

Combinations of the above matrices are also possible. Whatever matrix is used, however, relevant information on the product's performance (*e.g.* mechanical and chemical stability), supported by extensive testing, must be available on the final waste form prior to its acceptance.

Each country's waste management programme typically includes some type of waste-form evaluation programme, whereby selection of the most suitable waste form is achieved by a detailed experimental testing programme. Table 2.8 provides an example matrix of such a series of experiments in the U.K. (Heafield *et al.*, 1989).

The level of treatment/conditioning will be dictated by handling and transport requirements, length of interim storage, disposal options and regulatory requirements (IAEA, 1990b).

Concrete/cement

The most predominant matrix for immobilisation of radioactive waste is concrete or cement, although these two materials exist in a wide range of chemical compositions with a corresponding range of properties. Structural integrity is often enhanced using steel reinforcing rods. The long-term performance of concrete/cement has been studied in detail, and concrete has been used successfully to immobilise low-activity liquid wastes since the 1950's.

The term cement is often used generically to include cement pastes (cement-water mixtures), mortars (mixtures of cement, water and fine aggregate) and grouts (pastes or mortars). As a waste matrix, concrete may be considered as a porous solid with water-filled pores, the solid consisting mainly of:

- a continuous cement phase; *e.g.* ordinary Portland cement (OPC), which consists of calcium hydroxide and hydrates of calcium silicate and calcium aluminate;
- a particle phase: aggregates added to the cement, usually 65-75 % total volume.

In its prepared form, the concrete consists of aggregate material including waste, entirely surrounded by cement paste. Given the chemical constituents present, the pore water of fresh concrete is highly alkaline, *i.e.* has a high pH, close to 13.

Concrete is a major constructional component of deep repositories, but as a waste matrix, cement and concrete systems also serve several useful functions, such as controlling water flow (by having low permeability) and radionuclide migration (by having low diffusion coefficients and some sorption capacity for radionuclides) in the near-field (see Chapter 4). In addition, concrete as a container material has both mechanical and chemical functions. Mechanically, the containers must be strong enough to support the required loads so that individual waste packages do not collapse. The chemical function is to maintain the pH in the near-field at a high value for as long as possible. In the short term, this buffering minimises the corrosion of steel containers and, in the long term, ensures low aqueous solubility limits for any traces of long-lived radionuclides (again, discussed more in Chapter 4).

By suitable mixing of cement/concrete with additives, the permeability of the final product may be controlled. Thus, for example, gas-permeable cements are intended for use at the Swedish L/ILW repository in Forsmark (SFR) to allow release of gas produced by the waste and canister corrosion. On the other hand, by using polymer resins as additives, the permeability of the concrete/cement matrix may be lowered sufficiently that the release of radionuclides from waste containers is diffusion controlled. An additive such as vermiculite, a naturally occurring clay mineral, may be pre-mixed with cement to provide additional sorption capacity for cations.

Bitumen

Technical (man-made) bitumen is the name given to a series of complex organic compounds, which may be obtained as a distillation fraction of petroleum, and which are soluble in carbon disulphide. Strictly, the term bitumen includes both liquid and solid fractions, but only the solid fraction, which includes asphalt, is relevant to radioactive waste management. Bitumens consist predominantly of carbon and hydrogen, but also contain oxygen, sulphur, and nitrogen as minor constituents. The average composition of this class of material is: carbon (80-88 %), hydrogen (8-11 %), oxygen (1-12 %), sulphur (1-7 %), and nitrogen (trace to 1.5 %) (Eschrich, 1980).

Natural bitumens have been in use for more than 5000 years, and geological occurrences also point to their long-term stability (Helmuth, 1989). This material offers a suitable matrix for certain L/ILW types, including evaporator concentrates, reprocessing precipitation sludges, and ion exchange resins, and has been successfully applied on a large scale in a number of nuclear installations (*e.g.* the Eurochemic/Belgoprocess reprocessing plant). Different types of bitumen – direct distillation, blown, and cracked – are commercially available from the petrochemical industry.

In using bitumen as a matrix for L/ILW, the final waste form offers several advantages over cementation, including improved volume reduction and low permeability. On the other hand, the bituminisation process involves relatively high temperatures (150-230 °C). In addition, bitumens are susceptible to microbial attack, although the conditions which favour biodegradation in nature include a high oxygen content, *i.e.* aerobic conditions (Petit, 1991). Whether or not a microbial population will be viable depends on a number of factors, but mainly temperature of the repository environment and the presence of an adequate supply of nutrients. With respect to the latter, C, P, S, N, and O are expected to be the controlling elements.

While decomposition of organic materials ultimately yields gases (CO_2 and CH_4), some important intermediate products of bitumen degradation are high molecular weight humic acids, which offer sorption sites for some radionuclides, *e.g.* Ni. The resultant effect then depends more on the long-term behaviour of the organic species than on the radionuclide itself. In colloidal form, for example, the mobile organic species can increase the migration of a sorbed radionuclide. On the other hand, such a colloid may be retarded in some formations due to its size, in which case the overall mobility is correspondingly decreased.

For soluble organic degradation products, the possibility of forming soluble complexes with waste elements must be considered, particularly with respect to an enhanced mobility of radionuclides.

Polymers/resins

Different types of thermosetting and thermoplastic polymer matrices have been used to immobilise ion exchange resins (as an alternative to bitumen) and also liquid waste such as evaporator concentrate which is rich in borate and nitrate. These polymer matrices include polymethyl methacrylate (PMMA), phenol-formaldehyde (PhF), styrene divinylbenzene (for evaporator concentrates), polyester, polyurethane and epoxy (for spent ion exchange resins).

The main considerations in favour of using resins as a waste form are:

- industrial-scale production,

- good mechanical strength,
- inflammable,
- slow degradation rate; low water permeability,
- efficient incorporation of soluble salts, (*e.g.* nitrates, sulphates),
- good resistance to radiation,
- good thermal stability.

The solidification technology generally involves thorough mixing of the binder with catalyst, followed by the waste material, to the required viscosity. Addition of the initiator/promoter results in gelation within minutes and hardening in several hours. A cross-linking agent may also be added to the initial mixture if additional mechanical strength is desired.

Comparison of conditioning matrices

Table 2.9 shows an intercomparison of conditioning processes for LLW, ILW and alpha wastes (from IAEA, 1992).

2.4.3 Waste packaging

Conditioning of waste also involves packaging of the waste form, and this Section briefly addresses the types of container or canister in which the waste form is emplaced for disposal.

For HLW, interim storage of the waste form is likely to occur for at least 40 years prior to disposal in a deep repository, to allow the heat output to decay to an acceptable level. Thereafter, the primary function of the container is to delay the groundwater reaching the waste form, whether borosilicate glass, ceramic, or spent fuel itself. In some cases, an additional function of the canister may be to reduce the extent of irradiation of the surrounding near field, and the resultant production of H_2 and H_2O_2.

The two main approaches in canister design are to provide either a thick layer of metal which corrodes at a known slow rate (*e.g.* carbon steel), or a thinner layer of a corrosion-resistant material such as titanium (Hough and Marples, 1991). The principal canister materials under review for HLW are mild steel, stainless steel, copper and titanium. Where borosilicate glass is the waste form, molten glass is cast directly into stainless steel cylinders. In Sweden, the waste package involves encapsulation of spent fuel in copper canisters. In this case, the space between fuel rods is filled with copper or lead to permit good heat dissipation (KBS, 1983).

Steel (standard carbon steel) drums, both 200 litre and 400 litre volumes, and concrete/steel boxes/tanks, with volumes in the range 2 to 10 m^3, are the most common container materials for L/ILW waste, with concrete used whenever extra shielding is required for the ILW waste form. Such concrete

Table 2.9 Comparison of conditioning matrices for LLW, ILW and alpha-contaminated wastes (from IAEA, 1992). Reproduced by permission of IAEA.

Characteristic	Cementation	Bituminisation	Polymer fixation
Application	Variety of LLW/ILW	Chemical precipitates, low heat and low alpha wastes	Waste concentrates such as resins, sludges, evaporator bottoms and ashes
Process condition	Room temp.	High temperature	Room temperature Proper process control and understanding of chemical reactions necessary
Volume change	Increases by factor of 2	Overall reduction good, owing to removal from waste	Increases by factor of 1.2
Chemical compatibility	Waste with water can be fixed	Water removal before bituminisation required	Waste with water
Dimensional stability in presence of water	Very good	Very poor for certain wastes such as spent ion exchange resins and evaporator concentrates	Very good
Leaching rate	Satisfactory	Low	Very low
Radiation resistance	Very good	Satisfactory	Very good (radiation resistance up to 0.1 Gy)
Compatibility with environmental conditions	Many conditions	With most conditions	With most conditions

may be added as a liner, for example, inside a steel drum, or may serve as an overpack for the canister(s) during transportation. High-density polyethylene (HDPE) is also used as a 'high integrity container' material for certain types of ILW (non-organic), having low water permeability and a superior corrosion resistance to that of steel. Other candidates for high integrity container materials are specially coated (epoxy-coated) carbon steel and reinforced vinyl ester polymer.

2.4.4 Waste characterisation

Performance assessment requires a knowledge and understanding of the properties of the near-field and how they will evolve with time. This, in turn, requires corresponding information on the waste itself and on its conditioning prior to final disposal. In order to predict the behaviour of the waste (form) and its interactions with other constituents of the near-field, the basic characteristics to be established are:

- radionuclide inventory,
- physical and chemical properties of the waste package.

A radionuclide inventory, such as that shown in Table 2.3 or 2.4/2.5, provides the necessary information regarding specific radionuclides present and their activity levels. At a minimum, this information allows a waste stream to be evaluated in terms of maximum concentration of each radionuclide which is available at any time in the future. In addition, priority radionuclides can be identified as a function of timescale and exposure scenario, depending on their mobility and half-life.

For most repository concepts, some segregation of waste packages into different vaults or chambers is envisaged, separation being based on activity content and material content (characteristics) of the waste package. Thus, it is important to have as accurate an inventory as possible, and not just total inventory, but one for individual waste streams.

Radionuclide inventories are derived from a number of sources, *e.g.* based on:

- historical information provided by waste producers,
- direct activity measurement of waste streams,
- direct activity measurements of waste packages, combined with calculations using isotope ratios,
- comparison with analogous waste streams,
- standard radionuclide compositions,
- activity calculations performed for known fuel burnup.

In general, there is a lack of quantitative activity data in support of detailed waste radionuclide characterisation. Thus, given the above variety of sources

Table 2.10 Important waste elements and waste materials (bulk) for quantification (applicable to L/ILW waste disposal in the U.K.).

Waste Elements	**Waste Material**	
	Metals	**Organics**
Tin	Stainless steel	Cellulosics
Selenium	Other types of steel	Halogenated plastics
Nickel	Magnesium and Magnox	Non-halogenated plastics
Niobium	Aluminium	Organic ion-exchange resins
Zirconium	Zircaloy	Halogenated rubber
Molybdenum	Other metals	Non-halogenated rubber
	Other organics	

of data, it is not surprising that there are considerable uncertainties in many of the final 'numbers', and an in-depth review of such data should at least seek to identify inconsistencies, either in key isotope ratios quoted, or in terms of the known process(es) which generated each waste stream.

Information on the physical and chemical properties of the waste package gives an indication of its stability and long-term performance. In this respect, knowledge of the bulk material composition, including material masses, can be used to assess evolution of the near-field following waste package degradation *e.g.* subsequent gas generation or the release of organic species. In the U.K. waste management programme, for example, determination of accurate quantities of several bulk materials (see Table 2.10) in different waste streams is required as a basis for modelling gas generation and chemical evolution in the near-field. In addition, specific elements have been identified as important in limiting the solubility of associated radionuclides in groundwater, and these are also included in Table 2.10.

The above characteristics/properties will form the basis of waste acceptance criteria which should be developed for disposal at each repository on a site-specific basis, to ensure that the intended disposal route is compatible with the waste stream being considered. Several other criteria must be met before waste is accepted for final disposal, including quality assurance issues relating to the origin of the waste and its method of conditioning. Thus, a fully implemented quality assurance programme is central to establishing and meeting waste acceptance criteria requirements.

All steps in the waste handling sequence should be carefully monitored and reviewed, *viz.*:

- collection and sorting of raw waste,

- conditioning of waste,
- interim storage conditions,
- transport to disposal site,
- receipt at disposal site,
- handling and final positioning at disposal site.

For each of the above stages, significant properties (or changes in properties) of the waste should be identified and, whenever possible, quantified, so that preset (limiting) values on these properties can be used as waste acceptance criteria. Such limiting values can then be used to provide input data for assessing probable rates of release of individual radionuclides into the near field.

Integration of the above characterisation process will benefit from a computerised waste management system, developed according to acceptable quality assurance standards. For example, Electricité de France, the major waste generator in France, utilises a software package (DRA) which allows centralisation of waste characterisation data, facilitated by bar code labelling of waste packages (Communaux and Lantes, 1993). This system also allows real-time data exchange with ANDRA, the French national radioactive waste management agency. After two years of operation, implementation of the computerised system had allowed the tracking of more than 150,000 waste packages and 1,100 shipments (Communaux and Lantes, 1993).

2.5 References

Chapman, N.A. and McEwen, T.J. (1991) Geological aspects of the British programme for deep disposal of nuclear wastes. Chapter 18 in *Geological Problems in Radioactive Waste Isolation*, (P.A. Witherspoon, editor), *Lawrence Berkeley Laboratory Report* No. LBL-29703. Lawrence Berkeley Laboratory, Berkeley, California, USA.

Claiborne, H.C. (1972) Neutron-induced transmutation of high-level radioactive waste. *Oak Ridge National Laboratory Report* ORNL-TM-3964. Oak Ridge National Laboratory, USA.

Communaux, M. and Lantes, B. (1993) Radioactive waste computerized management. In *Environmental Remediation and Environmental Management Issues*, (eds. R. Baschwitz, R. Kohout, J. Marek, P.I. Richter and S.C. Slate), Volume **3**, pp. 471-473, Proc. 1993 International Conf., Prague, September, 1993. American Society of Mechanical Engineers, New York, New York.

Deer, W.A., Howie, R.A., and Zussman, J. (1982) *Rock-Forming Minerals*, 2nd edition, vol. 1A, Orthosilicates. Longmans, London and New York.

Eschrich, H. (1980) Properties and long-term behaviour of bitumen and radioactive waste–bitumen mixtures. *SKB Technical Report* TR 80-14. Swedish Nuclear Fuel and Waste Management Company, Stockholm, Sweden.

Heafield, W., Barlow, P., and Hunt, A. (1989) Equilibrium leach testing of intermediate level waste. In *Scientific Basis for Nuclear Waste Management Vol. XII*, (W. Lutze and R.C. Ewing, editors), Materials Research Society, Pittsburgh, Pennsylvania, pp 709-714.

Helmuth, K-H. (1989) Natural analogues of bitumen and bitumenized waste. *Finnish Centre for Radiation and Nuclear Safety Report* STUK-B-VALO 58.

Hough, A. and Marples, J.A.C. (1991) Repository source term for vitrified high level waste. In *Performance of High Level Waste Forms and Engineered Barriers under Repository Conditions.* Final Report of Co-ordinated Research Programme 1984-1989. International Atomic Energy Agency, Vienna, pp 191-225.

IAEA (1988) Nuclear Research Reactors in the World – *Reference Data Series No. 3.* International Atomic Energy Agency, Vienna, Austria.

IAEA (1989) Decontamination and decommissioning of nuclear facilities. *IAEA Technical Document Series* TECDOC-511. International Atomic Energy Agency, Vienna, Austria.

IAEA (1990a) Qualitative acceptance criteria for radioactive wastes to be disposed of in deep geological formations. *IAEA Technical Document Series* TECDOC-560. International Atomic Energy Agency, Vienna, Austria.

IAEA (1990b) Evaluation of low and intermediate level radioactive solidified waste forms and packages. *IAEA Technical Document Series* TECDOC-568. International Atomic Energy Agency, Vienna, Austria.

IAEA (1992) *Radioactive Waste Management.* An IAEA source book. International Atomic Energy Agency, Vienna, Austria.

KBS (1983) Final storage of spent nuclear fuel *KBS-3.* Chapter I General. Swedish Nuclear Fuel Supply Company, Stockholm, Sweden.

Lee, W.W-L. and Choi, J-S. (1992) Releases from exotic waste packages from partitioning and transmutation. In *High Level Radioactive Waste Management, Proc. 3rd International Conf.*, April, 1992, Las Vegas, Nevada. American Society of Civil Engineers, New York, New York, pp 1387–1396.

Lineberry, M.J. and Phipps, R.D. (1989) Preparations for the IFR fuel cycle demonstration. *Transactions of the American Nuclear Society* **60:** 170.

Moncouyoux, J.P., and Sombret, C.G. (1992) French high-level waste management research and development programme. In *High Level Radioactive Waste Management, Proc. 3rd International Conf.*, April, 1992, Las Vegas, Nevada. American Society of Civil Engineers, New York, New York, pp 2406-2409.

NEA/OECD (1986) *Decommissioning of Nuclear Facilities – Feasibility, Needs and Costs.* Nuclear Energy Agency, Paris.

Petit, J-C. (1991) Design and performance assessment of radioactive waste forms: what can we learn from natural analogues? In *Proceedings of 4th Natural Analogue Working Group Meeting*, (B. Côme and N.A. Chapman, editors), Pitlochry, June, 1990, *CEC Report* EUR 13014EN. Commission of the European Communities, Brussels, Belgium.

Pettersson, B.G. (1992) Improving the management of spent radiation sources. *IAEA Bulletin* **34**, No. 3, 19-23. IAEA, Vienna, Austria.

Rees, J.H. and Rodwell, W.R. (1988) Gas evolution and migration in repositories – current status. *U.K. Nirex Safety Series Report* No. NSS/G100. U.K. Nirex Ltd, Harwell, U.K.

Ringwood, A.E., Oversby, V.M., Kesson, S.E., Sinclair, W., Ware, N., Hibberson,

W. and Major, A. (1981) Immobilization of high-level nuclear reactor wastes in SYNROC: a current appraisal. *Nuclear and Chemical Waste Management* **2:** 287-295.

Robertson, D.E., Thomas, C.W., Wynhoff, N.L., and Hetzer, D.C. (1991) Radionuclide characterization of reactor decommissioning waste and spent fuel assembly hardware. *Pacific Northwest Laboratory, U.S. Nuclear Regulatory Commission Report* NUREG/CR-5343.

Roddy, J.W., Claiborne, H.C., Ashline, R.C., Johnson, P.J., and Rhyne, B.T. (1986) Physical and decay characteristics of commercial LWR spent fuel. *Oak Ridge National Laboratory Report* ORNL/TM-9591/V1&R1.

Skålberg, M and Liljenzin, J.-O. (1992) Partitioning and transmutation. A review of the current state of the art. *SKB Technical Report* TR 92-19. Swedish Nuclear Fuel and Waste Management Company, Stockholm, Sweden.

SKB (1992) Treatment and final disposal of nuclear waste. RD & D Programme 92. Swedish Nuclear Fuel and Waste Management Company, Stockholm, Sweden.

Smith, G.M. and Charles, D. (1991) Neutron activation of elements and the radiological implications for development and use of low activation materials in DT fusion reactors. *AEA Technology Report* AEA-FUS-122. AEA Technology, Harwell, U.K.

U.S. Department of Energy (1988) 1987 State-by-state assessment of low-level radioactive wastes received at commercial disposal sites. National Low-Level Radioactive Waste Management Program, DOE/LLW-69T.

U.S. National Research Council (1990) Rethinking high-level radioactive waste disposal. *Commission on Geosciences, Environment, and Resources, National Research Council.* U.S. National Academy Press, Washington D.C.

White, T.J. (1986) The extended defect as a mechanism for the immobilization of HLW species in zirconolite, perovskite and hollandite. In *Scientific Basis for Nuclear Waste Management*, **Vol. IX**, (L.O. Werme, editor), Materials Research Society, Pittsburgh, Pennsylvania, pp 283-290.

White, I.F. and Smith, G.M. (1984) Editors, Assessment of management modes for graphite from reactor decommissioning. *Euratom Report* EUR-9232.

3

Repository and barrier concepts

3.1 Introduction

The functional requirements for geological disposal of nuclear waste include the following key points:

- the waste must not be released to the biosphere at concentrations deemed to present an unacceptable hazard,
- the waste must be removed and isolated from the effects of human activity or catastrophic natural events,
- the technology to implement disposal must be readily available as well as achievable at a reasonable cost,
- in some countries, the retrieval of some types of disposed nuclear wastes must be technologically and economically feasible, if so desired by future generations,
- the processes which control safe performance of nuclear waste disposal must be well-characterised and accepted for modelling and that sufficient, relevant data be obtained and used in such models to demonstrate predicted performance reliably.

There are two fundamental approaches to the geological disposal of nuclear waste. The first is to attempt to *contain* the waste in place for as long as necessary to prevent unsafe release to the environment. It is clearly difficult to prove such containment for the extremely long time periods over which some types of nuclear waste may be hazardous. The second approach is a *controlled release* and dispersal of the nuclear waste within the environment through reliance on well-understood natural processes. Effective mobilisation and dispersal causes the concentrations of the nuclear waste to be reduced to levels which present no unacceptable hazard.

The purpose of this chapter is to describe the types of geological disposal which have been proposed, with emphasis on the leading concepts involving a combination of natural and engineered barriers. Such 'repository' concepts are defined here as systems of engineered structures emplaced within a well-characterised natural setting which will provide safe disposal of nuclear waste.

Conceptual and functional requirements for individual barriers are described. Several design options or deep geological repositories are summarised.

3.2 Geological disposal systems

At least four types of geological disposal schemes have been proposed for nuclear waste:

- deep (>100 metres) injection of liquid wastes into porous strata,
- ultra-deep (> several kilometres) burial in boreholes drilled from the surface,
- shallow (~ several metres) burial in trenches or bunkers covered with engineered barriers,
- deep (50 to 1000 metres) burial containing an engineered barrier system.

The latter two options involving a combination of natural and engineered barriers are referred to here as *nuclear waste repositories*. These concepts have received the principal amount of research and development internationally and will be the focus of further discussion.

Most repository designs rely on elements of both containment and controlled release approaches. Containment requires that the nuclear waste be encapsulated into a matrix, or waste form, that is physically isolated from the environment. During containment, the radionuclides within the waste form are undergoing radioactive decay. For all nuclides, containment for a sufficient number of half-lives will effectively reduce their inventory to extremely low levels. For example, a containment time period equal to 10 half-lives leads to a reduction in inventory by a factor of 1024. The optimal duration for a containment period depends on the type of waste and its radionuclide inventory. For LLW, having low concentrations of long-lived (half-lives > 100,000 years) nuclides, containment for a period of 300 years may be sufficient to result in a waste form with inventory levels of radionuclides at acceptably safe levels. Assuming proper institutional controls for this time period, a LLW could be safely emplaced within a shallow-land burial repository. No controlled-release phase would be necessary, permitting a wide selection of geological formations that were suitably conducive to complete containment for the specified duration.

Demonstrating containment for a time period equivalent to 10 times or more of the half-life of long-lived (> 1000 years) nuclides is a more difficult task. For this reason, geological repository systems for HLW and long-lived ILW are typically designed also to promote controlled release and dispersal of radionuclides upon containment failure. In concept, deep geological formations provide an ideal setting into which nuclear waste can be released and dispersed without adversely impacting the biosphere. Indeed, the projected efficiency in dispersal of released nuclear waste by many types of host rocks is so great that any containment requirements, other than for initial emplacement, may seem unnecessary and unjustified. As will be discussed in later chapters, however, the tremendous dispersal capacity of natural geological formations must be tempered with a significant degree of uncertainty regarding the spatial variability of such formations. This spatial uncertainty is further compounded by temporal uncertainty in what will happen to the natural system over time periods of 1,000 years or greater.

3.3 Multiple-barrier repository designs

Uncertainty inherent in the properties and performance of natural systems is a fundamental reason that repository designs have embraced a *multiple-barriers concept* to nuclear waste disposal, based on the combined performance of natural and engineered barriers. In this concept, there is a series of engineered and natural barriers that are nested one inside the other. Initially, this concept postulated that each barrier by itself should be capable of ensuring safe isolation of nuclear waste. This overly ambitious and unnecessary constraint has evolved to a more realistic assumption that the barriers will act in concert. Furthermore, the barriers should not have common failure modes. That is to say, a given set of postulated future conditions that may cause one barrier to fail in its expected performance should not lead to other barriers also failing.

A conceptual set of multiple-barriers for a repository is shown in Figure 3.1. Starting from the innermost set of barriers, these include the waste form or matrix, a container or canister, a backfill or buffer and finally the encompassing geological formation or 'geosphere'. In addition, the underground access tunnels and shafts must be backfilled and sealed where they intersect zones of high permeability/flow within the rock. The following sections describe the functions and, where applicable, design goals for such barriers. Leading candidate materials for engineered barriers are also noted.

3.3.1 Waste form

The main functions of the waste matrix or waste form are derived from pre- and post-closure requirements. During the pre-emplacement period, the waste form provides a convenient solid form to facilitate handling and transportation of nuclear wastes (see Chapter 2). After emplacement, the function of the waste form is to limit the rate of release of radionuclides following containment failure. This last topic is a main focus of near-field performance assessment, described in Chapter 4.

LLW is typically reduced in volume (distillation, incineration) and compacted within steel drums. The higher the waste activity and the longer

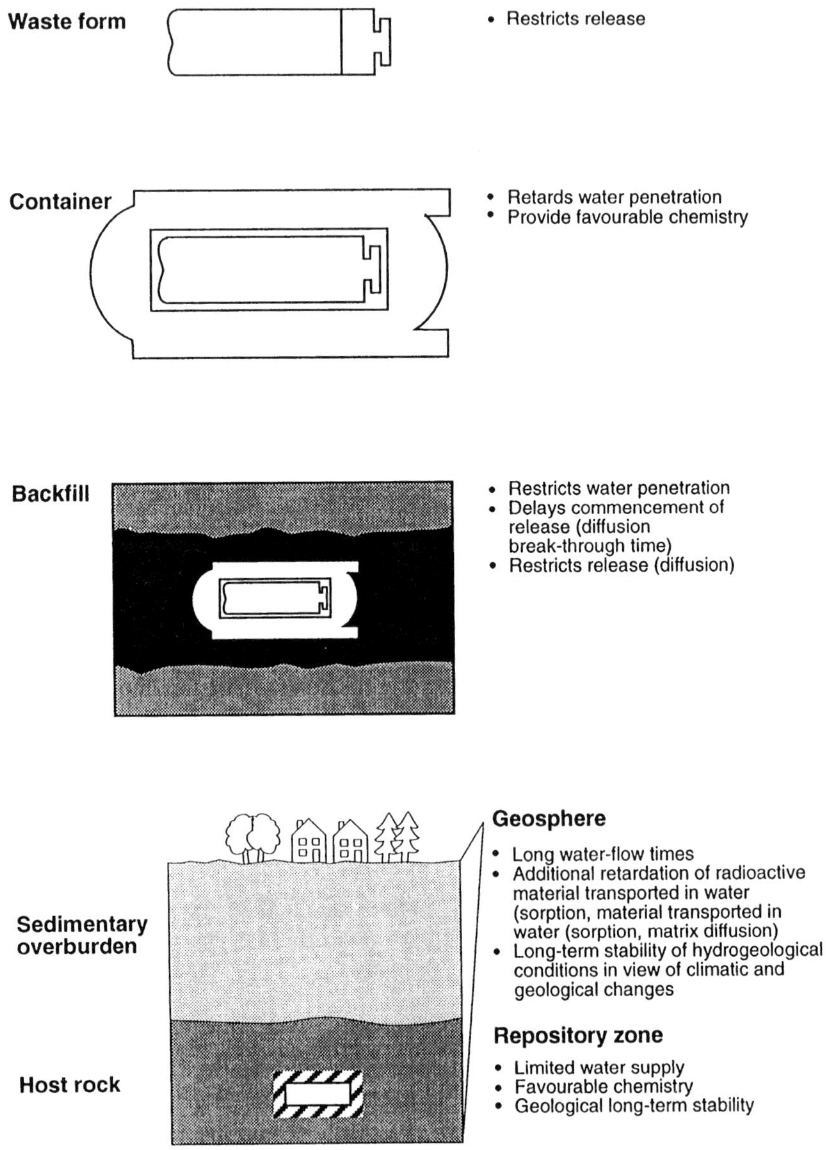

Figure 3.1 Schematic illustration of the multi-barrier system of waste containment; in this instance the Swiss concept for disposal of high-level waste (from Chapman and McKinley, 1989). Reproduced by permission of Nagra.

the duration of this activity, the more effort is put into conditioning the waste. ILW is encapsulated into cement, resin, polymer or bitumen matrices, sometimes requiring mechanical reductions in size of fragments.

HLW, derived from reprocessing of spent nuclear fuel, is in the form of a liquid. To produce a final waste form, this liquid is first reduced to a calcine powder or concentrated slurry. These forms of the waste can then be chemically or physically incorporated within a variety of cementitious, ceramic, glassy or other non-radioactive matrices. Such matrices must be selected on the basis of chemical compatibility with the waste components, ease and cost of fabrication and, perhaps most importantly of all, release performance under repository conditions.

In some countries, the direct disposal of spent nuclear fuel is the preferred option. Other than possible separation of fuel pins from the reactor assembly, no conditioning of spent fuel prior to disposal is envisaged. The characteristics of spent fuel relevant to release performance are outlined in Chapter 4.

3.3.2 Container

The function of the container (also called canister or overpack) is to isolate nuclear waste from groundwater for a designated period of time. For HLW, containment strategies range from long-term containment ($> 10^6$ years), assuring the elimination of most radioactivity by radioactive decay, to relatively short-term containment (500 to 1000 years), isolating waste from groundwater until short-lived fission products (and their associated heat production) decay. For LLW and ILW containing short-lived radionuclides, containment considerations may be reduced to the time period covering transportation, handling and emplacement of the wastes.

Three types of containers can be identified:

- *stable materials* under expected repository conditions, for which the mass-transfer rate of reactants (or products) rather than interface kinetics can be used to establish extremely low corrosion rates,
- *corrosion-resistant materials* that have an extremely low general-corrosion rate because of the formation of passive/insoluble surface layers, and
- *corrosion-allowance materials* that have a moderate general-corrosion rate but are inexpensive enough to fabricate into relatively thick-walled containers.

The reliance on mass-transport rates, rather than corrosion rates, for stable container materials provides a high predictive reliability to expected containment times. It must be demonstrated for the site, however, that the favourable chemical conditions for container stability will persist during this period. For corrosion-resistant and corrosion-allowance materials, one of two basic failure modes will determine the time to loss of containment

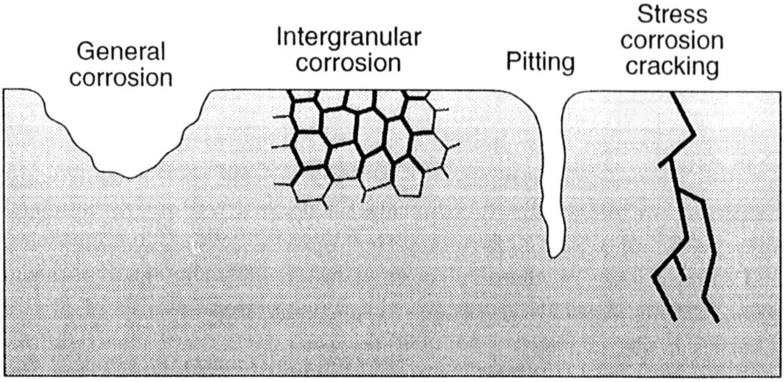

Figure 3.2 Schematic views of the four intrinsic modes of corrosion: general corrosion, intergranular corrosion, pitting and stress corrosion cracking.

(Figure 3.2):

- *general corrosion* until the container thickness has been reduced to a mechanical yield threshold or
- *localised corrosion* of the bulk container, occurring by one of several possible mechanisms (*e.g.* pitting, stress corrosion cracking, intergranular corrosion).

Failure of corrosion-resistant materials is usually assumed to occur by one of the first corrosion modes, while failure of corrosion-allowance materials is assumed to occur by the second mode. Preferential corrosion (by any mode) at the heat-affected zone of welds may be the dominant failure mechanism for either type of material, in the absence of annealing.

3.3.3 Backfill/buffer

The design functions of the backfill (also called buffer or packing) span both the *containment* and *controlled release* periods. These design functions of the backfill may include some or all of the following:

- reduce possible tectonic shearing forces on containers,
- prevent the container from settling within the emplacement hole/tunnel,
- conduct heat from the engineered barriers (*i.e.* keep near-field temperatures low),
- filter fine particles and colloids that may form during waste package reactions,
- provide a low-hydraulic conductivity, diffusional-transport barrier between the waste form and rock,
- retard the diffusional-transport rate of many dissolved radionuclides,

- chemically buffer the composition of intruding groundwater.

The first four of these functions are achieved by physical properties of the backfill; plasticity, bearing capacity, thermal conductivity and filtration coefficient. The hydraulic conductivity, filtration, retardation and chemical buffering functions are described in detail in Chapter 4.

Swelling smectitic clays, notably sodium- and calcium-montmorillonite, display favourable characteristics for all these properties. Because these clays swell upon contact with water, compacted blocks of montmorillonite have extremely low hydraulic conductivities and are effective barriers to advective flow. In addition, swelling of the clay acts to self-heal gaps and to penetrate small fissures in the adjoining rock. Cement grouts and concretes also meet many of the functional requirements for backfills and for certain cementitious wastes, provide a more chemically compatible near-field environment.

Volatile species (*e.g.* H_2, CH_4) may be generated within the backfill from container corrosion, microbial degradation of organic materials and, to a lesser extent, radiolytic decomposition of water and organic materials (NEA, 1992; also, see Chapter 4). For low permeability, saturated backfill materials, generation of such volatiles may lead to the formation of a free gas phase within the backfill. This free gas phase may, in turn, physically disrupt or impede the performance of the backfill. As described in Chapter 4, certain repository designs incorporate gas vents/locks to prevent the localized build-up of generated gases.

For repositories located in unsaturated rock or in the vadose zone near the surface, a highly conductive backfill may be incorporated into the design. The function of such a barrier is to divert infiltration of surface water or episodic flows of water (*e.g.* rainfall, flash flooding) which might otherwise come into contact with the waste packages. These conductive barriers are shown in more detail later in Section 3.4.2.

3.3.4 Shaft seals/grouts

Zones of high permeability/flow that are intersected by underground tunnels and shafts may be sealed to prevent formation of preferential pathways for radionuclide migration. In addition, exploratory boreholes used to develop three-dimensional characterisation of the repository site might be expected to penetrate the repository rock volume, although not the underground excavations themselves. Such boreholes also must be sealed to prevent them from becoming preferential pathways for release to the surface. Desirable performance characteristics for sealants include:

- low hydraulic conductivity,
- tight bonding to the host rock,
- long-term physical and chemical stability under repository conditions,

- resistance to erosion/failure under high hydraulic heads.

A variety of sealing materials have been studied, including cementitious materials, polymers, resins, chemical grouts, clay and fracture-filling mineral precipitates, ceramics and metals. Performance, cost, type of host rock and techniques for emplacement are key factors in the overall analysis and selection of sealants. At this time, cementitious materials and clay materials are the leading candidates for sealants. Intersected zones of weak rock may need strengthening to assure that underground excavations remain open. Grout is the most commonly proposed material for this function.

Both bedded salt and salt dome formations are under investigation for disposal of various types of nuclear wastes in several countries (IAEA, 1993). In repositories located in salt, underground excavations will be backfilled with crushed salt. In addition, low permeability dams may be emplaced along the length of underground excavations. The purpose of these dams is to minimize possible convective flow of water through the relatively high porosity of crushed-salt backfill. This convective flow might arise from rapid inital influx of brine followed by the compaction of the salt backfill from the inevitable convergence of the surrounding rock salt.

3.3.5 Natural barriers

Natural 'barriers' are derived from the conditions and features of the geological formations in which a repository is to be constructed. A number of types of geological formations have been investigated regarding their suitability for a nuclear waste repository (Chapter 5). These include generic 'granites' (also called 'crystalline' rock), salt domes, bedded salt formations, sedimentary formations (particularly mudstone or shale), basalt, tuff and unconsolidated ocean sediments. It is required that the portion of such formations within which the repository will be constructed maintains long-term stability and mechanical integrity. Other favourable characteristics of geological formations can be categorised by hydrogeological and chemical functions that are conducive to safe, long-term isolation of nuclear waste.

Categorisation by hydrogeological function or conditions includes:

- formations having limited access of water at 'dry' sites (*e.g.* salt formations, sites located above the water table) or,
- hydrogeologically-saturated formations in which radionuclide migration is restricted to diffusion-limited mass transport of nuclides (*e.g.* mudstone or other impermeable sedimentary formations), or
- saturated geological formations in which migration and dispersal of released radionuclides will be controlled by flow in fractures (*e.g.* basalt, granite and other igneous rocks).

For repositories within such saturated geological formations, additional favourable conditions include long-flow times to reach the biosphere and additional retardation of migrating radionuclides by the rock matrix.

Alternatively, geological formations may be categorised by their chemistry, particularly their capability and capacity for buffering the pH and Eh (redox potential) of co-existing pore waters (see Chapter 5). As will be discussed in Chapter 4, both containment and controlled-release performance of engineered barriers, nuclear waste forms and radionuclides are sensitively related to these chemical variables. In particular, the solubility of many radioelements are extremely low under low Eh and/or high pH conditions. Furthermore, retardation of radionuclides within the geological formations is strongly affected by these same chemical variables. Division of geological formations by chemistry includes:

- rock types that actively impose low Eh, typically controlled by the abundance of reactive ferrous (Fe^{2+})-bearing mineral phases (*e.g.* basalt, some types of granite) or
- rock types that only moderately control the pore water chemistry at neutral to mildly alkaline pH values and slightly oxidising to slightly reducing Eh values (*e.g.* tuff, mudstone, some types of granite) or
- rock types that impose no control or buffering of pH and Eh (*e.g.* salt domes and bedded salt).

Suitable repository sites can be established based on favourable hydrogeological features, chemical features or both. It is worth noting, however, that the long-term (>1000 years) persistence of favourable chemical features for a given rock type can be more reliably predicted than can the long-term persistence of favourable hydrogeological features.

3.4 Basic repository designs

3.4.1 Deep geological repositories

Several designs have been developed for deep (> 50 metres) waste repositories. All are based on the multiple barriers concept including an engineered barrier system (EBS) which is emplaced within a suitable geological formation. The EBS is emplaced within underground excavations that are connected to the surface via vertical or inclined shafts (Figure 3.3). Figure 3.4 diagrammatically shows all the engineered and natural barrier components within an integrated repository system.

Three principal designs for a repository can be identified:

- in-tunnel emplacement,
- borehole emplacement,
- cavern emplacement.

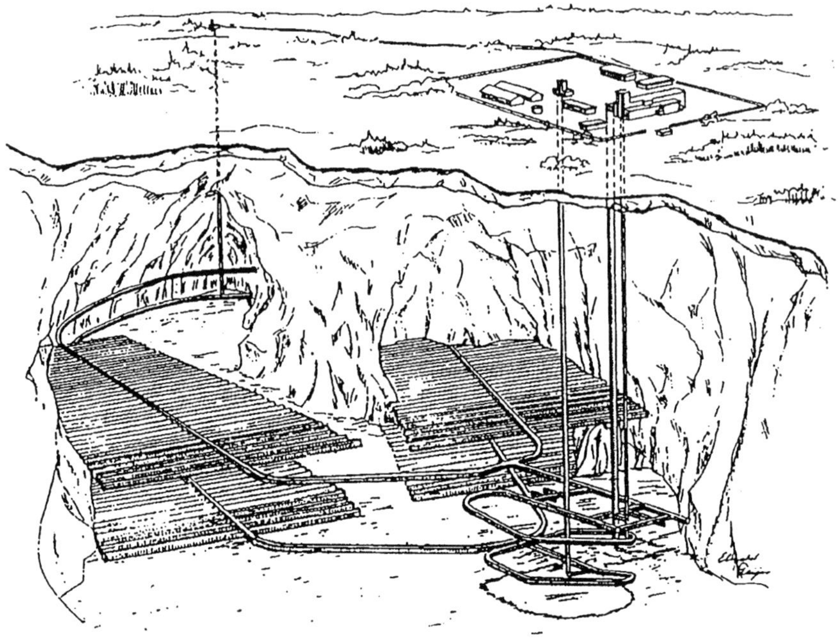

Figure 3.3 Artists view of tunnels and shafts in the final repository for spent fuel (SKB, 1983). Reproduced by permission of SKB.

For in-tunnel emplacement (shown schematically in Figures 3.5 and 3.6), waste containers are emplaced in tunnels excavated in the rock. As each waste container is emplaced, the container is backfilled and the tunnel sealed. Alternatively, borehole emplacement (Figure 3.7) involves excavating additional boreholes within the floor or walls of tunnels. One or more waste containers would be loaded into these holes and backfilled. For repository closure, the tunnels themselves would be backfilled and sealed. Geometric variations on these two types of emplacement are shown in Figure 3.8.

A different option is cavern emplacement, in which multiple waste containers or boxes are stacked together within one large silo of engineered barriers. In the 'WP-cave', concept (Figure 3.9), HLW containers are placed within a radial set of tunnels connected by inner and outer shafts. Individual containers are not specifically backfilled. This central storage excavation would, in turn, be surrounded by a thick (several metres) bentonite-sand backfill. Outside this flow barrier, a large number of vertical boreholes would be drilled to form a hydraulic cage. This cage is expected to conduct most of the water past the central storage excavation, thereby reducing the local hydraulic gradients.

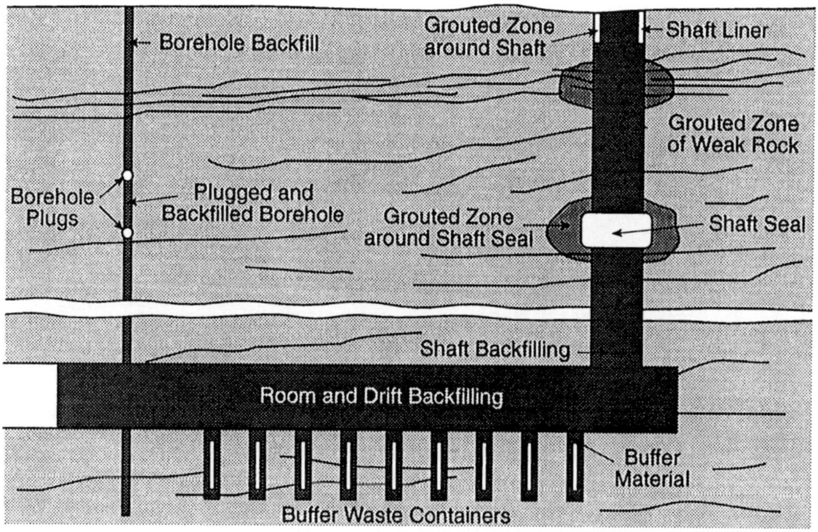

Figure 3.4 Engineered barriers and components of a nuclear fuel waste disposal vault (Cheung *et al.*, 1991). Reproduced by permission of Materials Research Society.

Figures 3.10 and 3.11 show moderately deep (~ 50 metres) silo repositories for the disposal of low- and intermediate-level wastes (L/ILW). A common design characteristic is the gas vents/gas locks to permit the escape of gases which may be generated from container corrosion and microbial degradation. A horizontally configured L/ILW repository emplaced into the side of a mountain (Figure 3.12) also incorporates gas vents as a design element.

The suitability of each of these repository types (or variants of these types) would be based upon the waste type to be disposed, as well as the thermal and mechanical properties of the geological formation selected for disposal. In particular, the in-situ stress state of the rock, the response of the rock to thermal heating from HLW (see Chapter 4), the regional hydraulic gradients and the occurrence and flow characteristics of fracture zones may set important constraints on the design. The need for (and cost of) drilling or boring tunnels, excavation supports, ventilation of underground workings, shield transportation and emplacement machinery, removal of mined rock and waste packaging (containers, backfill) also impact upon repository designs. Examination of these factors is, however, outside the scope of this book.

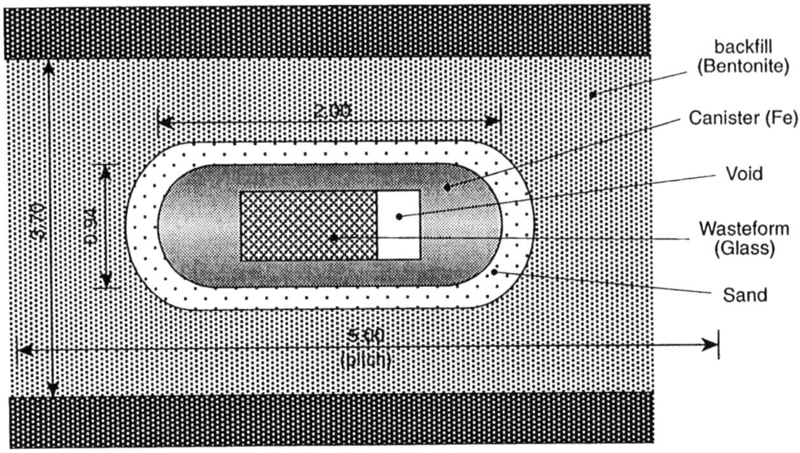

Figure 3.5 Diagrammatic representation of the Nagra (Switzerland) HLW near-field (from McKinley, 1991). Reproduced by permission of Nagra.

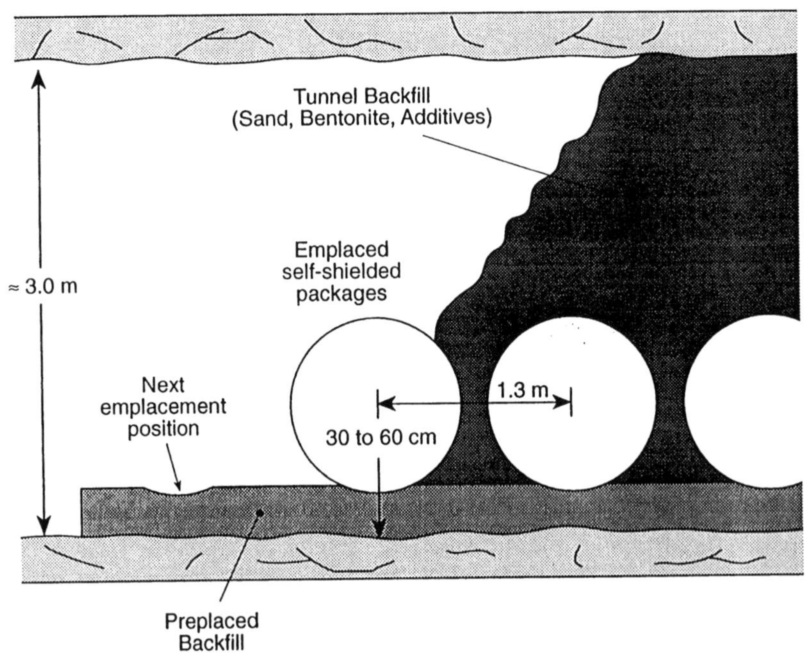

Figure 3.6 Possible emplacement configuration for USDOE (USA) HLW waste package design concept.

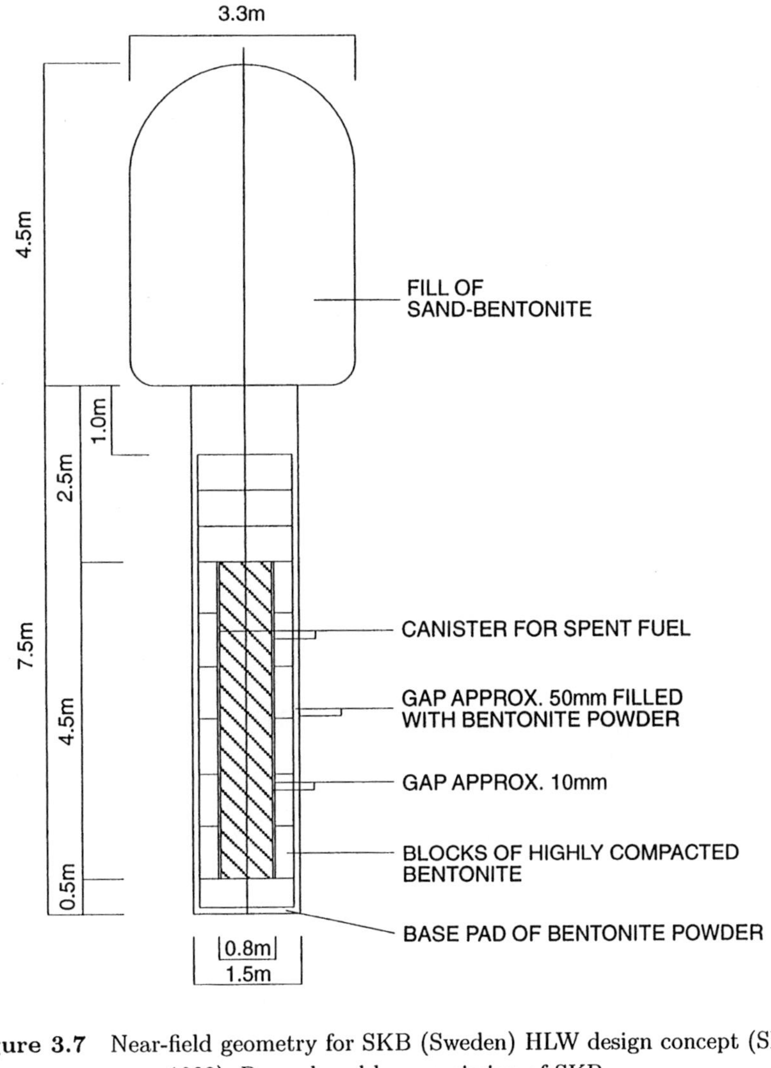

Figure 3.7 Near-field geometry for SKB (Sweden) HLW design concept (SKB, 1983). Reproduced by permission of SKB.

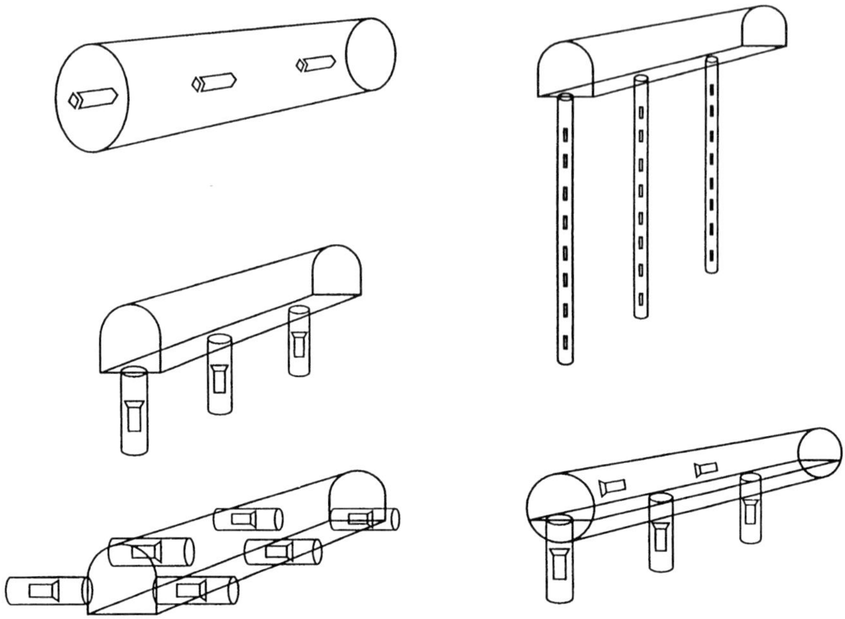

Figure 3.8 Geometric variations for waste package emplacement. From Hara, K. *et al.*, *Proceedings of 2nd International Conference on High-Level Radioactive Waste Management,* 1991, pp 347-354. Copyright 1991 by the American Nuclear Society, LaGrange Park, Illinois, U.S.A.

3.4.2 Near-surface/shallow land disposal

Near-surface disposal of LLW is being investigated and implemented in several countries. Early designs incorporated a simple soil 'cap' or cover over shallow trenches/bunkers containing the encapsulated wastes (see Figure 3.13). Especially in humid areas, these soil covers have generally proven less than satisfactory, often serving as the principal pathway for water entry into the waste. Reduction of water infiltration rate has become the principal focus of engineered barriers for near-surface disposal. Three basic types of barriers, or covers, are being investigated:

- resistive-layer barrier,
- conductive–layer barrier,
- bioengineering management.

The resistive-barrier concept (Figure 3.14) is based on a low permeability layer such as compacted clay. This low permeability assures low flow rates of water through the layer that can be monitored.

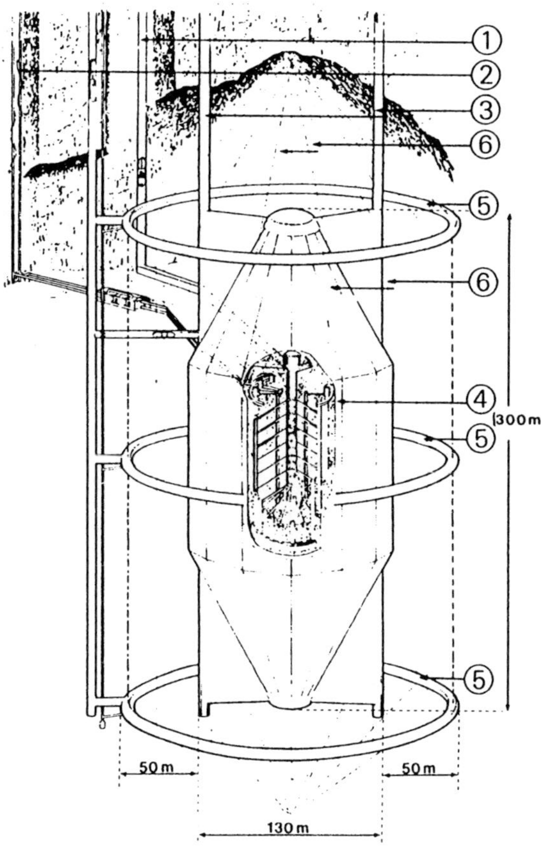

1. *Transportation shaft*
2. *Ventilation shaft*
3. *Main shaft for excavation and refilling of slot*
4. *Bentonite-quartz barrier with thickness of 5 m*
5. *Drift for hydraulic cage*
6. *Drillhole for hydraulic cage*

Figure 3.9 Artist's view of a WP-Cave structure. The dimensions of the bentonite-sand barrier are 300 metres in height and 130 metres in width (SKB, 1989). Reproduced by permission of SKB.

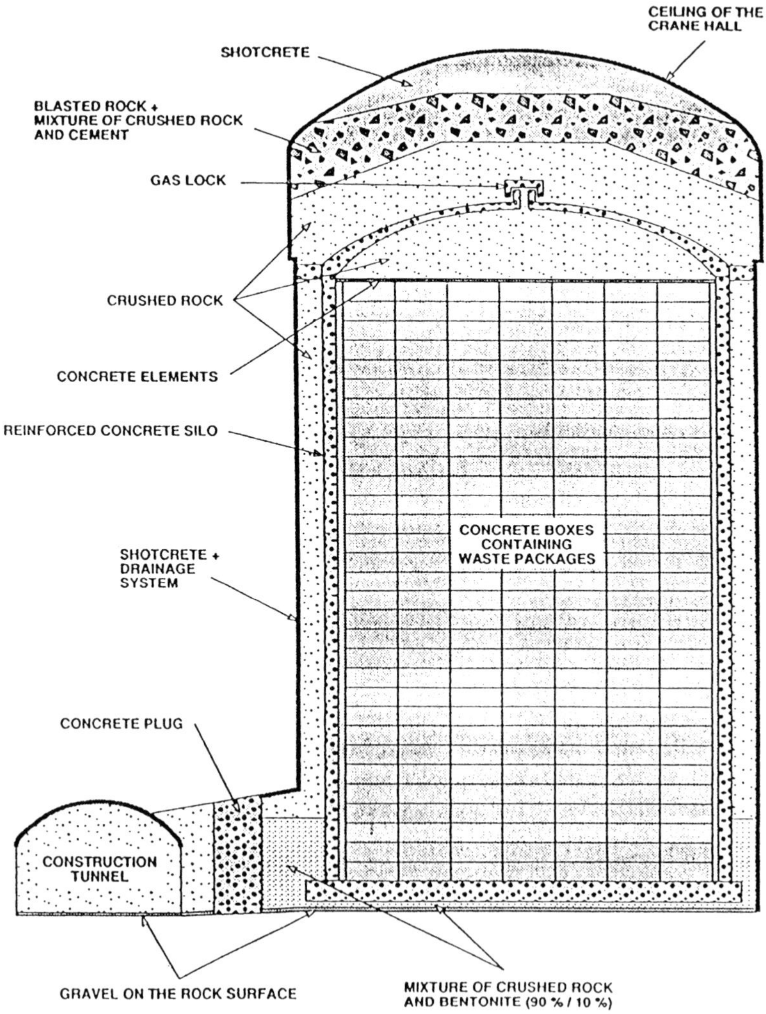

Figure 3.10 Schematic of Finnish VLJ Repository for ILW (NEA, 1992). Reproduced by permission of NEA/OECD.

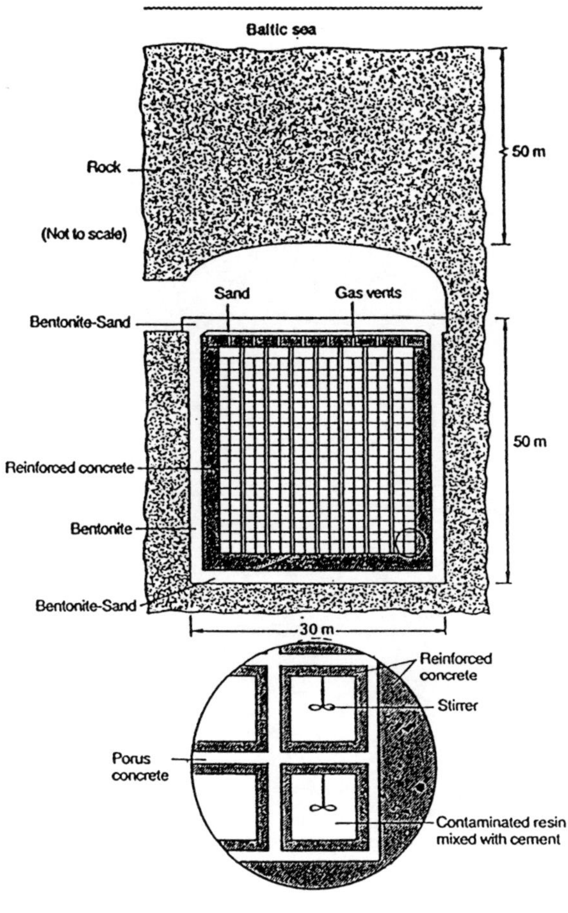

Figure 3.11 Schematic of Swedish SFR Repository for L/ILW Disposal (NEA, 1988). Reproduced by permission of NEA/OECD.

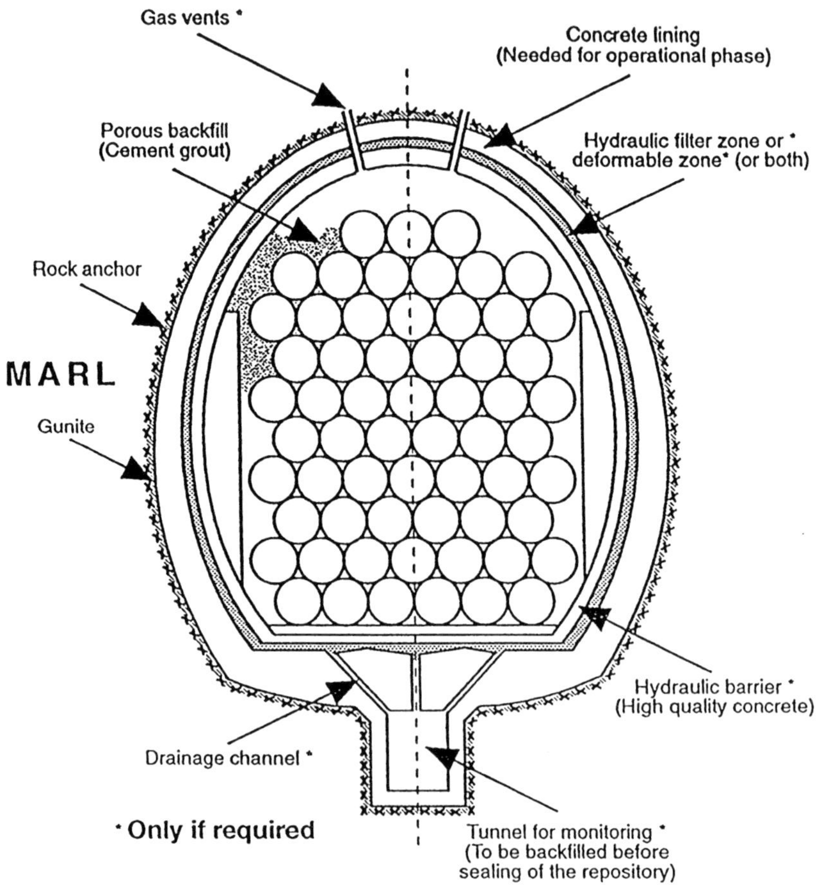

Figure 3.12 Schematic of Swiss repository for L/ILW (NEA, 1988). Reproduced by permission of NEA/OECD.

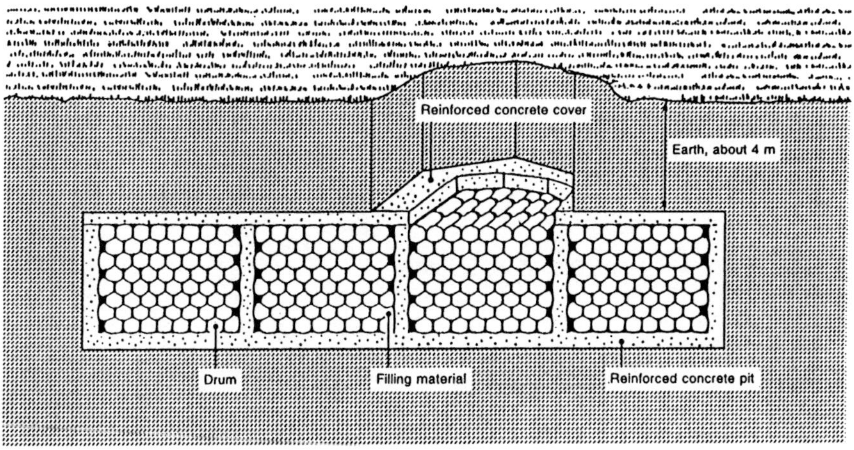

Figure 3.13 Conceptual view of reinforced concrete pit, Rokkasho repository, Japan (from IAEA, 1993). Reproduced by permission of IAEA.

The conductive barrier concept makes use of capillary forces to divert water away from and around the waste (Figure 3.15). A highly permeable layer, such as sand, is emplaced between the overlying rock and the wastes. Infiltrating water is diverted into this conductive layer. The diverted water is at all times at negative capillary potential or under tension and there can be no leakage of water across this zone to contact the waste.

The bioengineering management approach incorporates a combination of engineered geometry to enhance run-off with moisture-stressed vegetation to control deep water percolation through the disposal cover. This is shown schematically in Figure 3.16.

Various combinations of all these barriers and covers can be incorporated into highly engineered near-surface disposal designs. Figure 3.17 shows how several engineered barriers have been combined with a disposal mound located above the original ground surface for L/ILW disposal in France (IAEA, 1993). A more elaborate example is shown in Figure 3.18 (Schulz *et al.*, 1994). The combined vegetation and soil cover minimizes the percolation of water to the engineered barriers. The topmost gravel layer serves to divert any episodic event of high flow. The resistive layer acts to limit the infiltration rate of water received by the underlying conductive layer. The conductive layer diverts all this limited water flow around the concrete waste bunker. The gravel capillary-break layer prevents any drips of water from reaching the concrete. As a consequence, the concrete is exposed only to a stagnant film of water which will greatly retard the degradation rates of the concrete and the waste containers.

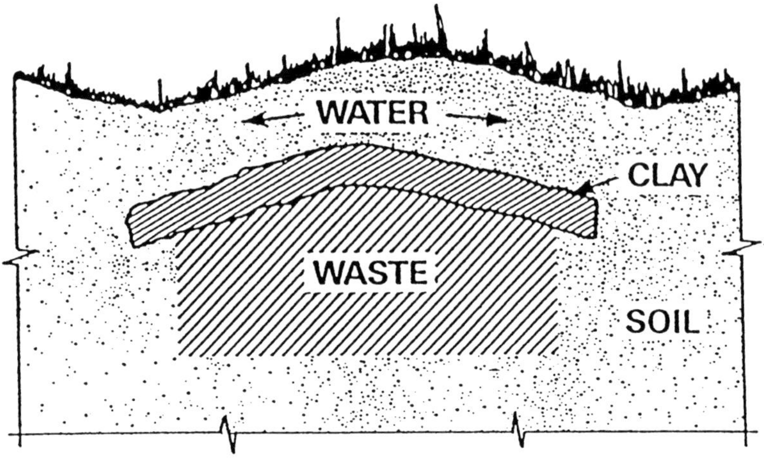

Figure 3.14 Concept of resistive layer barrier (from Schulz *et al.*, 1994). Reproduced by permission of U.S. Nuclear Regulatory Commission.

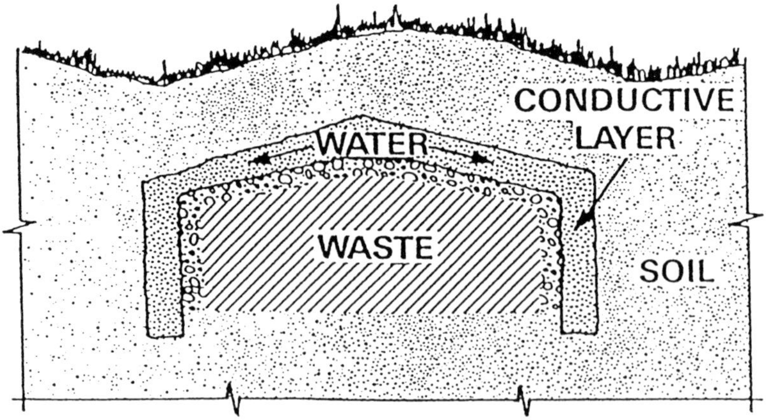

Figure 3.15 Concept of conductive layer barrier (from Schulz *et al.*, 1994). Reproduced by permission of U.S. Nuclear Regulatory Commission.

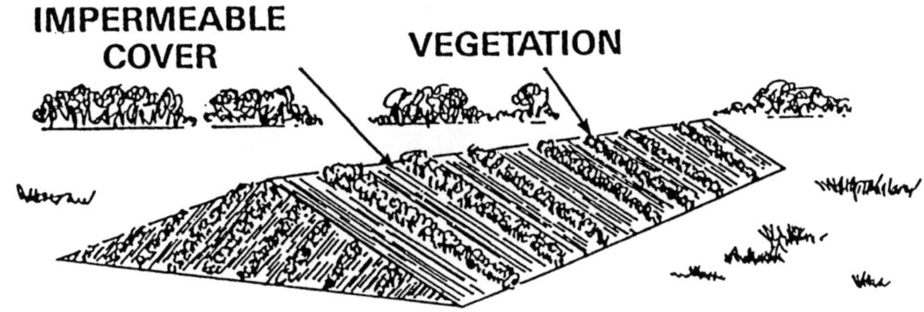

Figure 3.16 Concept of bioengineering management (from Schulz *et al.*, 1994). Reproduced by permission of U.S. Nuclear Regulatory Commission.

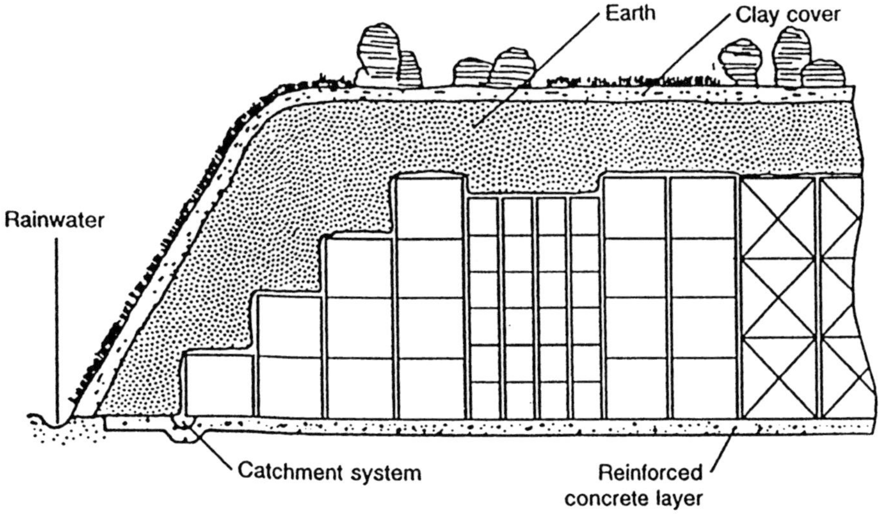

Figure 3.17 Schematic of LLW disposal concept at the Centre de la Manche, France (from IAEA, 1993). Reproduced by permission of IAEA.

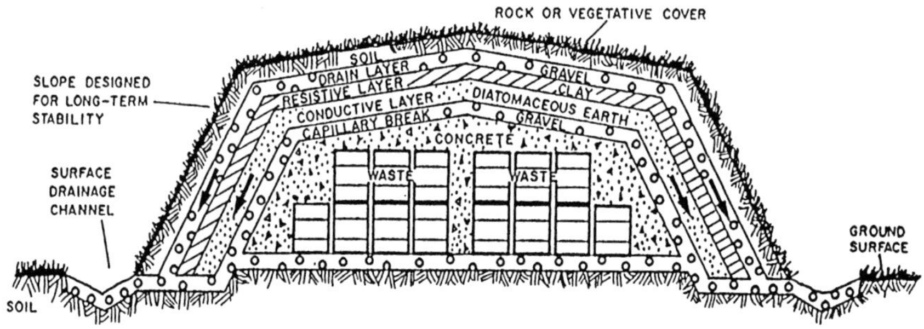

Figure 3.18 Schematic of resistive layer barrier overlying a conductive layer barrier as it might be used with an earth-covered concrete bunker (from Schulz *et al.*, 1994). Reproduced by permission of U.S. Nuclear Regulatory Commission.

3.5 References

Chapman, N.A. and McKinley, I.G. (1987) *The Geological Disposal of Nuclear Waste*, John Wiley & Sons.

Cheung, S.C.H., Gray, M.N., Young, R.N. and Mohamed, A.M.O. (1991) The effects of moisture content, salinity and temperature on the load-bearing capacity of a dense clay-based backfill. In *Scientific Basis for Nuclear Waste Management XIV*, (eds. T. Abrajano and L. Johnson), Materials Research Society, Pittsburgh, USA, 491-498.

Hara, K., Fujita, T., Saotome, A., Uchida, M. and Sasaki, N. (1991) Conceptual design of geological disposal system of high-level radioactive waste management. *American Nuclear Society*, La Grange Park, Illinois, USA, 347-354.

IAEA (1993) Report on radioactive waste disposal. *Technical Report Series* No. 349, International Atomic Energy Agency, Vienna, Austria.

KBS (1983) Final storage of spent nuclear fuel. Volumes 1-5, in KBS-3, Swedish Nuclear Fuel Supply Co./Division, Stockholm, Sweden.

McKinley, I.G. (1991) Integrated near–field assessment. In *Proceedings from the Technical Workshop on Near-field Performance Assessment. SKB Technical Report* 91-59, SKB, Stockholm, Sweden.

NEA (1988) *Near-Field Assessment of Repositories for Low and Medium Level Radioactive Waste.* Nuclear Energy Agency, Paris, France.

NEA (1988) *Feasibility of Disposal of High-Level Radioactive Waste into the Seabed*, 8 Volumes. Nuclear Energy Agency, Paris, France.

NEA (1992) *Gas Generation and Release from Radioactive Waste Repositories.*

Nuclear Energy Agency, Paris, France.

Schulz, R.K., Ridky, R.W. and O'Donnell, E. (1994) Control of water infiltration into near surface LLW disposal units – progress report on field experiments at a humid region site, Beltsville, Maryland. *U.S. Nuclear Regulatory Commission Report* NUREG/CR-4919 Volume 7. U.S. Nuclear Regulatory Commission, Washington, D.C., U.S.A.

SKB (1989) WP-Cave assessment of feasibility, safety and development potential. *SKB Technical Report* 89-20, SKB, Stockholm, Sweden.

4

The near-field

4.1 Introduction

The near-field, comprising the engineered barrier system of wasteform, backfill and canister (schematically shown in Figure 4.1), provides confidence in the reliability of radioactive waste isolation through:

- **Uniformity** – EBS components can be designed and constructed with well–described, uniform properties;
- **Testability** – evaluation of the performance of individual and assembled components of the EBS can be tested readily over a wide range of controlled conditions;
- **Design Flexibility** – components of the EBS can be designed to provide acceptable performance over a wide range of conditions.

Together, these factors provide the capability to demonstrate a *robust* EBS performance with high *predictive reliability* for the near-field sub-system. *Predictive reliability* is the assurance that the actual performance of the system (or its components) will be as good or better than the predicted performance. These favourable factors for the EBS do not obviate the need for performance evaluation of natural barriers. They do, however, argue for a balanced isolation strategy based on both sub-systems.

The purpose of this chapter is to identify and characterise the processes that affect the performance of the near-field sub-system. Particular emphasis will be placed on the calculation of a 'source-term', or release rates of radionuclides, from the EBS to the geosphere. General design and functions of the EBS have been briefly outlined in Chapter 3. The dominant processes affecting the isolation of nuclear waste within the near-field are described here. Based on these processes, conceptual and mathematical models for predicting the source-term release of radionuclides from the near-field to the far-field are described. Finally, performance assessment models and codes that have been developed to model the behaviour of the EBS and near-field are reviewed and summarised.

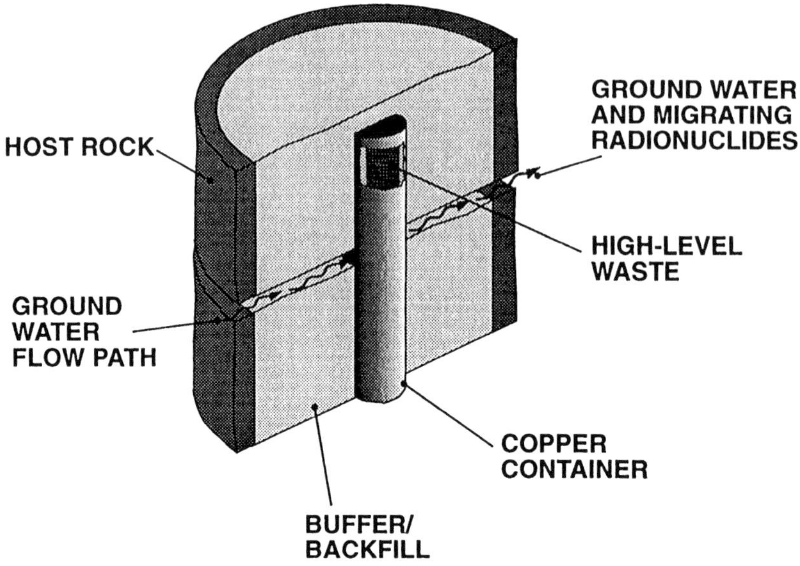

Figure 4.1 Schematic diagram of an engineered barrier system (EBS) for radioactive waste isolation.

4.2 Design and function of the EBS

As outlined in Chapter 3, an EBS is designed to:

- contain radionuclides for an initial period following emplacement of nuclear waste in a repository (*containment period*), and
- assure acceptably low release rates of radionuclides to the far-field following the loss of containment (*controlled-release period*).

The *containment period* is met by encapsulating the waste in an inert or slowly corroding metal or ceramic container. A surrounding layer of clay backfill or concrete liner may be included in EBS designs. This has the primary performance function to stabilise mechanically the waste containers and to prevent groundwater flow through the emplacement hole. In addition, this surrounding backfill or liner material will chemically react with and change the composition of the groundwater that intrudes into this region.

Even with good engineering control, however, containment will eventually fail and groundwater will contact the waste form. Demonstration of acceptably low releases of radionuclides from the EBS during the *controlled-release period* is typically achieved by either:

- limited access of water at 'dry' sites (*e.g.* salt formations, sites located above the water table) or
- diffusion-limited mass transport of nuclides through a layer of low-permeability backfill material for groundwater saturated sites.

The primary focus here is on quantitative calculations for controlled release ('source-term') of radionuclides from the near-field to the far-field. This topic has recently been reviewed as part of an International Workshop on near-field performance assessment (NEA, 1993).

The ability of the near-field, and the EBS in particular, to isolate nuclear waste safely can be divided into four categories of processes:

- Environment
- Containment
- Release
- Transport

Performance assessment modelling of the near-field is based on the modelling of these processes, as well as the linkage of these processes together. The following sections briefly describe the processes involved in each of these categories.

4.3 Near-field environment

Nuclear waste will be emplaced in excavated underground or near-surface workings. The site for waste emplacement will have undergone considerable site characterisation, including determination of:

- thermal,
- mechanical,
- chemical, and
- hydrogeological characteristics.

Depending on the type of excavation and nuclear waste to be disposed, waste emplacement will perturb the ambient characteristics. Furthermore, the near-field environmental conditions will continue to evolve after emplacement, affecting the long-term performance of the near-field. Because of this, considerable resources have been devoted to experiments and modelling to evaluate these perturbations. In many instances, the changes among these environmental conditions are linked or coupled. Coupling means that perturbations in temperature, for example, may lead to additional changes in chemical, hydrogeological, and/or mechanical conditions. In this section the important conceptual features in modelling thermal-chemical-hydrogeological-mechanical conditions of the near-field are qualitatively described with emphasis on deep geological repositories.

4.3.1 Thermal conditions

Temperature, and associated temperature gradients, are particularly important parameters in near-field performance assessment, particularly for heat-producing nuclear waste. This is because changes in hydrogeological, mechanical, and chemical conditions will be sensitively coupled to spatial and temporal variations in temperature.

Representative temporal pressure-temperature (P-T) histories of the EBS for HLW in a hydrogeologically saturated site and an unsaturated (*i.e.* above the water table) site are shown in Figure 4.2. The P-T trajectory of a typical (30 °C/km) geothermal gradient within the Earth is also included for comparison. The initial, post-excavation pressure in the near-field will be equal to the atmospheric pressure of 0.1 MPa (1 bar). There will be a region of transitional pressures between the atmospheric pressure of the excavated emplacement hole and the hydrostatic pressure within undamaged rock. Emplacement of HLW may locally increase rock temperature to >100 °C, and water within the depressurised region will vapourise. Assuming ventilation of the underground workings, vapour-only conditions may be maintained within the EBS, as well as the depressurised portion of the surrounding rock.

Following repository closure at a hydrogeologically saturated site above 100 °C, the fluid pressure within the depressurised region of the near-field and EBS will increase as re-intruding groundwater continues to be vapourised. Eventually, the increasing fluid pressure reaches the liquid-vapour curve, and fully saturated conditions under hydrostatic pressure and elevated temperature will be obtained. The timing for this process depends on a number of factors, including heat-generation rate and permeability of the surrounding host rock. For an unsaturated site, the pressure remains at about 0.1 MPa; the decreasing heat-generation rate causes a gradual decrease in temperature until the liquid-vapour curve is encountered at approximately 100 °C and 0.1 MPa.

There may be three important consequences of groundwater vapourisation. The first is that aqueous transport of radionuclides is effectively prevented during this initial period. Second, the chemical composition of the water that does eventually enter the EBS may be significantly altered from the ambient composition. Third, mineral precipitation may occur at the vapourisation front within the near-field rock, resulting in changes of porosity, hydraulic conductivity, and groundwater chemistry.

There are three fundamental modes of heat transfer; conduction, convection, and radiation. Repository heat transfer is normally dominated by conduction, but may involve significant contributions from convection by groundwater and radiation, depending on site-specific hydrogeological conditions and EBS design considerations. For unsaturated conditions, the loss (or gain) of latent heat from vapourisation (or condensation) of water can also significantly affect

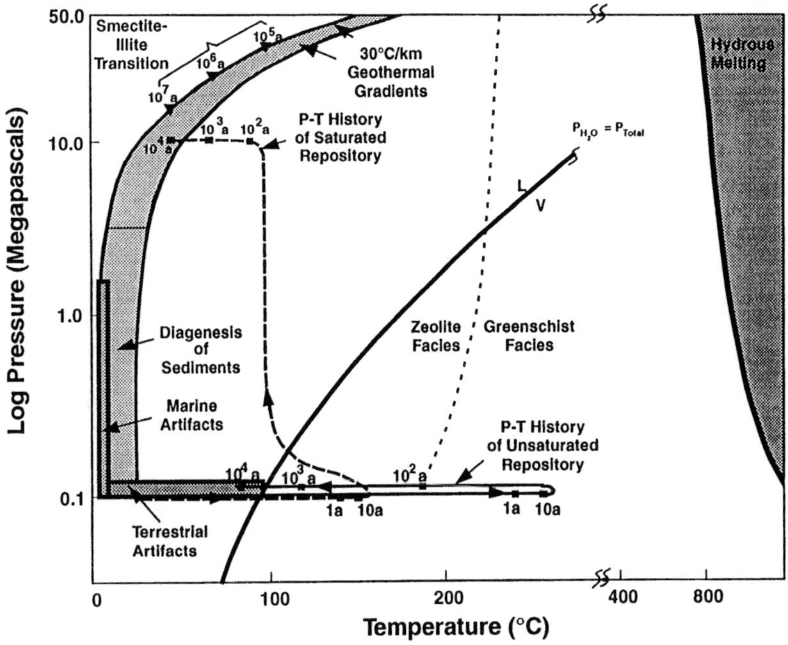

Figure 4.2 Pressure-Temperature (P-T) diagram showing conditions within a HLW EBS over time. P-T trajectories shown in years (a).

heat transfer between the waste container and surrounding engineered and natural barrier materials.

Repository heat-transfer calculations are typically performed at a variety of scales; for an individual waste package, for an emplacement drift (or panel) of waste packages, and for the entire repository system. Scaled analyses are needed to evaluate fully the impact of heating on hydrogeological, mechanical, and chemical conditions. These conditions can be expected to vary spatially and temporally as more and more of the rock mass with heterogeneous properties is included in the heat-transfer calculations. Figure 4.3 portrays the thermal evolution in the near-field for a representative repository containing reprocessed HLW glass.

4.3.2 Mechanical conditions

The mechanical conditions of the near-field rock, including induced stresses and structural damage, will also be perturbed by nuclear waste emplacement in deep geological formations. Excavation techniques create damaged rock zones of variable extent around the rock openings. Excavation also produces

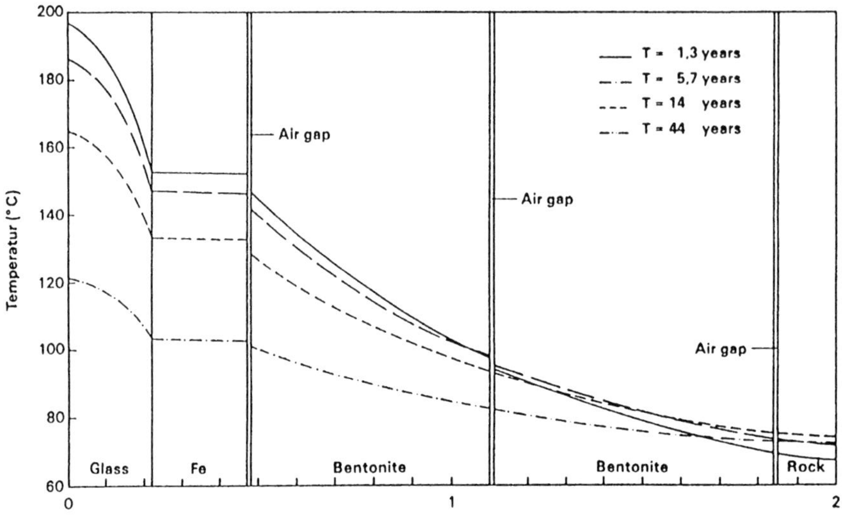

Figure 4.3 Temperature profiles in the near-field of a HLW repository in hard crystalline rocks. The profiles (at various times after disposal) are calculated for the glass waste form, its iron container, the bentonite buffer in the disposal tunnel, and the host rock (Nagra, 1985). Reproduced by permission of Nagra.

a concentration in the mechanical tangential stress around the boundary of the openings. Temperature increases arising from heat-producing waste will give rise to a thermal tangential compressive stress. Changes in rock stresses are shown graphically in Figure 4.4 for a representative underground excavation. Depending on the elastic, expansive, and strength properties of the surrounding rock, as well as *in situ* stresses and formation of cracks/fractures, the openings may remain intact, creep, or fail. Failure of the rock could compromise or reduce the life expectancy of the nuclear waste containers.

In addition to changed rock stresses induced by waste emplacement, EBS materials in a deep geological repository will be subjected to loads ranging between hydrostatic and lithostatic pressures. Hydrostatic pressure is the pressure exerted on the system by a hypothetical column of water equal in height to the depth of the system, and the lithostatic pressure is the analogous pressure exerted by a column of rock of the same height. The more relevant pressure that affects chemical interactions between the groundwater and natural or engineered barrier materials is the fluid pressure. The fluid pressure (P_f) within the near-field will evolve over time and is sensitively related to temperature and degree of hydrogeological saturation in the host

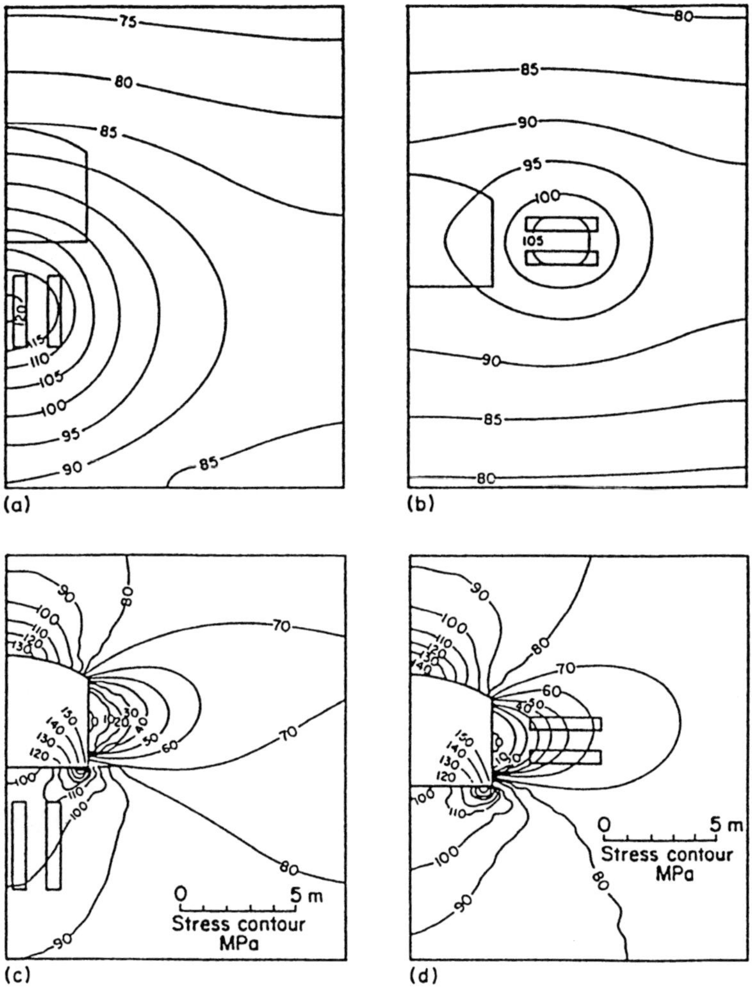

Figure 4.4 (a) and (b) show the estimated rise in temperature around a high-level nuclear waste room 30 years after emplacement of the waste. Temperature in °C with 5 °C contours. (a) shows the situation for a vertical-borehole concept, (b) describes the horizontal-borehole concept. (c) and (d) show the combined horizontal stress around the repository after the same period. (c) relates to the vertical-borehole concept and (d) to the horizontal-borehole concept. From Roxburgh, I.S., *Geology of High Level Nuclear Waste Disposal*, 1987. Reproduced by permission of Chapman and Hall.

rock.

The final fluid pressure can vary between the hydrostatic and lithostatic pressures, depending on the degree of transfer of the lithostatic load from the rock to the groundwater. Under certain unusual conditions, such as long-range artesian effects, the fluid pressure could conceivably exceed the lithostatic pressure. A positive pressure differential between fluid and lithostatic pressures can be maintained by the repository system up until the yield strength of the rock (typically on the order of 10 MPa) is surpassed. Fluid pressures exceeding this limit will result in fracturing of the rock, and consequent reduction of fluid pressure.

The load on near-field materials, in the absence of strain, will eventually readjust to the lithostatic pressure. The potential difference between fluid and lithostatic pressures will result in anisotropic stresses within the near-field. For a repository located at 1 km depth or less, however, the effect of such non-uniform stresses on solution chemistry is expected to be negligible.

4.3.3 Chemical conditions

Perturbation of the ambient chemical composition of groundwater by emplacement of EBS materials is an important factor in the long-term performance of a nuclear waste repository. Although the EBS may modify certain characteristics of the intruding groundwater, gross characteristics such as salinity may remain unaltered (Chapter 5). Groundwater thus becomes the link between performance assessment of the near- and far-fields. The three most important factors affecting near-field performance are: (1) pH; (2) redox conditions, characterised by redox potential (Eh, volts), or the negative logarithm of the activity of the aqueous electron (pe, dimensionless) or oxygen fugacity (f_{O_2}, pascals); and (3) anion concentrations in the pore water.

Interactions with backfill materials such as clay can cause the pH of intruding groundwater to increase to values between 9 and 10; interaction with concrete can result in even greater pH changes to values up to 13. These alkaline to hyper-alkaline fluids will display significantly different reactivity toward container and waste constituents than the near-neutral pH of ambient groundwaters. Figure 4.5 illustrates changes in near-field pH and Eh as a function of time that have been calculated for hypothetical repositories of HLW in a copper canister surrounded by bentonite, and LLW encapsulated in a mild steel container surrounded by a cement backfill.

Depending on the amount and availability of polyvalent aqueous species (*e.g.* Fe, S, C) in EBS materials, the redox characteristics of intruding groundwaters may also change. This can profoundly affect the performance of the EBS; for example, the solubility of several multi-valent radioelements (Tc, U, Pu) are sensitively related to redox, and the stability and corrosion mechanisms of container material varies as a function of redox and pH.

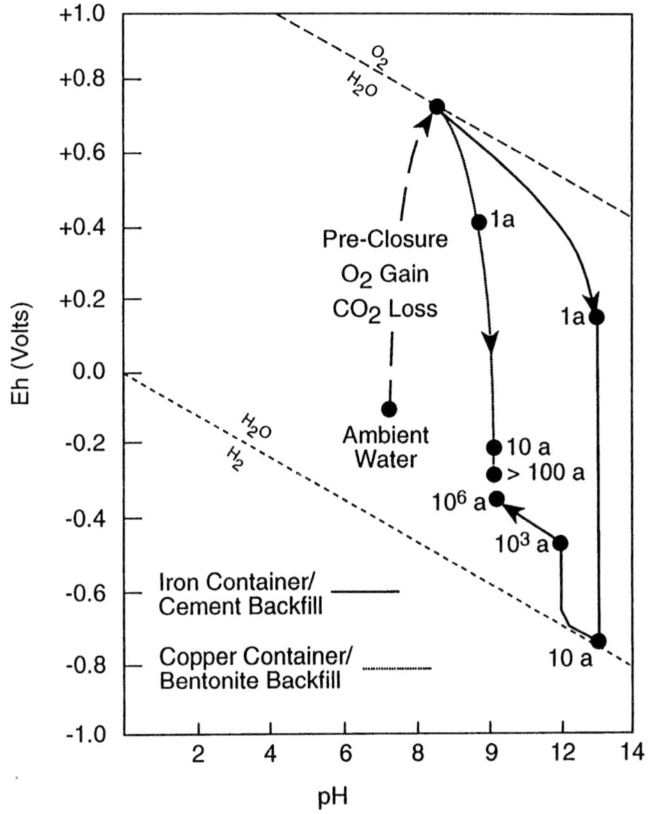

Figure 4.5 Evolution of near-field chemistry.

Lastly, the dissolution of minor solids (carbonates, sulphates, phosphates, organics) contained within EBS materials will result in an initial, local increase in the concentration of dissolved complexants. These complexants can lead to deleterious effects on the performance of the EBS, particularly leading to increased solubility and mobility for dissolved radioelements that will eventually be released from the nuclear waste.

The following sections briefly review special issues with regard to the chemical environment of the near-field.

Bentonite

Bentonite is a common choice of backfill in both HLW and I/LLW disposal concepts. Bentonite consists primarily of calcium or sodium montmorillonite

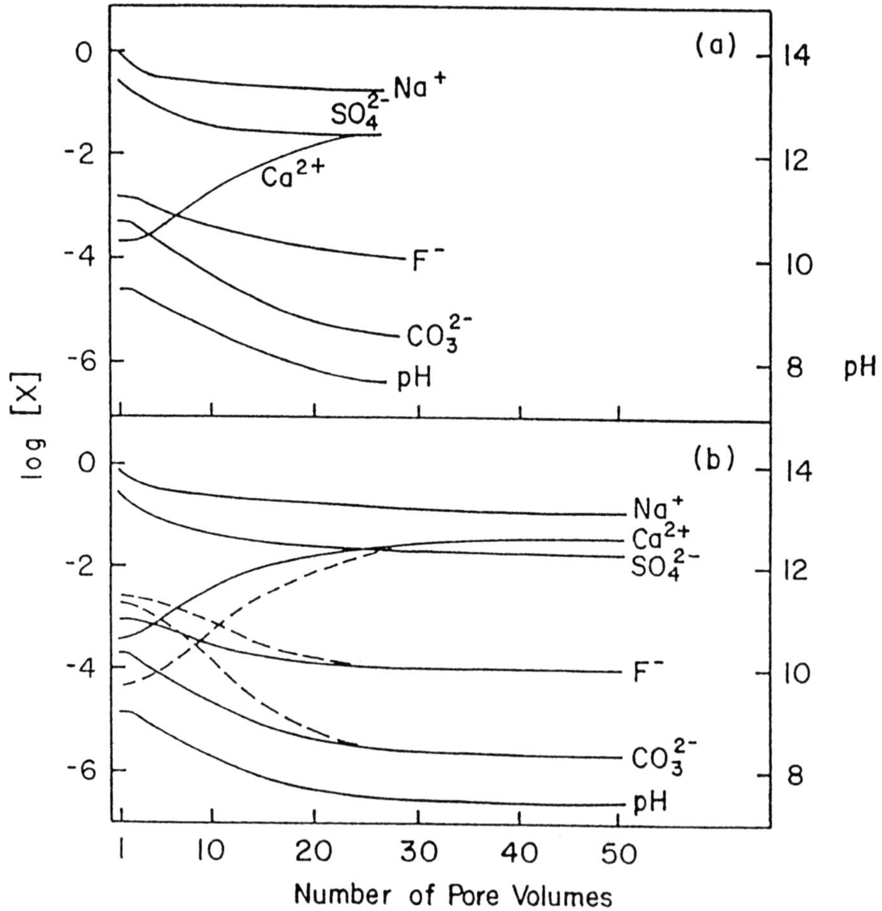

Figure 4.6 Evolution of bentonite pore fluid chemistry on mixing and reaction with ambient groundwater. From Lemire and Garisto (1989). Reproduced by permission of Atomic Energy of Canada Ltd.

clay, with minor (but chemically important) amounts of quartz, feldspar, kaolinite, carbonates, sulphides, sulphates and organic matter. Interaction of bentonite with groundwater is complex, but may be considered to consist of:

- rapid ion-exchange of alkali and alkaline-earth cations,
- relatively slower dissolution of carbonates, sulphates and organics,
- slow dissolution-precipitation reactions of the montmorillonite and other silicates.

In general terms, these reactions lead to an increase in pH and increases in the carbon, sulphur and organic content of the pore fluid in comparison with the ambient groundwater. Although montmorillonite clays are thermodynamically unstable with respect to non-swelling micas and sheet silicates, alteration processes occur exceedingly slowly, on geological timescales. The alteration of the bentonite and the concomitant change in the chemical composition of the pore fluid is controlled by the rate of groundwater throughput, which in turn, is controlled by diffusion (Figure 4.6). Many of the issues regarding clay stability and behaviour are discussed in more detail by Snellman *et al.* (1987), Higgo (1987), Meunier (1993) and Arthur *et al.* (1993).

Cement and concrete

Cement and concrete are composed of calcium hydroxide ('portlandite'), sodium and potassium hydroxides, hydrates of calcium silicate and calcium aluminate, and a siliceous aggregate which may occupy 65-75 % of the total volume (Chapter 2). The dissolution of sodium, potassium and calcium hydroxides and calcium silicate hydrates within the cement leads to the development of a hyperalkaline pore fluid (pH 12.5-13). The elevated pH conditions help minimise canister corrosion and radioelement solubility. The passage of groundwater through the EBS with time will dissolve components from the cement/concrete, sequentially leaching the alkali hydroxides, followed by portlandite then calcium silicate hydrates. Simple modelling of the behaviour of a cement/concrete EBS system by application of typical groundwater flow conditions to be expected at a repository site enables prediction of the evolution of chemical conditions with time (Figure 4.7). The evolution of cement and entrained pore fluids in a repository environment with time has been discussed *inter alia* by Atkinson (1985), Berner (1987) and Reardon (1992). Although cement and concrete are tried and tested constructional materials, there may be concerns with regard to the suitability of this material in a geological disposal environment (see Chapter 5).

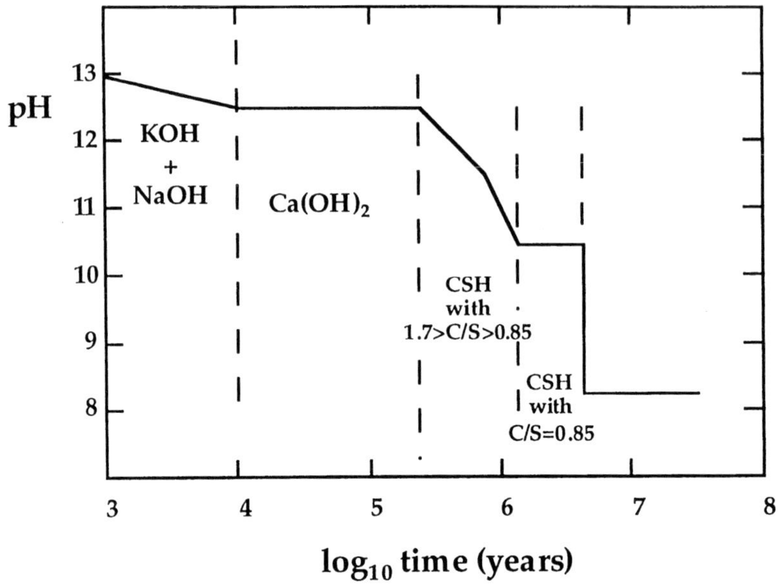

Figure 4.7 Estimated evolution of cement pore fluid chemistry in a repository of diameter 20 m containing 185 kg m^{-3} cement situated in a groundwater flux density of 10^{-10} m s^{-1} at 25 °C. After Atkinson (1985). Reproduced by permission of AEA Technology.

Temperature

Thermal conditions may have an indirect effect upon near-field chemical conditions. The temperature dependence of the dissociation constant of water tends to decrease the pH of near-field pore fluids with increasing temperature. 'Neutral' pH of pure water at 25 °C and atmospheric pressure is 7, but is 6.1 at 100 °C. Since the solubilities of silicates, hydroxides and carbonates are also temperature dependent, and these solids may act as near-field pH buffers, then calculation of *in situ* pH may be complex. For example, the solubility of portlandite, $Ca(OH)_2$ (an important pH buffer in cement and concrete) decreases with increasing temperature. The pH of a fluid saturated with portlandite at 25 °C is 12.5, but will decrease to 11.1 at 70 °C (at 0.1 MPa). The precise conditions of geothermal gradient, waste heating and depth of burial will therefore have a large effect upon *in situ* near-field pH, and may be considerably different from earth surface temperature and pressure.

Pressure

As previously noted in the analysis of fluid pressure evolution within the near field, groundwaters will initially flow into the depressurised region of the EBS and surrounding damaged rock. Depressurisation of the water may cause degassing of the groundwater, particularly CO_2. This may result in attendant precipitation of calcite ($CaCO_3$) in open pore space. In addition, dissolved ferrous iron (Fe^{2+}) will be quickly oxidised to insoluble amorphous oxides and hydroxides within this transitional region of the near-field. Coupled with potential dissolution/precipitation of silica compounds (SiO_2) such chemical changes arising from temperature and pressure gradients in the near-field may, in turn, lead to changes in hydrogeological properties (porosity, hydraulic conductivity) within the near-field.

Organics

The large amounts of organic materials in LLW (paper, rubber, plastics, textiles), and to a certain extent in ILW (ion exchange resins) pose additional problems to those associated with the disposal of HLW. Microbially-mediated degradation of these organic materials with time under repository pressure-temperature conditions will generate significant amounts of organic acids, carbon dioxide, methane, water and hydrogen sulphide. Organics may influence radionuclide mobility in four ways:

- complexation with cationic radionuclides,
- generation of colloidal organic material upon which radionuclides may sorb,
- change the chemical environment through redox reactions with polyvalent elements,
- generate gas via degradation reactions.

The most important organic component of LLW is probably cellulose, a polysaccharide or condensation polymer of sugar. Degradation of cellulose proceeds through hydrolysis, which liberates various monomeric alcohols:

$$[C_6H_{10}O_5]_n + nH_2O \rightarrow nC_6H_{12}O_6 \qquad (4.1)$$

and a variety of microbially-mediated oxidation reactions giving rise to species such as saccharinic acids.

Studies on the effects of cellulose degradation products upon solubility and sorption indicate detrimental effects with regard to actinide mobility (Atkinson *et al.*, 1993). Although these effects are documented, what is less clear is the rate of organic degradation and the potential for the existence of cellulose degradation products in near-field pore fluids over the long term (*i.e.* times greater than 1000 years).

Ion exchange resins and bitumens (used for encapsulating wastes in some concepts) are less reactive than cellulose and are unlikely to be major sources of organic complexants in a radioactive waste repository (Wiborgh *et al.*, 1986).

The removal or destruction of organic materials prior to disposal is thus beneficial and can be achieved by methods such as incineration, wet oxidation, electrochemistry, and biological techniques. Clearly, there are penalties in terms of costs for this waste management option.

Gas

Considerable effort is being expended internationally to understand potential problems associated with the generation of gases in a radioactive waste repository (NEA/OECD, 1992). Gas generation in a waste repository is of concern for the following reasons:

- gases may be generated at rates and volumes sufficient to cause physical disruption of the waste form and backfill,
- groundwater may be expelled from the near-field and could potentially lead to gas-driven groundwater flow in the far-field,
- gases may react chemically with the pore fluid and EBS to degrade performance of near-field barriers,
- gas may de-saturate regions of the repository to promote other reactions,
- radioactivity may be transported from the repository in gaseous media,
- the concentration of methane or other flammable gases may reach levels that make fires or explosions possible.

The principal gas generation mechanisms are the following:

- the anaerobic corrosion of metals, principally steel/iron,
- the degradation of organic materials,
- the radiolytic decomposition of water,
- helium production by alpha decay,
- release of gas from the host rock,
- volatilisation of organic materials.

Metal corrosion mechanisms are well understood (Romanoff, 1989; Section 4.4), but the rates of gas generation are less well known. Anaerobic corrosion of steel will yield hydrogen as a product:

$$3Fe + 4H_2O \rightarrow Fe_3O_4 + 4H_2 \quad (4.2)$$

A typical steel corrosion rate of 0.3 μm/a will yield 1.3 l/a of hydrogen per m^2 of steel surface area. Other metals (zircaloy *etc.*) will also corrode to produce hydrogen, albeit on a minor scale compared to steel (Agg *et al.*, 1993).

Gas may be generated via microbially-mediated organic degradation reactions according to a sequence of energetic favourability (Table 4.1).

Table 4.1 Typical sequence of bacterially-mediated redox reactions, after Stumm and Morgan (1981).

	Reaction	ΔG° (kJ) (pH=7)
(1)	Aerobic respiration $CH_2O + O_2 \rightarrow CO_2 + H_2O$	-502
(2)	Denitrification $CH_2O + 0.8NO_3^- + 0.8H^+ \rightarrow CO_2 + 0.4N_2 + 1.4H_2O$	-477
(3)	Mn(IV) reduction $CH_2O + 2MnO_2 + 4H^+ \rightarrow 2Mn^{2+} + 3H_2O + CO_2$	-340
(4)	Nitrate reduction $CH_2O + 0.5NO_3^- + H^+ \rightarrow CO_2 + 0.5NH_4^+ + 0.5H_2O$	-327
(5)	Fermentation $3CH_2O + H_2O \rightarrow 2CH_3OH + CO_2$	-118
(6)	Fe(III) reduction $CH_2O + 4Fe(OH)_3 + 8H^+ \rightarrow 4Fe^{2+} + 11H_2O + CO_2$	-116
(7)	Sulphate reduction $CH_2O + 0.5SO_4^{2-} + 0.5H^+ \rightarrow 0.5HS^- + H_2O + CO_2$	-105
(8)	Methanogenesis $CH_2O + 0.5CO_2 \rightarrow 0.5CH_4 + CO_2$	-92.9

Predicting the rate at which these reactions may occur is difficult because of their dependence upon a complex interplay of pH, temperature and the nature of the organic material. It is conceivable that 90 % of cellulose in the waste may be converted to carbon dioxide and methane with each kg giving rise to 0.2 kg of methane and 0.66 kg of carbon dioxide. Complete degradation may take several hundred years. Carbon dioxide released during organic degradation will dissolve in the near-field pore water, reducing the pH and thus become available to react with the portlandite component of the cementitious backfill to produce calcite. Similarly, hydrogen may react with carbon dioxide to produce methane. The composition of gas generated by anaerobic degradation of organics (about 50 % methane) can present a flammable hazard in air even when mixed substantially with other gases (flammable range of methane is 5-15 %).

Radioactive gases may be produced where organic materials containing radionuclides volatilise or undergo chemical or microbial degradation. Tritium and ^{14}C may be significant in this respect. Other gases include ^{85}Kr and a wide range of volatile inorganic and organic compounds which may incorporate radionuclides.

Allowances must be made for the escape of gas from the repository by both the installation of gas vents in waste canisters and the use of gas-permeable backfill. For deep disposal of LLW/ILW in the UK, a backfill permeability of 10^{-19} m^2 for a 10 cm diameter vent (or 10^{-18} m^2 for a 1 cm vent) would be sufficient to preclude canister and backfill damage, assuming a maximum gas generation rate of 4 m^3 y^{-1} m^{-3} waste at STP (Nash *et al.*, 1989; Nash and Rodwell, 1990). Typical cementitious backfill materials can withstand internal pressures in the range 0.2-3 MPa before failure (Wiborgh *et al.*, 1986).

Gas migration is considered later (Section 4.6.1).

Microbiology

The presence or introduction of microbes may also impact the near-field chemical environment, notably Eh, pH, and concentrations of anions and organic complexants. In addition to accelerating the degradation of organic materials (discussion above), bacteria may also assimilate CO_2, resulting in the formation of organic compounds. Also, microbially-mediated corrosion of container materials and concrete may occur. Viability of such bacteria may be set by extremes in temperature, salinity, pH, Eh, radiation, and pressure within a repository, or alternatively, nutrient and energy availability could be limiting factors (see West and McKinley [1985] for a review of the importance of microbial activity in the repository environment).

Radiolysis

Radiation from nuclear waste can also affect the chemical environment in the near-field. Alpha (α), beta (β), gamma (γ) and neutron (η) radiation can be produced during decay of radioactive wastes. The α and β particles are not penetrating, and are stopped by relatively thin layers of materials. Their effect on near-field chemistry will be restricted to radiolysis of groundwater that might eventually contact the surface of the waste. Both γ and η radiation are more penetrating, and could interact with engineered and natural barrier materials surrounding the waste. The effect of radiation on material properties is anticipated to be negligible because of low energies imparted during interactions with solids and, for the case of neutron radiation, its low intensity. Radiolysis (either α, β or γ) of water may occur, however, leading to production of a variety of radiolytic radical species. Most of these species are extremely reactive or unstable, and quickly convert to more stable and inert species. Radiolysis leads to production of equal amounts of oxidising and reducing species; however, the relative reactivity of these species can vary considerably. In particular, H_2, a dominant reducing species produced by radiolysis, is far less reactive than most of the co-generated oxidising species. The net effect is that the water becomes a more oxidising reactant with respect to reaction with engineered and natural barrier materials. Radiolysis reactions in groundwater are further complicated by the presence of dissolved 'scavenger' ions, as well as the presence of reactive surfaces of materials. As yet, comprehensive data for analysis of radiolysis in compositionally complex groundwaters do not exist, although there are a number of design and modelling approaches that can be implemented to minimise or eliminate deleterious radiolysis effects on near-field performance.

4.3.4 Hydrogeological conditions

The near-field hydrogeological conditions may be significantly changed due to deep waste emplacement. Under saturated conditions, emplacement of backfill materials with permeabilities lower than the host rock will cause the water to flow around the emplacement hole rather than through it. This simple picture is complicated by the damaged zone of rock that typically occurs near the surface of rock excavations. This damaged zone will have higher permeability than the intact rock, leading to preferential flow in the zone between the intact rock and impermeable backfill. This behaviour is shown graphically in Figure 4.8.

For unsaturated conditions, introduction of backfill materials that have higher permeabilities (a gravel layer or an air gap) will allow capillary suction by the rock to retain water in the pores, preventing contact with the waste container. Prediction of groundwater movement in the near-field for both saturated and unsaturated situations may be further complicated by the natural heterogeneities/anisotropies in the properties of the host rock.

Changes in thermal, chemical, and mechanical conditions may also contribute to changes in the near-field hydrogeology. Creation of a low-pressure/high-porosity region from rock excavation will create the initial perturbations to groundwater flow. Formation of new cracks/fractures within the surrounding rock during excavation will further perturb the near-field flow regime. Thermal buoyancy effects may also induce changes in flow, as could possible changes in porosity attributed to formation of new, secondary minerals attributable to significant temperature gradients induced in the rock.

Of key concern is the saturation of clay layers in backfilled tunnels and around waste packages. Several illustrative studies (*e.g.* Pruess and Bodvarsson, 1982; Andrews *et al.*, 1986; Grindrod and Takase, 1992) have been made, with estimates of resaturation time ranging from 1 to 1000 years. For impermeable host rocks, the log of the resaturation time for backfill is predicted to be inversely proportional to the log of the permeability of the host rock (Pruess and Bodvarsson, 1992), and depends less upon backfill porosity and other factors. Alternately, Andrews *et al.*, (1986) found resaturation times to be a sensitive function of the characteristic capillary pressure versus water saturation relationship of compacted bentonite.

Grindrod and Takase (1992) have developed a conceptual 'capillary bundle' model for resaturation based on the different wetting behaviour of large and small capillaries of compacted clay. They note that compacted clay is not typical of most porous media. The small size of individual platelets means

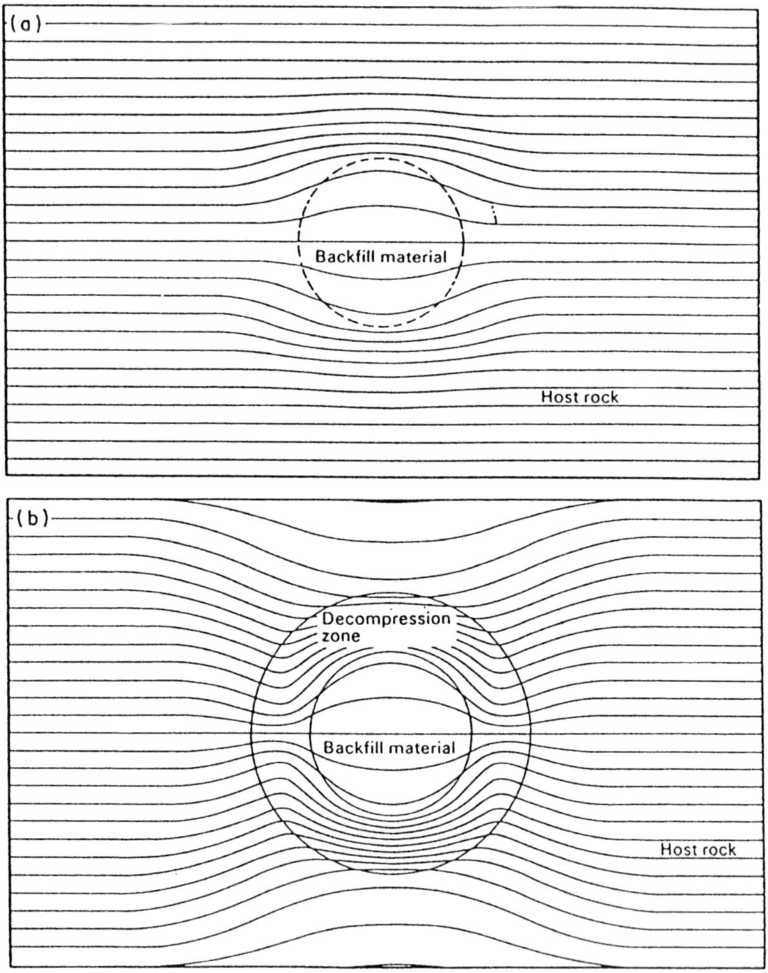

Figure 4.8 (a) Shows theoretical flow lines, in an homogenous isotropic medium, around a repository tunnel containing a low hydraulic conductivity backfill. (b) Shows the same, but with a more permeable decompressed zone in the rock caused by tunnelling (Nagra, 1985). Reproduced by permission of Nagra.

that a porosity of perhaps 20-30 % by volume is distributed through pores measuring less than a few nanometers. Consequently, capillary pressures are large (10-100 MPa) for gas-water interactions in compacted clay. This is due to the strong affinity of clay surfaces for water. In addition, the wetting interfaces will be driven by large potential forces. Large capillaries are more amenable to wetting. Therefore wetting interfaces will initially move most rapidly along them. Conversely, smaller capillaries will be slower to wet, although water will ultimately move along these interfaces as the initial high drag forces are overcome. Thus, initially, wetting is controlled by the most permeable conduits for flow in the clay, and supported by the less permeable, though less-reversibly saturated smaller pores. This will be followed by the desaturation of the most permeable pores in the creation of preferential pathways for the release of the gas phase from within the inner clay-canister boundary. Successful comparisons have been made between the results of this model and laboratory studies of resaturation of bentonite (Grindrod and Takase, 1992).

4.4 Containment

There are three key aspects of containment that affect the ability of the near-field to isolate nuclear waste in a safe fashion. These aspects are:

- mean time to containment failure,
- distribution of containment failure over time,
- mode of containment failure.

Although container failure modes and models are highly material- and design-specific, the effect of containment on subsequent source-term release from the near-field is well recognised. Longer total containment times permit radioactive decay of relatively short-lived radionuclides. For example, strontium-90 and caesium-137 will undergo 10 half-life decays within the first 300 years after disposal. Demonstration of total containment for this duration would effectively reduce the initial inventory by over a factor of 1000. Likewise, a mean time to containment failure of 10,000 years would be effective at reducing the activity of all radionuclides with half-lives less than 1,000 years to negligible values. It should be noted, however, that most radionuclides of concern from a dose perspective have half-lives greater than 10^5 years (*e.g.* ^{129}I, ^{237}Np, ^{99}Tc, ^{135}Cs). Therefore, extended containment beyond 300-1000 years to reduce radionuclide inventories in nuclear waste is not a particularly effective isolation strategy, unless mean time to containment failure approaches 10^6 years or greater.

Even with good engineering control, however, total containment will eventually fail and groundwater will contact the waste form. For a repository composed of multiple waste packages, it is likely that the individual containers of a repository will fail over a range of times. This distribution of containment failure times would arise from differences in local environmental conditions across the repository, as well as slight differences in fabrication of containers. This expected, yet unquantified, distribution in containment failures over time can have a significant effect on the peak release (or dose) rates of radionuclides. Certain long-lived, high-solubility, non-sorbing radionuclides can rather quickly escape from the near-field (as well as the far-field). If all waste containers were to fail at the same time, the resulting superposition of releases from all waste packages could result in an unacceptably high peak release of radionuclides. If, however, there is a distribution of releases attributable to a distribution in containment failures over time, the peak releases of rapidly released radionuclides will be attenuated (*e.g.* SKB, 1983).

Finally, the mode of containment failure can set an important constraint on the subsequent release of radionuclides from the container. Near-field, source-term calculations typically assume that a container is physically removed upon containment failure (although chemically present), a conservative, yet wholly unrealistic assumption. Different modes of containment failure, however, lead to a range of conditions that may significantly affect the subsequent release of radionuclides from the near-field. For example, pitting or other localised corrosion modes may lead to small penetrations of the container (*e.g.* SKB, 1992). Alternatively, the general corrosion mode may lead to the formation of corrosion products that may represent a significant mass-transport or chemical barrier to the release of radionuclides from the waste form to the far-field. Calculations on partially failed containers show that, under certain circumstances, partially failed containers provide a more effective 'barrier' to release of radionuclides than engineered clay layers or impermeable host rock that may surround the waste container. Perhaps even more important, the effect of release through partially failed containers under these conditions effectively de-couples the performance of the engineered barrier system from the hydrogeological regime of the far-field.

4.5 Release

Upon loss of containment, intruding groundwater will contact the nuclear waste forms. Subsequent release of radioelements to the groundwater is controlled by a number of physical and chemical characteristics of the waste form and the radioelements themselves. These characteristics set important limitations on the source-term performance of the near-field that are identical for all nuclear waste forms. These characteristics include:

- radionuclide inventory of waste forms,
- possible inhomogeneous distribution of radionuclides within waste forms,

- variability of characteristics of waste forms,
- chemical constraints on release (waste-form reaction rate, radionuclide solubilities).

These topics are briefly reviewed in the subsequent sections.

4.5.1 Radionuclide inventory

Of prime concern is the abundance of radionuclides contained within the waste form. Information on radionuclide inventories for reference pressurised water reactors (PWR) and boiling water reactors (BWR) have been calculated using codes such as ORIGEN (Roddy *et al.*, 1986) for different assumed burn-ups. A problem with using such sources of information is that radionuclides that occur at minor and trace abundances in fuel are often not reported; while such nuclides may not be of importance for calculation of initial heat generation and radiation fields, they may be extremely important in long-term (> 1,000 year) safety analyses. This problem is compounded by the potential, yet generally unreported, uncertainties in all of the nucleonic calculations leading to the predicted inventories.

For HLW borosilicate glass, the predicted radionuclide inventories are based on ORIGEN-type calculations on expected inventories derived from the reactor history of reprocessed fuel, combined with information on the efficiency in radioelement separations during reprocessing. This information may be checked based on limited measurement of the concentration of major radionuclides in the final waste glass. As for spent fuel, the calculation of radionuclide abundances, as well as confirmation from analytical measurements, will be uncertain.

For LLW, fission product concentrations are low, with extremely low or zero concentrations of actinides. Further categories of non-HLW have been variously developed in many nations, as determined by operational or regulatory criteria. In an important sense, these differences in international definitions of waste types are irrelevant to performance assessment of the near-field. What is important is that every nuclear waste type can, and should, be evaluated by application of the same, general approach to near-field performance assessment.

Chapter 2 presents representative emplacement-inventories for nuclear waste forms. Note that the abundances of individual radionuclides will change over time from radioactive decay and in-growth. The abundances at the time of emplacement will be affected by programmatic decisions regarding temporary surface storage of nuclear waste, as well as by design criteria on thermal loading of containers that may seek to achieve certain thermal conditions for set periods of time within the underground repository.

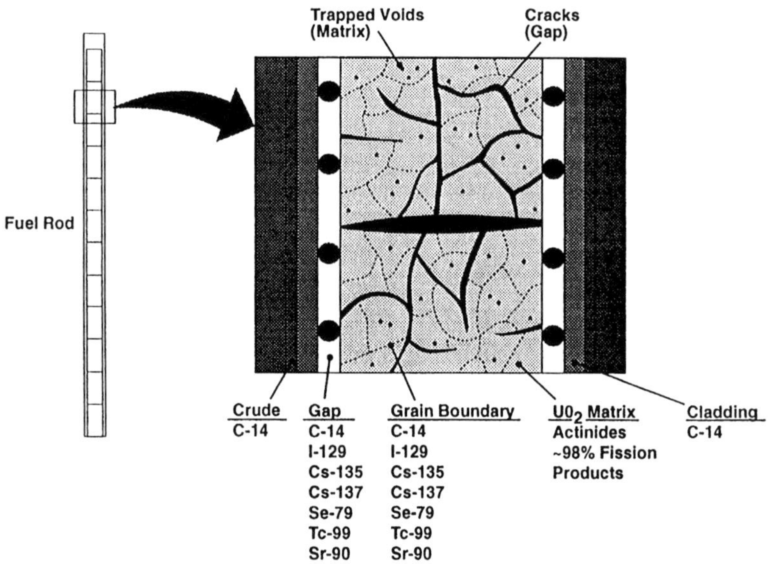

Figure 4.9 Schematic diagram of spent nuclear fuel, showing locations of various radionuclides.

4.5.2 Distribution of radionuclides within waste forms

Abundances of radionuclides may not provide sufficient information for near-field performance assessment. The location and relative 'accessibility' of these radionuclides with respect to reaction with groundwater may also be needed. For example, reprocessed HLW borosilicate glass is a relatively uniform, homogeneous source of radionuclides, while spent fuel is not (see Figure 4.9). Spent fuel includes both the reactor assemblies ('hardware') and the individual spent fuel rods. The spent fuel rods can, in turn, be subdivided into separate source-regions based on physicochemical characteristics and inventories:

- UO_2 matrix,
- zircaloy cladding,
- grain boundary phases, and
- phases of volatile fission products that form at the matrix-cladding 'gap'.

The zircaloy cladding, in turn, may develop an additional surface layer, or 'crud', that has characteristics and inventories different from the bulk cladding material.

LLW is particularly complex with respect to the distribution of radionuclides among different solid components. Contaminated components of LLW can include dry solids, glassware, plastics, animal carcasses, medical wastes, metals, ion exchange resins, paper, clay, and cement. One useful approach to overcome the problem of inventory distribution is centralised conditioning of the waste. Conditioning can be achieved by mechanical homogenisation and encapsulating the resulting LLW mixture into cement, resin, polymer, or bitumen blocks. This encapsulation has the additional advantage of providing shielding of the radiation, especially for cement matrices. In less centralised disposal programmes, LLW from multiple sources are received with stated maximum activities of radionuclides, but having no detailed description of types or composition of materials. Characterisation is further confounded by indiscriminate blending of these multiple sources. This arbitrary approach produces a final waste form which makes long-term performance difficult to assess.

In addition to the possible inhomogeneities in waste forms, the expected variability in characteristics, particularly inventories, for these different sources must be assessed. For processed waste forms, such as HLW glass and encapsulated LLW, this variability should be low. For spent fuel, however, there will be a range of types, reactor conditions, and power histories all contributing to a potentially significant variability in characteristics and inventories in what is collectively called spent fuel. In particular, the readily accessible and soluble fission products that migrate to the inner cladding surface may be rapidly mobilised upon container failure; hence, the variability in the amount of this 'instant release' gap fraction is of critical importance to performance assessment. Correlations between gap inventories and reactor operating characteristics are needed to determine such variability (Garisto *et al.*, 1989).

4.5.3 Chemical constraints on release

The release of radionuclides from nuclear waste forms will be controlled by one of two limiting chemical processes. The first is the dissolution (or alteration or reaction) rate, typically expressed in unit of mass of waste form dissolved per surface area of waste form per unit time (Bourcier, 1991; Mouche and Vernaz, 1988; Shoesmith and Sunder, 1992). Continued dissolution of the waste form leads to the second process for controlling the release of radionuclides in the near-field, solubility-limits imposed by radionuclide-containing solid phases (Apted, 1982; Lemire and Garisto, 1989; see Figure 4.10). Such solubility limits may be imposed either by the primary waste form (Garisto and Garisto, 1985) or through formation/precipitation of new alteration phases from the groundwater solution (Savage, 1986; see Figure 4.11). Considerable research on waste form dissolution and radioelement solubilities have been conducted. Both processes are sensitively dependent on environmental conditions of the near-field, particularly temperature and groundwater composition, and much of this has been recently reviewed (NEA, 1993).

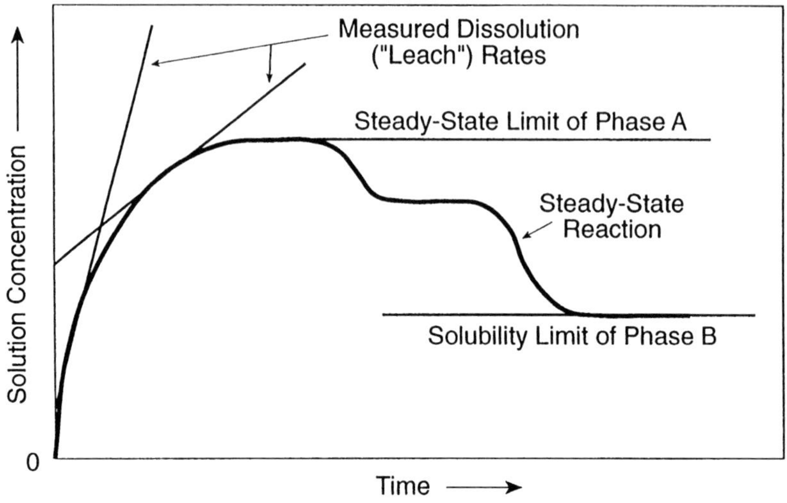

Figure 4.10 Solid dissolution behaviour as a function of time. Phases A and B refer to secondary solids incorporating radionuclides.

The relative importance of these two processes in controlling radionuclide release from the near-field may be determined from the relative rate of dissolution (release) at the waste form surface (R_{rxn}) versus the rate at which dissolved radionuclides migrate away from this surface (R_{mig}). If $R_{rxn} > R_{mig}$, then the concentration of the dissolved radionuclides will continue to increase at the waste-form surface until a solubility limit is exceeded, thereafter restricting the release. Conversely, if $R_{rxn} < R_{mig}$, then the interface concentration of dissolved radionuclides remains low and release is restricted by the reaction rate of the waste form.

In general, release models for both bounding processes must be developed to assure that realistic release conditions are modelled. The release of the most soluble radionuclides, such as carbon-14, iodine-129, and selenium-79, are expected to be controlled by the reaction rate of waste forms as well as by adsorption on barrier materials (Figure 4.12). Releases of the remaining, insoluble radionuclides are expected to be limited by solubility constraints. Radionuclides are often one of several isotopes for a radioelement, therefore these isotopes 'share' the solubility of the solid solubility-limited phase for that radioelement.

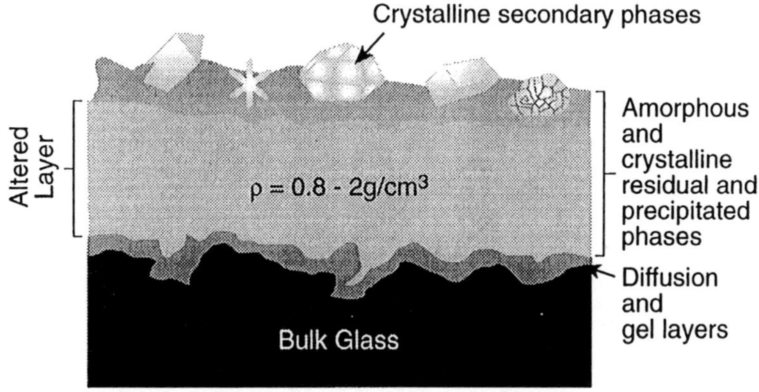

Figure 4.11 Features observed on reacted glass surfaces (from Mendel, 1984). Reproduced by permission of Battelle Pacific Northwest Laboratory.

4.6 Transport

The transport of dissolved, as well as gaseous or colloidal, radionuclides from the near-field to the far-field will be determined by the following processes and factors (Pigford and Chambré, 1988):

- transport mechanisms (diffusion, advection),
- physical properties of engineered and natural materials (hydraulic conductivity, porosity, diffusion coefficient, fractures *etc.*),
- retention/retardation properties of engineered and natural barriers (sorption, filtration),
- coupling of processes, such as flow and chemical reaction.

4.6.1 Transport mechanisms

Groundwater

The primary mechanisms for radionuclide transport within engineered and natural barrier materials are groundwater flow (advection) in a pressure ('head') gradient, or diffusion in a concentration gradient, or both. In the absence of a flow-limiting engineered material, transport in the near-field will be controlled by the dominant transport mechanism in the host rock, and may be assumed to be advection except for impermeable mudstone and salt formations. For advection, flow is governed by Darcy's law that relates the total volumetric flow rate of water through a unit cross-sectional area normal to the direction of flow ($\mathbf{q}$), to the permeability (k) and pressure gradient

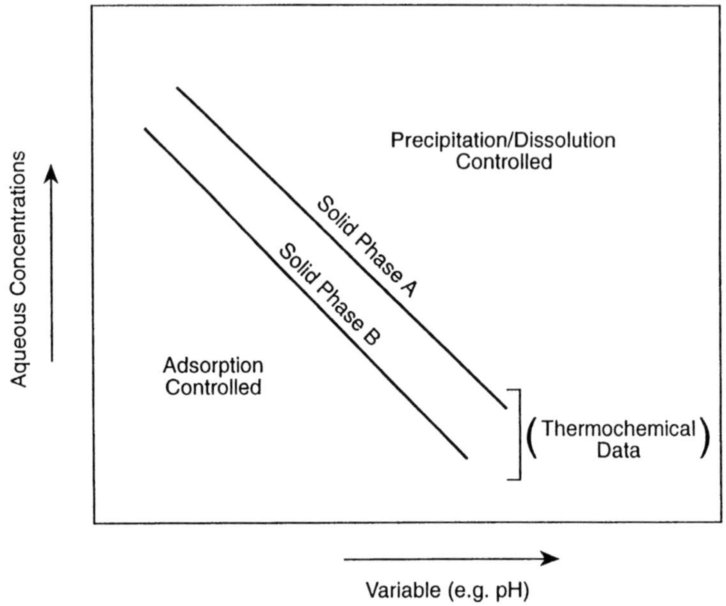

Figure 4.12 Schematic diagram illustrating how to set upper limits of radionuclide concentrations. From Rai, D. and Kittrick, J.A. 'Mineral equilibria in the soil system', in *Minerals in the Soil Environment*, Soil Science of America Book Series No. 1, 1989, pages 161-197. Reproduced by permission of the Soil Science Society of America.

(∇P):

$$\mathbf{q} = -k\nabla P \tag{4.3}$$

$$\overline{\mathbf{v}} = \mathbf{q}/\theta \tag{4.4}$$

where $\overline{\mathbf{v}}$ is the average pore velocity and θ is the effective porosity (total porosity actively involved in transport). Most groundwater flows, and all expected repository flows, occur well within the laminar flow range, where the linear Darcy law is applicable. There will also be a lower limit to the validity of Darcy's law, requiring some threshold gradient to initiate flow. For a waste container with radius of R, a general guide to this transitional region is based on dimensionless Peclet numbers (Pe), where:

$$Pe = \frac{|\overline{\mathbf{v}}|R}{D_e} \tag{4.5}$$

and D_e is the effective diffusion coefficient in the porous medium.

For repository systems in which the Peclet number is below values of 0.1 to 1, diffusional transport will dominate. Note that because hydraulic gradients may be expected to vary over the operational lifetime of a nuclear waste repository (> 10,000 years), demonstration of low hydraulic conductivities of near-field materials provides the most reliable basis for assuring low, diffusion-limited transport of radionuclides.

Diffusional transport of aqueous species is represented through Fick's second law of diffusion:

$$\frac{\partial c}{\partial t} = \nabla . \left(\frac{D_e}{\theta} \nabla c \right) \tag{4.6}$$

where the time dependent change in concentration c is a function of the effective diffusion coefficient and the concentration gradient. Solution of this equation must be based on geometric and boundary conditions for the specific near-field configuration of interest.

Introduction of flow-limiting barrier materials, such as compacted-clay or concrete backfills, can separate the performance of the near-field from the far-field. Flow-limiting barriers restrict transport from the EBS to diffusion only, regardless of changes in the advective flow rate in the host rock. This greatly simplifies the analysis of transport, as well as assuring a reliably low rate of mass transfer from the EBS (Pigford, 1987). Mass-transfer rates from the EBS are not completely decoupled from advection in the host rock, however. Increasing flow rate in the host rock can reduce the concentration of released radionuclides at the backfill-rock interface. This maintains a steep concentration gradient at this interface, resulting in a slightly elevated diffusional mass-transport rate compared to the case for low flow in the rock. This effect can be readily incorporated into diffusional transport models by applying a zero concentration ('swept away') boundary condition at the backfill-rock interface.

Gas

The migration of gas through a saturated porous medium is influenced by a large number physical phenomena. In the context of gas and water transport, the most important of these are:

- two-phase flow;
- relative permeability and capillary pressure effects;
- solubility limited dissolution of the gas phase in the liquid phase;
- reaction-diffusion-advection behaviour of the dissolved species;
- adsorption of the gas phase by the medium;
- mixing or diffusion of vapour in the gas phase;

- viscous fingering (instability of simple geometrical configurations);
- dynamic response of the medium to pressure conditions.

The extent to which each of these phenomena influences transport is very much dependent on the gas generation rate, pressure conditions and material properties of the medium through which the gas may move. Thus it is advisable to scope the effects of each phenomenon in a number of disposal concepts, using both analytical and numerical mathematical models. The results of such calculations provide estimates of the physical conditions likely to be encountered in the repository. Sensitivity analyses could also give a clear indication of the model parameters that had most bearing on the predicted conditions. This could provide input into experimental work, in terms of deciding what properties it was desirable to test and how to measure them. Ideally, modellers and experimentalists can work interactively to design experiments for calibration and validation, and interpret the results. The models that might be considered include:

- analytical gas reaction-diffusion models;
- analytical gas flux models, possibly incorporating the elastic response of the porous medium to changes in gas pressure;
- analytical gas-water interface models for a partially saturated or fully saturated porous medium;
- a numerical model for a full two-phase flow system with an elastic response of the porous medium to changes in gas pressure;
- a fractal model for the simulation of viscous fingering.

4.6.2 Physical transport properties

As noted in the previous section, the primary properties that influence the magnitude of advection and diffusion in the near-field, as well as the far field, are the hydraulic conductivity, the kinematic porosity, and the effective diffusion coefficient. Representative matrix hydraulic conductivities and diffusion coefficients for compacted bentonite are described in the following paragraph. It is assumed that large regional fracture systems will be avoided during repository siting, and waste packages will not be emplaced in boreholes which intersect large fractures.

Measured effective diffusion coefficients in compacted bentonite range from 6.5×10^{-9}cm^2 s^{-1} to 1.1×10^{-5}cm^2 s^{-1}. Under saturated conditions, diffusion coefficients in bentonite will not be below 10^{-6}cm^2 s^{-1}. Permeability measurements on compacted bentonite have proven difficult because of the non-Darcy response of the material and the high hydraulic gradients that are typically needed to obtain measurable flows in the material. The most recent studies conducted independently by ultracentrifugation and low-gradient methods both indicate hydraulic conductivities lower than 10^{-14} cm s^{-1}, up to three orders of magnitude lower than previously reported values (Conca *et al.*, 1993; Dixon *et al.*, 1992).

Transport rates are affected by the cross-sectional area through which radionuclides are migrating. For aqueous radionuclides, limited apertures in partly failed containers or into small fractures within surrounding rock may impose severe constraints on their transport rate from the EBS (SKB, 1992).

4.6.3 Retention/retardation of radionuclides

Transport of radionuclides by both advection and diffusion will be modified by various modes of retention/retardation. The retention/retardation capability of engineered and natural barrier materials is controlled by a number of physical and chemical processes. Several of these processes are shown diagrammatically in Figure 4.13. For most radionuclides, the principle retention/retardation processes involve direct interaction of solute with solid phases, which is generally termed 'sorption'. Retardation processes can be subdivided into two fundamentally distinct categories:

1. physical restriction of solute transport (*e.g.* filtration, ion-exclusion), and
2. direct uptake with solids.

In the strict sense, only the latter process should be considered as 'sorption'. Direct uptake of radionuclides can, in turn, be further sub-divided into processes that are reversible (*e.g.* ion-exchange, physical absorption) or irreversible (*e.g.* precipitation). Typically, several or all of these retardation processes may occur simultaneously under near-field conditions, and it is important to separate out physical from chemical processes, as well as reversible from irreversible processes.

A common approach to quantification of sorption is the distribution coefficient (K_d, in $m^3\ kg^{-1}$), which is operationally defined as the ratio of the concentration of radioelement l sorbed on a solid (S_l in moles/kg) and the concentration of radioelement l in the coexisting solution (C_l, in moles/m^3):

$$K_d = \frac{S_l}{C_l} \tag{4.7}$$

with the important condition that the interaction between solution and solid surface be in equilibrium. The conceptual model further assumes that this relationship is independent of solute concentration, the radionuclide behaves ideally (*i.e.* its behaviour is independent of other solution species), the reaction is reversible, and the temperature is constant or irrelevant. Figure 4.14 shows *in situ* measurements of K_d for selected radioelements in bentonite as a function of compaction density (or more accurately, volumetric water content).

In reality, sorption is often observed to show strong dependence on solute concentration. Accordingly, more complex 'isotherm' relationships,

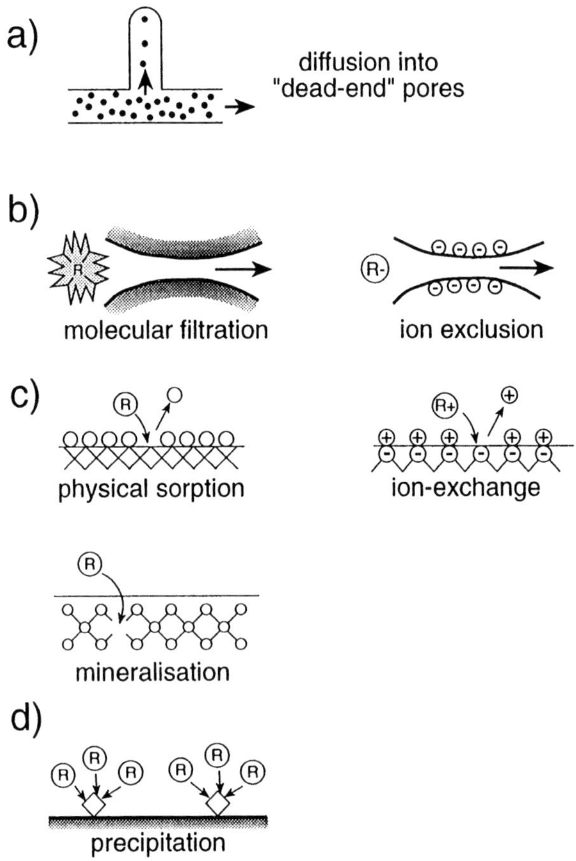

Figure 4.13 Retardation mechanisms (from McKinley and Hadermann, 1984). Reproduced by permission of Nagra.

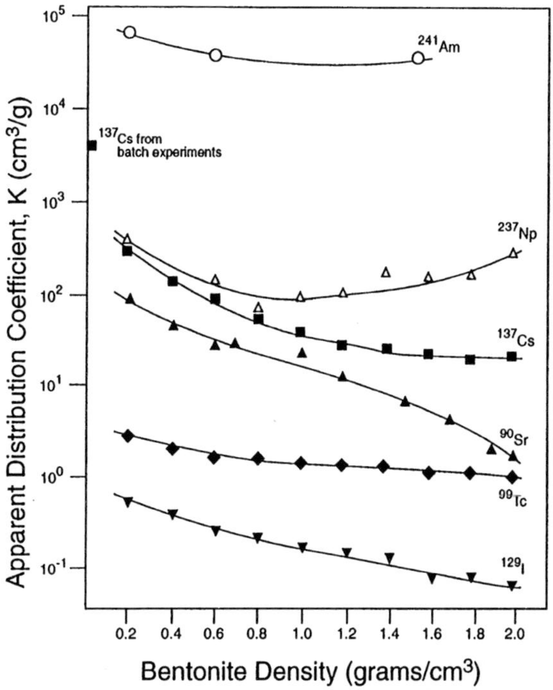

Figure 4.14 Apparent distribution coefficients for some nuclides as a function of bentonite compacted density. From Conca, J.L. *et al.*, *Proceedings of the 2nd International Conference on High-Level Radioactive Waste Management,* 1991, pp 1382-1389. Copyright 1991 by the American Nuclear Society, LaGrange Park, Illinois, U.S.A.

both theoretically and empirically based, have been derived to incorporate such behaviour. Beyond these approaches is a thermodynamic approach, in which the energetics of chemical and physical forces for attraction between solute and solid surface are explicitly modelled. The thermodynamic approach to sorption is limited by a paucity of thermodynamic data on species and solids relevant to systems for nuclear waste disposal, and sluggish kinetics in such systems at low temperature that may lead to disequilibrium conditions, thereby invalidating this approach.

Dissolved chemical species are retarded through three processes:

1. sorption and other interactions with surface functional groups exposed along mineral edges;
2. exchange with ions within the lattice itself, especially the interlayer positions; or

3. adsorption/exchange with other minerals or colloids along the pathway, such as phosphates that interact with most species in solution.

Species display different degrees of retention/retardation, depending on the number and magnitude of sorption processes in which they participate. For example, iodine and pertechnetate form anionic complexes that are not significantly retarded, resulting in high mobilities within bentonite and porous media in general. Uranyl ions do not exchange with interlayer positions, but do complex with surface functional groups and the uranium mobility is intermediate. Americium complexes are highly retarded by both sorption and exchange and their mobilities are very low.

Exchange of a species with an interlayer cation or group in the solid requires that it move from its solvated state into the interlayer position and must be accompanied by a favourable free-energy change. However, movement from site to site within the interlayer space is much less favourable and is hindered by the energy requirements necessary to break bonds within the interlayer sites. Therefore, diffusion through the interlayer space is very slow, being closer to solid diffusion rates ($10^{-18}$to 10^{-15}cm^2 s^{-1}). Diffusion coefficients between $10^{-10}$and 10^{-5}cm^2 s^{-1} in bentonite, or any porous material, reflect aqueous diffusion through bulk water films and channels, and not through interlayer spaces.

Physical retention/retardation processes may also operate in the far-field (Chapter 5). Such physical retention/retardation mechanisms are particularly important in evaluating the potential transport of radionuclide-bearing colloids or suspended particles. Colloids may form from the dissolution of waste forms, organic degradation, radiolysis reactions, or precipitation at the boundaries between different chemical environments, as well as occurring naturally in many groundwaters. Radionuclide-bearing colloids/particles in groundwater represent an additional, transportable constituent that can, under certain circumstances, vastly exceed the solubility-limited concentrations for dissolved radionuclides. The small-scale pore size and tortuosity in compacted bentonite are assumed to physically entrap particulate and even colloid-size (> 1 μm) particles. Modelling colloid migration in the far-field is discussed in Chapter 5.

4.6.4 Coupled processes

There may be strong coupling between any number of processes in the near-field, *e.g.* thermo-mechanical, flow and chemical reaction, *etc.* Some of the issues relevant to understanding coupled processes will be addressed by consideration of those associated with coupled flow and reaction.

Considerations of chemical reaction and migration concerned in the evolution of the near-field environment usually involve some modelling simplifications. For example, chemical systems are often equilibrated instantaneously (by calls to large equilibrium databases); transport may be represented by reaction path models, where a parcel of water is moved through the near-field towards the steel overpack; diffusive effects may be ignored; and the back-coupling of temperature gradients and flow effects with the chemistry may also be neglected.

In some instances, the equilibration times necessary for a chemical equilibrium to pertain between the incoming saturating water, and the minerals of an initially dry clay buffer zone, may be such that transport processes must be coupled to the reactive chemistry. This may be true for both the near-field and the far-field evolution. Moreover, the effect of diffusive transport of solutes out of the near-field, or inwards, towards corroding canister materials, may be important even during the resaturation phase. For example, aerobic corrosion must be limited by the diffusive supply of oxygen.

For transport in clays, for example, fluid flow may be driven by chemical gradients producing anomalies not modelled by a standard Darcy's Law approach. There is now a large literature covering the effects of coupled non-linear processes within specific geochemical systems. In solving such problems it is necessary to model the interactions between the chemical reactions, chemical transport, fluid flow and the temperature distribution. The state of the art of reaction-diffusion theory (*e.g.* Grindrod, 1991) provides a number of qualitative and quantitative techniques enabling verification of numerical solutions, even in the presence of such non-linear coupling. A twin approach to such problems is to make a detailed numerical investigation of a large coupled system, supported by qualitative analysis aimed at verifying the calculated solutions.

Non-linear models and fingering

This issue may relevant to understanding the behaviour of redox-fronts in the near-field. The reactive transport of an aqueous species, with concentration c, through a saturated porous medium can be described as follows:

$$\frac{\partial}{\partial t}(\theta c) = \nabla.(D\nabla c - c\mathbf{q}) - \frac{R(c,d)}{\varepsilon},$$

$$\frac{\partial d}{\partial t} = \frac{R(c,d)}{\varepsilon},$$

$$\frac{\partial \theta}{\partial t} = -\nabla.\mathbf{q}, \quad \mathbf{q} = -k(\theta)\nabla P \tag{4.8}$$

$$\theta = g(c, d, \ldots).$$

Here θ is the porosity, D is the dispersion coefficient, $\mathbf{q}$ is the Darcy velocity, d represents the density of the solid and/or adsorbed phase of the species, assumed immobile, ∇P is the pressure gradient, R is the reaction term and $1/\varepsilon$ is the reaction rate. Notice that since the composition of the medium is altered by the reaction, θ depends on the state of the system in order to account for any consequent volumetric changes. For reversible reactions there exists a continuously differentiable function f such that $R(c, f(c)) = c$, for all $c \geq 0$, so that $d = f(c)$ represents that chemical equilibrium. $k = k(\theta)$ denotes the permeability of the altering medium.

If the reactive chemistry equilibrates on a faster time scale than that associated with the transport processes ($\varepsilon \to 0$ in (4.8)), a single non-linear transport equation results:

$$\frac{\partial}{\partial t}(\theta c + f(c)) = \nabla.(D\nabla c - c\mathbf{q}) \tag{4.9}$$

$$\frac{\partial \theta}{\partial t} + \nabla.\mathbf{q} = 0$$

$$\theta = g(c).$$

It can be shown (analytically) that systems like (4.8) and (4.9) yield unique planar front solutions when the inlet fluid is in disequilbrium with the host medium. The one dimensional (planar) travelling waves for such systems are non-linear, so that solutions are obtainable analytically, allowing verification of numerical codes designed to solve more general problems. However they may also give rise to scalloping or fingering effects (Figure 4.15). This may happen when the reacted medium has a lower permeability than the unreacted medium. Hence if perturbations to the plane wave geometry form fingers, the flow can become channelled towards the finger tips, and the solute delivered preferentially to these points. This results in further growth of the structure; however, such instability will be diffusion limited. Note that the change in permeability due to the reaction plays no role in the construction of the planar front. However it must play a key role in the mechanism symmetry, breaking instabilities. Under certain assumptions such problems may be reduced to a moving boundary formulation, having a sharp reaction front.

A similar phenomenon can occur where reaction fronts migrate up temperature gradients (Figure 4.16). The increase in reaction rates due to increased ambient temperatures may allow reaction fingers to outrun the reaction zones nearby leading to instability. These fingers depend crucially on the existence of the thermal gradient: if the gradient diminishes then the planar waves recover stability. Again such structures are diffusion limited.

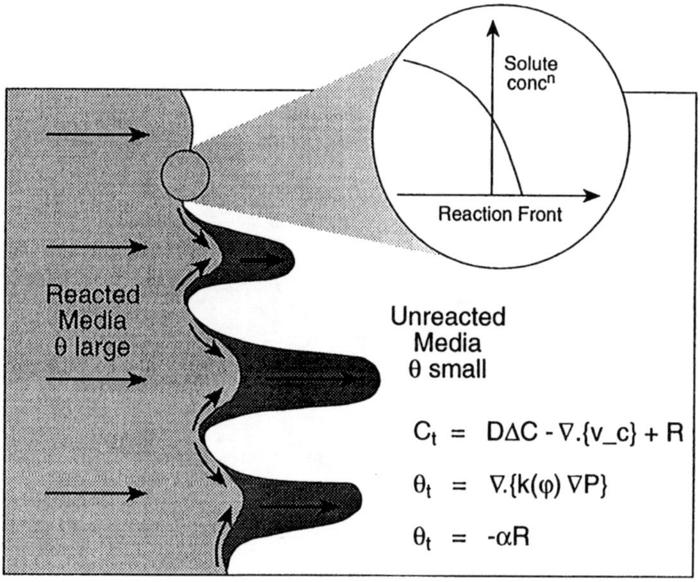

Figure 4.15 Coupling of flow and chemical transport may result in frontal instabilities (fingering).

4.7 Source-term modelling

4.7.1 Overview

The release of radionuclides from the near-field to the far-field is referred to here as the 'source-term'. Source-term models were recently summarised and reviewed as part of an international workshop (NEA, 1993). The migration is controlled by the transport mechanisms, radioactive decay, groundwater chemistry, and the geometry of the system. The transport mechanisms include advection, diffusion, and dispersion. In advection, a dissolved species is carried by the flowing groundwater. In molecular diffusion, the species migrates under its concentration gradient by colliding with carrying fluid molecules. The concentration of a dissolved species may also spread in space due to dispersion effects. Many factors contribute to dispersion in a porous medium, such as heterogeneity, variation of flow velocity in space, and surface and chemical effects (Bear, 1972). Groundwater flow velocity, diffusion coefficient, backfill or rock hydraulic properties are among the factors influencing the transport of a dissolved radionuclide.

In an unsaturated disposal site, the scenario for the migration of dissolved species is similar to that in a saturated medium, provided that a continuous groundwater pathway exists. In addition to dissolved radionuclides, gaseous

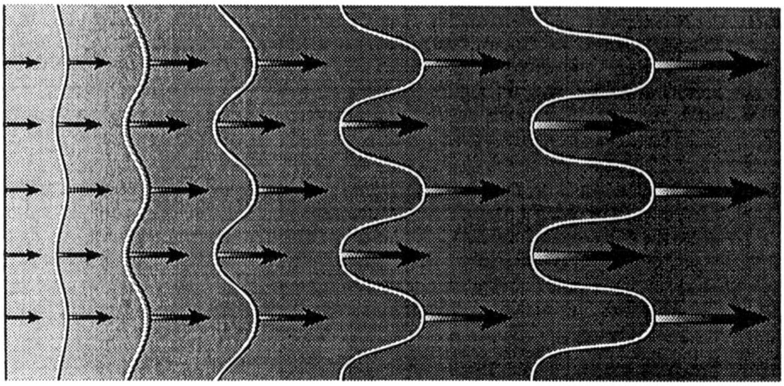

Figure 4.16 Frontal instability driven by a thermal gradient: structures are diffusion limited.

radionuclides can migrate from the waste to the host rock once the canister fails.

To simulate near-field transport, one needs to solve coupled equations for:

- flow (including continuity of the carrying fluid and Darcy's law),
- mass-conservation of the radionuclide in the transport region,
- mass-conservation inside the canister (including dissolution and transport effects),
- energy-conservation, and
- chemical reactions.

The conservation equations contain second-order differentials of spatially independent variables. There are three kinds of boundary conditions for solving such kind of partial-differential equations (Zwillinger, 1992):

- First kind: dependent variable (*e.g.* concentration of a radionuclide) on the boundary is prescribed;
- Second kind: normal derivative of dependent variable (*e.g.* radionuclide diffusive flux) on the boundary is prescribed;
- Third kind: linear combination of dependent variable and normal derivative (*e.g.* summation of diffusive and advective fluxes of a radionuclide) on the boundary is prescribed.

Treatment of the transport phenomena at the waste-backfill interface and the backfill-rock interface is more problematical. Generally, the continuity of concentration and mass-transfer flux is imposed at these interfaces. If any interfacial or transitional mechanisms exist at an interface, the interface may be treated as a separate region where general conservation laws with all possible mechanisms included are applied.

Solutions for such kinds of general problems are limited not only by computing techniques, but also by lack of data for the coupling phenomena and poor understanding of some coupling mechanisms. Accordingly, simplified models and methods are used to solve an idealised system. For example, in the case of a reasonably constant or very low groundwater flow velocity, flow equations can be eliminated. The energy equation disappears if the temperature gradient is small or if the temperature dependence of hydraulic, transport, and chemical reaction parameters can be neglected.

Problems can also be simplified by decoupling these equations. For example, if the groundwater convection induced by the waste heating is not important, the flow equations can be decoupled from the energy equation. Dealing with the coupling of transport and chemical reactions, the system equations include the chemical speciation equations for the aqueous, sorbed, and precipitated radionuclides. Solubilities of radionuclides depend on pH and redox which vary along the radionuclide migration path. These equations are based on the concepts used to describe an aqueous chemical system: chemical equilibrium; mass balance; mass action; and charge balance. In the time scale of the near-field performance assessment, equilibrium sorption can be assumed due to the fast reaction rate compared to the slow rate of migration. Furthermore, because sorption of most radionuclides will occur at relatively low solution concentrations, the assumption of linear equilibrium isotherms for sorption using the equilibrium distribution coefficient is often used to simplify the sorption model. These topics have been discussed in the previous chapters.

The solution method can also be simplified. For example, using conservative boundary conditions, one can decouple or avoid solving the transport equation in the host rock or geosphere. Constant concentration boundary conditions will be a good source condition for solubility limited elements. A quasi-steady-state approach can be employed for a finite region with a characteristic time (a time for a system to reach steady state) much less than the time scale used in the model.

The solution technique employed by a model is important to the accuracy and efficiency of modelling. Analytical methods are accurate and fast but limited to a system where the number of processes is reduced and the geometry is simplified. Numerical methods, on the other hand, are capable of modelling a relatively comprehensive system but the code of the numerical scheme needs to be tested and benchmarked. One can manoeuvre the solution techniques between analytical and numerical methods to get the optimum combination of accuracy and efficiency. Examples can be seen later in the detailed model review.

A model, whether employing analytical or numerical methods, should be tested to examine how close the modelling results are to the reality. This is and will remain challenging to the nuclear waste modellers who predict the distant future. Experimental data and observations from natural analogues are needed to test and calibrate models.

4.7.2 Available models

Currently, there are many models developed worldwide to simulate the near-field transport of radionuclides. They have been developed for different sites or the same sites but using different approaches. The most comprehensive and technically defensible of these have been developed for deep geological disposal of HLW and spent fuel. Such codes and models are generally suitable for analysis of other nuclear waste forms and near-surface disposal designs.

A recent review (Apted, 1993; NEA, 1993) has been conducted on the following source-term codes: AECL's Vault Model used in SYVAC (AECL, 1991), STRENG (Grindrod *et al.*, 1990), MESHNOTE (Robinson *et al.*, 1992), CALIBRE (Robinson and Worgan, 1992), STAG (Robinson, 1992), AREST (Apted *et al.*, 1989), QUADC (Chuang, 1991), SOTEC (Sagar *et al.*, 1992), SKB's Compartment Model (Romero *et al.*, 1991) and Resistance-Network Model (Nilsson *et al.*, 1991), MELODIE (Baudoin *et al.*, 1993), and MARNIE (Martens *et al.*, 1993). Emphasis has been placed on codes that are being used for source-term analysis, either for individual waste packages or for entire repository plus EBS.

The review (Apted, 1993; NEA, 1993) details the solution method of each model and comparisons among models are made. Differences in geometric and transport factors are also compared among models. Possible confusing nomenclature and terminology over transport parameters are clarified using unified symbols and definitions. Tables 4.2 and 4.3 summarise this review.

The source-term models can be classified as the following four types:

1. Simple models with complete solutions. These models start with a simple (*e.g.* a reduced number of processes, or simplification of geometry) system relative to other models. They solve the equations completely. The obtained results reflect the real physical picture described by the governing equations, but this is an inherently incomplete picture. The solution techniques can be either analytical (*e.g.* AREST code) or numerical (*e.g.* SOTEC code).
2. Comprehensive models with approximate solutions. These models start from a more complete and realistic system description compared to the first-type models. Some approximations are made in solving equations. The obtained results partially reflect the physical picture described by the governing equations. Examples of codes that adopt this approach are the Vault Model, STAG, STRENG, CALIBRE, MESHNOTE, and MELODIE. Under certain conditions, the results are very closest to the solutions without the approximation.
3. Comprehensive, coupled models with complete solutions. These models

Table 4.2 Summary of review of source-term models (from NEA, 1993). Reproduced by permission of NEA/OECD.

	Vault Model	STRENG	MESHNOTE	CALIBRE	STAG	AREST	QUADC	SOTEC	SKB Comp.	MELODIE	MARNIE
Geometry	1-D planar	1-D cylinder	1-D cylinder	2-D cylinder	1-D cyl, planar	1-D sphere, planar	2-D planar	1-D sphere	2-D	2-D quasi 3-D	1-D planar
Transport regions	buffer, back-fill, rock	buffer	buffer	can, back-fill, rock	backfill, rock	backfill, rock	backfill, rock	rock	buffer, rock	rock	backfill, rock
Time dependent	yes	yes	yes	yes	yes	yes	yes	yes	yes	yes	yes
Mechanisms	diff, adv, disp[1]	diff	diff	diff, adv[2]	diff, disp, adv[3]	diff	diff, adv, disp	diff	diff, adv[2]	diff, disp, adv	diff, disp, adv
Phases	aqueous, solid	aqueous, solid	aqueous, solid	aqueous, solid	aqueous, solid	aqueous, solid	aqueous, solid	aqueous, solid	aqueous, solid	aqueous, solid	aqueous, solid
Sorption model	linear equilib.	linear equilib.	non-linear isotherm	linear equilib.	linear equilib.	linear equilib.	C-S,I-C, S-C[4]	linear equilib.	linear equilib.	linear equilib.	linear equilib.
Precipitation front	no	no	yes	yes	no	no	yes	no	no	no	yes
Radioactive decay	yes	yes	yes	yes	yes	yes	yes	yes	yes	yes	yes
In-growth	yes	yes	yes	yes	yes	no	yes	no	yes	yes	yes
Rock region	finite	n/a	n/a	finite	finite	infinite	finite	finite	finite	finite	n/a
Rock boundary	zero conc	zero conc, mix. tank	mixing cell	zero conc	zero conc	zero conc	conc, flux	zero conc	zero conc	conc, flux	conc, flux
Hetero-geneity	no	no	no	no	no	no	yes	no	yes	no	yes
Fracture	no	no	no	yes	no	no	yes	no	yes	yes	no
Limited can. surface	no	no	no	yes	no	no	yes	no	yes	no	no
Coupled regions	yes (approx.)	n/a	n/a	yes	yes (approx.)	yes	n/a	n/a	n/a	n/a	yes (approx.)
Dissolution coupled	yes (approx.)	yes	no	yes (approx.)	yes (approx.)	no	yes	yes (approx.)	no	no	yes
Multi decay chains	no	yes	yes	yes	yes	no	yes	no	no	no	yes
Chemical coupling	no	no	no	redox front	no	no	yes	no	no	no	no
Solution method	analytical	numerical, analytical	finite diff.	finite diff.	numerical, analytical	analytical	finite element	finite diff.	compart-ment	finite element	finite diff.

[1]dispersion is considered in evaluating mass-transfer coefficients. [2]advection is considered in fractures only. [3]advection and dispersion are considered in rock region only. [4]complexation-sorption, ion exchange, and surface-complexation. Approx.: approximated solution methods are employed. n/a not applicable.

attempt to simulate a system as closely as possible. Numerical methods have to be employed to accomplish the complete modelling. The computation takes more time than the second-type models. The results may be the closely approximated solution to the governing equations. QUADC and MARNIE are examples of codes in which this approach has been used.

4. Screening- or scoping-type models: SKB's Compartment and resistance-network models use simple solution methods to model a system with a complex geometry. The computation would be very fast. The results may not be as accurate as others.

Note, every model simplifies a real system more or less. Therefore, 'complete' and 'simple' are relative terms.

No attempt has been made to judge which approaches or codes are better or poorer than others. The choice of modelling method depends on performance assessment criteria imposed by each country, physical, chemical, and hydrogeological conditions of a specific site, and types of waste form. Generally, a simplified analytical model can be used to benchmark a numerical code which can model a more complete system. A model which attempts to include many mechanisms also requires significantly more amounts of data from experiments or natural analogues.

4.8 References

AECL (1991) The vault model. *Atomic Energy of Canada Limited Draft Environmental Impact Study Report.* Atomic Energy of Canada Limited, Pinawa, Manitoba, Canada.

Agg, P.J., Moreton, A.D., Rees, J.H., Rodwell, W.R. and Sumner P.J. (1993) NSARP Reference Document: Gas generation and migration: January 1992. *UK Nirex Report* NSS/B142. United Kingdom Nirex Ltd, Harwell, U.K.

Andrews, R.W., LaFleur, D. and Pahwa, S. (1986) Resaturation of backfilled tunnels in clay. *Nagra Technical Report* 86-27. Nagra, Baden, Switzerland.

Apted, M.J. (1982) Overview of hydrothermal testing of waste package barrier materials at the Basalt Waste Isolation Project. In *MCC Workshop on Leaching Mechanisms of Nuclear Waste Forms*, (ed. J.E. Mendel). *Pacific Northwest Laboratory Report* PNL-4382. Pacific Northwest Laboratory, Richland, Washington, U.S.A.

Apted, M.J. (1993) Survey and review of near-field performance assessment. *Nagra Internal Report* 93-60. Nagra, Wettingen, Switzerland.

Apted, M.J., Liebetrau, A.M. and Engel, D.W. (1989) The analytical repository source term (AREST) model: analysis of spent fuel as a nuclear waste form. *Pacific Northwest Laboratory Report* PNL-6347. Battelle Pacific Northwest Laboratory, Richland, Washington, U.S.A.

Arthur, R.C., Apted, M.J. and Conca, J.L. (1993) Dealing with uncertainty in the chemical environment in bentonite backfill. In *Scientific Basis for Nuclear Waste*

Management Vol XVI, (ed. C. Interrante). Materials Research Society, pp 389-394.

Atkinson, A. (1985) The time dependence of pH within a repository for radioactive waste disposal. *United Kingdom Atomic Energy Authority Report* AERE-R11777. United Kingdom Atomic Energy Research Establishment, Harwell, U.K.

Atkinson, A., Williams, S.J. and Wisbey, S.J. (1993) NSARP Reference Document: The Near Field: January 1992. *UK Nirex Report* NSS/B139. United Kingdom Nirex Ltd, Harwell, U.K.

Baudoin, P., Bourgeois, M., Certes, C., Escalier des Orres, P., Gomit, J.M. and Greneche, P. (1993) MELODIE: a code for radiological consequences assessment associated with radioactive waste repositories in deep geological formations. In *Proceedings of Safewaste 93*, Avignon, France.

Bear, J. (1972) *Dynamics of Fluids in Porous Media.* Dover Publications Inc., New York, U.S.A.

Berner, U. (1987) Modelling porewater chemistry in hydrated portland cement. In *Scientific Basis for Nuclear Waste Management Vol X*, (eds. J.K. Bates and W.B. Seefeldt). Materials Research Society, pp 319-330.

Bourcier, W.L. (1991) Overview of chemical modelling of nuclear waste glass dissolution. In *Scientific Basis for Nuclear Waste Management Vol XIV*, (eds. T. Abrajano and L. Johnson). Materials Research Society, pp 3-18.

Chuang, G.M. (1991) A two-dimensional finite element model study of the near-field of a high level radioactive waste repository. Unpublished PhD Thesis, University of Washington, Seattle, U.S.A.

Conca, J.L., Apted, M.J. and Arthur, R. (1993) Direct determination of transport parameters. In *Scientific Basis for Nuclear Waste Management Vol XVI*, (ed. C. Interrante). Materials Research Society, pp 839-844.

Conca, J.L., Ashida, T. and Sato, H. (1991) Apparent and simple diffusion coefficients in compacted bentonite. In *Proceedings of the Second International Conference on High-level Radioactive Waste Management*, American Nuclear Society, La Grange Park, Illinois, U.S.A., pp. 1382-1389.

Dixon, D.A., Ranjan, R.S. and Graham, J. (1992) Applicability of Darcy's Law in laboratory measurement of water flow through low permeability clays. *48th Canadian Geotechnical Society Meeting*, Toronto, Ontario, Canada.

Garisto, N.C. and Garisto, F. (1985) A UO_2 solubility function for the assessment of used nuclear fuel disposal. *Atomic Energy of Canada* Report AECL-10009. Atomic Energy of Canada Ltd., Pinawa, Manitoba, Canada.

Garisto, N.C., Vance, E., Stroes-Gascoyne, S. and Johnson, L.H. (1989) Instant release fractions for the assessment of used nuclear fuel disposal. *Atomic Energy of Canada* Report AECL -9584. Atomic Energy of Canada Ltd., Pinawa, Manitoba, Canada.

Grindrod, P. (1991) *Patterns and waves: the theory and applications of reaction-diffusion equations.* Oxford University Press.

Grindrod, P. and Takase, H. (1992) The saturation behaviour of clay buffers. *Intera Internal Report.* Intera Information Technologies Corporation, Henley-on-Thames, U.K.

Grindrod, P., Williams, M. and Grogan, H. (1990) STRENG: a source term model for vitrified high level waste. *Intera Internal Report* 12124-1, Version 3. Intera Information Technologies Corporation, Henley-on-Thames, U.K.

Higgo, J.J.W. (1987) Clay as a barrier to radionuclide migration. *Progress in Nuclear Energy* **19**: 173-207.

LeMire, R.J. and Garisto, F. (1989) The solubility of U, Np, Pu, Th and Tc in a geological disposal vault for used nuclear fuel. *Atomic Energy of Canada Report* AECL-10009. Atomic Energy of Canada Ltd, Pinawa, Manitoba, Canada.

McKinley, I.G. and Hadermann, J. (1984) Radionuclide sorption database for Swiss safety assessments. *Nagra Technical Report* NTB 84-40. Nagra, Baden, Switzerland.

Martens, K-H., Romstedt, P. and Fischer, H. (1993) Description of the MARNIE Code. Auftrags-Nr: 85511.

Mendel, J.E. (1984) Final Report of the defense high-level waste leaching mechanisms program. *Pacific Northwest Laboratory Report* PNL-5157. Battelle Pacific Northwest Laboratory, Richland, Washington, U.S.A.

Meunier, A. (ed.) (1993) *Clays and Hydrosilicate Gels in Nuclear Fields.* European Materials Research Society Symposium Proceedings 26, North Holland.

Mouche, E. and Vernaz, E. (1988) Aqueous corrosion of the French LWR solution reference glass: first generation model. In *Scientific Basis for Nuclear Waste Management Vol XI*, (eds. M.J. Apted and R.E. Westerman). Materials Research Society, pp 703-712.

Nagra (1985) Project Gewähr 1985: Nuclear waste management in Switzerland: feasibility studies and safety analyses. Volumes 1-8, Volume 9 (English Summary). *Nagra Gewähr Report Series* NGB 85-01/09. Nagra, Baden, Switzerland.

Nash, P.J., Nash, G. and Rodwell, W.R. (1989) Gas escape from vented waste canisters grouted into a repository vault. *UK Nirex Report* NSS/R201. United Kingdom Nirex Ltd, Harwell, U.K.

Nash, P.J. and Rodwell, W.R. (1990) Effects of gas overpresssurisation on the geological environment of a deep repository. *UK Nirex Report* NSS/R208. United Kingdom Nirex Ltd, Harwell, U.K.

NEA (1993) *Workshop on Near-Field Performance Assessment*, Cadarache, France, May 11-13, 1993. Nuclear Energy Agency, Paris, France.

NEA/OECD (1992) Workshop on: *Gas Generation and Release from Radioactive Waste Repositories.* Nuclear Energy Agency, Paris, France, pp 437.

Nilsson, L., Moreno, L., Neretnieks, I. and Romero, L. (1991) A resistance network model for radionuclide transport into the near-field surrounding a repository for nuclear waste. *SKB Technical Report* TR 91-30. Swedish Nuclear Fuel and Waste Management Company, Stockholm, Sweden.

Pigford, T.H. (1987) Performance assessment of coupled processes. In *Coupled Processes Associated with Nuclear Waste Repositories*, (ed. C-F. Tsang), Academic Press, New York.

Pigford, T.H. and Chambré, P.L. (1988) Near-field mass transfer in geologic disposal systems: a review. In *Scientific Basis for Nuclear Waste Management Vol XI*, (eds. M.J. Apted and R.E. Westerman). Materials Research Society, pp 125-141.

Pruess, K. and Bodvarsson, G. (1982) Hydrothermal conditions and resaturation times in underground openings for a nuclear waste repository. *Lawrence Berkeley Laboratory Report* LBL-18451. Lawrence Berkeley Laboratory, Berkeley, California, U.S.A.

Rai, D. and Kittrick, J.A. (1989) Mineral equilibria in the soil system. In *Minerals in the Soil Environment*, Soil Science of America Book Series No. 1, Soil Science Society of America, New York, U.S.A, pp 161-197.

Reardon, E.J. (1992) Problems and approaches to the prediction of the chemical composition in cement/water systems. *Waste Management* **12:** 221-239.

Robinson, P. (1992) STAG Guide for Version 2. *Intera Internal Report* IM2795-2, Version 1. Intera Information Technologies Corporation, Henley-on-Thames, U.K.

Robinson, P. and Worgan, K.J. (1992) The CALIBRE source term code: Technical Documentation for Project 90. *SKI Technical Report* SKI-TR 91:18. Swedish Nuclear Inspectorate, Stockholm, Sweden.

Robinson, P., Williams, M.J. and Worgan, K.J. (1992) A description of MESHNOTE. *Intera Internal Report* IM3479-1, Version 1. Intera Information Technologies Corporation, Henley-on-Thames, U.K.

Roddy, J.W., Claiborne, H.C., Ashline, R.C., Johnson, P.T. and Rhyne, B.T. (1986) Physical and decay characteristics of commercial LWR spent fuel. *Oak Ridge Laboratory Report* ORNL/TM-9591. Oak Ridge National Laboratory, Oak Ridge, Tennessee, U.S.A.

Romanoff, M. (1989) *Underground Corrosion.* NACE, Houston, Texas.

Romero, L., Moreno, L. and Neretnieks, I. (1991) A compartment model for solute transport in the near-field of a repository for radioactive waste. *SKB Technical Report* TR 91-48. Swedish Nuclear Fuel and Waste Management Company, Stockholm, Sweden.

Roxburgh, I. (1987) *Geology of High level Nuclear Waste Disposal.* Chapman and Hall.

Sagar, B., Codell, R.B., Walton, J. and Janetzke, R.W. (1992) SOTEC: a source term code for high level nuclear waste geologic repository, user's manual: Version 1.0. *CNWRA Report* 92-009. Southwest Research Institute, San Antonio, Texas, U.S.A.

Savage, D. (1986) The geochemical interactions of simulated borosilicate waste glass, granite and water at 100°-350°C and 50 MPa. *Nuclear and Chemical Waste Management* **6:** 15-39.

Shoesmith, D.W. and Sunder, S. (1992) The prediction of nuclear fuel (UO_2) dissolution rates under waste disposal conditions. *Journal of Nuclear Materials* **190:** 36-45.

SKB (1983) Final disposal of spent nuclear fuel - KBS-3, Vols 1-5. Swedish Nuclear Fuel and Waste Management Company, Stockholm, Sweden.

SKB (1992) SKB 1991: Final disposal of spent nuclear fuel. Importance of the bedrock for safety. *SKB Technical Report* TR 92-20. Swedish Nuclear Fuel and Waste Management Company, Stockholm, Sweden.

Snellman, M., Uotilla, H. and Rantanen, J. (1987) Laboratory and modelling studies of sodium bentonite groundwater interaction. In *Scientific Basis for Nuclear Waste Management Vol X*, (eds. J.K. Bates and W.B. Seefeldt). Materials Research Society, pp 781-790.

Stumm, W. and Morgan, J.J. (1981) *Aquatic Chemistry.* John Wiley and Sons, New York.

West, J.M. and McKinley, I.G. (1985) The geomicrobiology of nuclear waste disposal. In *Scientific Basis for Nuclear Waste Management Vol VII*, (ed. G.L. McVay). Materials Research Society, 487-494.

Wiborgh, M, Höglund, L.O. and Pers, K. (1986) Gas formation in a L/ILW repository and gas transport in the host rock. *Nagra Technical Report* NTB 85-17. Nagra, Baden, Switzerland.

Zwillinger, D. (1992) *Handbooks of Differential Equations, 2nd Ed.* Academic Press Inc., Boston, U.S.A.

5

The far-field

5.1 Geological nature of the far-field

The far-field, or geosphere, or geological barrier, can be defined as all the rock mass surrounding the repository as far as the surface. Some definitions of the far-field make no real distinction between the rock mass and the overlying biosphere. The rock mass surrounding a deep repository in any type of rock will be affected, out to a certain distance, by the presence of the repository. This affected zone, which will vary with rock type, repository design, heat loading from the repository, *etc.* is likely, at least in part, to be included within the near-field. Thus the near-field may be considered by certain waste disposal organisations to include the excavation damage zone (EDZ). The best location for the geosphere-biosphere interface is a matter for debate and is related more to modelling the transport of radionuclides across this boundary than to any other factor. For deep disposal concepts its precise location hardly influences the extent of the far-field.

The first questions that need to be asked when considering the far-field are:

- what is the role of the geological barrier and what can, or should, be expected of such a barrier?
- what are the desirable characteristics of the geological barrier?
- where are we most likely to find such a barrier?
- how necessary is a good geological barrier and could long-lived waste be disposed of almost anywhere at depth?

Some of the requirements of a geological barrier are intuitively obvious, such as the fact that the rock mass isolates the waste from man's environment and provides adequate shielding from radiation emitted from the waste. Beyond these lie more significant aspects of the barrier, in particular, for deep disposal:

- deep disposal places the waste out of reach of major disruptive events;
- deep disposal can provide an environment where the rates of natural processes affecting the waste are slow.

Recent work by SKB, as part of their Safety Assesment Review (SKB,

1992a, b) has come to the, perhaps rather provocative, conclusion that most sites which have been investigated within Sweden as part of their deep disposal research programme would prove to be suitable in terms of their safety-related requirements (all these sites are in crystalline rock). SKB also considered that the

> 'primary function of the rock is to provide stable mechanical and chemical conditions over long periods of time, so that the long-term performance of the engineered barriers is not jeopardised' (SKB, 1992a, b).

These conclusions are specifically related to SKB's KBS-3 disposal concept (SKB, 1983) in which the waste canister and the engineered components of the near-field are assumed to prevent absolutely for considerable periods of time, or at least very greatly minimise, the release of radionuclides from any part of the near-field.

These statements by SKB were made because performance assessment calculations had shown that in some situations the return times for radionuclides from a repository at 500 m depth were extremely short and the retardation offered by the geological barrier was minimal (SKB, 1992a). In other types of geological environment, such as in argillaceous rocks and evaporites, radionuclide transport from a repository at depth is usually far more restricted, with much longer return times for radionuclides to the biosphere (*e.g.* Nirex, 1989; Billington *et al.*, 1989; Bonne, 1991, Chapman and McEwen, 1991). Nevertheless, even in these environments there are likely to be some fast pathways for radionuclide transport which could effectively bypass the geological barrier (*e.g.* PAGIS, 1984; Nagra, 1988).

5.1.1 Potentially suitable geological environments

The necessary requirement for demonstrating the long-term safety of a repository system, particularly for a deep repository, implies that suitable geological and hydrogeological conditions are important. The general requirements for a suitable environment for deep disposal are presented in Chapman *et al.* (1986), and it is the following features of the *geological and hydrogeological environments*, and not the specific properties of particular rocks that are most important in this regard. It is also important when considering the environment in which a repository is placed not to limit consideration only to the geological environment but to consider the geological and hydrogeological environments together. The relevant features of the geological and hydrogeological environments are:

- predictable groundwater flow paths, preferably long and resulting in progressive mixing with older, deeper waters or leading to discharge at sea;

- predictable and suitable physical and chemical dispersion properties;
- ease of construction to allow for economic repository development;
- meeting the many accepted caveats regarding seismicity, depth, *etc.*

The environments considered to be most suitable for deep disposal, and which are likely to be generally applicable to most European countries, were then defined as (Figure 5.1):

- inland basinal environments;
- seaward dipping and offshore sediments;
- low permeability basement under sedimentary cover;
- hard rocks in low relief terrain, and
- small islands.

Evaporites, in particular salt domes, were not included in this list, since it was prepared with reference to the geology of the U.K. In other European countries, most notably Germany, the Netherlands, Spain and Denmark, evaporites may offer some of the most suitable disposal environments. The rationale behind the choice of these environments is given in Chapman *et al.* (1986).

It is important at this stage to appreciate that the flow of groundwater and the transport of radionuclides in the far-field will probably be the area of greatest uncertainty in any performance assessment. Many of the uncertainties associated with the far-field will be intractable and it may not be possible to increase the confidence levels of far-field structures and properties to those initially believed desirable. This fact does not imply, however, that no confidence or weight should be placed on far-field properties and that all emphasis should, as a consequence, be placed on the near-field engineered barriers (see discussion on these matters in SKB (1992a), SKI (1991) and Nagra (1994)). Another important factor to consider in this regard is that a robust performance assessment is probably only possible with relatively simple geological and hydrogeological environments (see, for example, McCombie *et al.*, 1991). The heterogeneity and spatial variability of the system is important here, especially for crystalline rocks, in which the heterogeneity of the hydraulic system can be particularly marked (Hodgkinson and McEwen, 1991).

5.1.2 Hydrogeological properties

Rocks potentially suitable for the deep disposal of long-lived waste will have low hydraulic conductivities. Great effort is expended to characterise these hydraulic properties. Although this work deserves considerable attention, it can also be argued that understanding dispersion and retardation behaviour deserves similar attention. Unfortunately, this is often not the

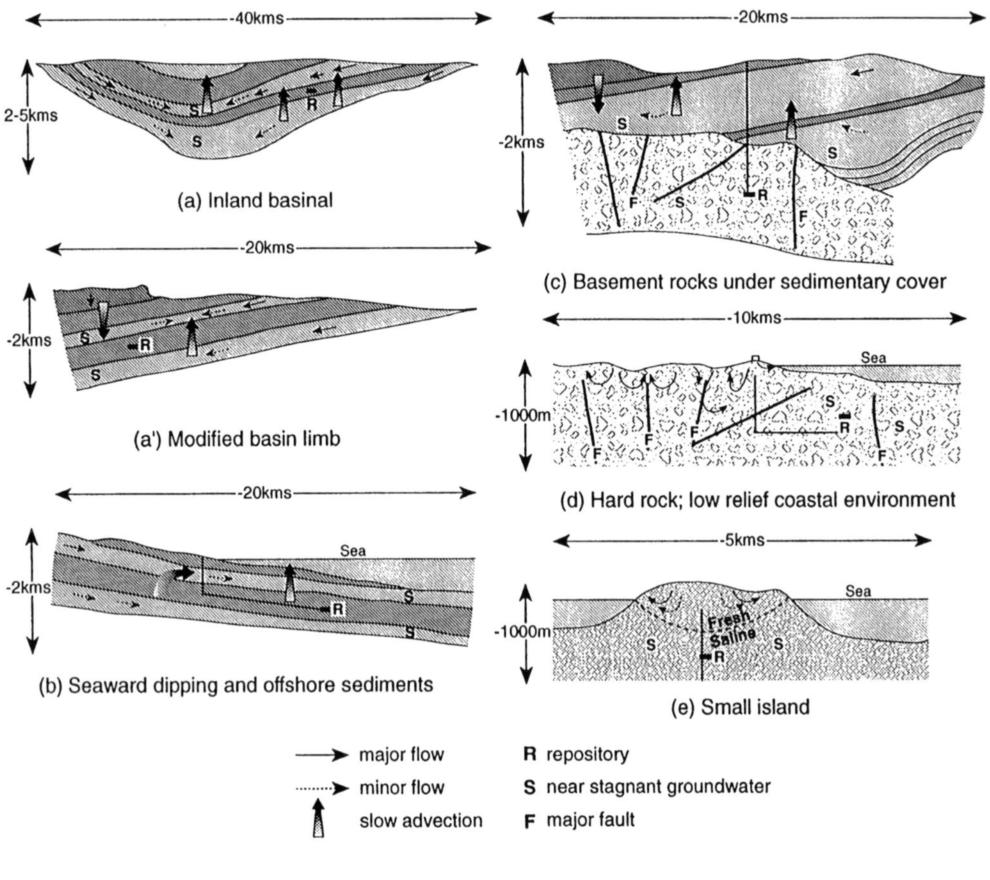

Figure 5.1 Potentially suitable environments for the deep disposal of long-lived radioactive waste (from Chapman *et al.*, 1986). Reproduced by permission of IAEA.

case (see Section 5.3). This imbalance in information required for modelling radionuclide transport should be borne in mind when reading the remainder of this section.

Values of the hydraulic properties of the types of rocks that might be considered for radioactive waste disposal are presented in Table 5.1. This table also includes two calculated values: (flux (the volume of water flowing through a unit section in a given time) and velocity):

$$\text{flux} = \text{hydraulic conductivity} \times \text{gradient} \tag{5.1}$$

$$\text{velocity} = \frac{(\text{hydraulic conductivity} \times \text{gradient})}{\text{porosity}} \tag{5.2}$$

The values of hydraulic conductivity which are presented in Table 5.1 represent the gross averages for a cube of rock of at least 100 m per side. In argillaceous formations considered for the disposal of radioactive waste hydraulic conductivities towards the lower value are most appropriate. In such circumstances fluxes may be so low that diffusion is the dominant transport process in some argillaceous rocks, especially at depth. In general, for $k < 10^{-10}\ ms^{-1}$ diffusional transport is likely to be dominant. These averages, however, do not include the potentially faster pathways formed by fracture zones, *etc.* which can effectively provide bypasses to the bulk of the rock mass. These faster pathways are discussed later. The path lengths and the associated travel times are, therefore, the maximum likely in these types of rock.

Table 5.2 illustrates the maximum and minimum values of groundwater travel times for the same rocks and environments as presented in Table 5.1. These pathlengths and travel times can be applied to a potential repository at a depth of 500 m or more. They illustrate that there can be very large ranges of travel times that are possible for all the rock types, though it needs to be emphasised that these represent extreme ends of the spectrum and that sites selected for the deep disposal of waste are more likely to lie nearer the maximum end of the range.

The hydraulic properties and in particular the transport properties of fracture zones and faults can vary very considerably. An example of the various types of conductive zone that can exist is provided by the site characterisation work that has been carried out by SKB over the past 15 years. SKB have erected a hierarchy of such zones which is presented in Table 5.3 (based on Birgersson *et al.*, 1992).

It can be seen that the porosity of the rock mass plays an important part in determining the groundwater flow velocities. Crystalline rocks normally have very low porosities. The porosity in these types of rock is made up of three components. These rocks are, however, often described as having dual porosities, as one of the components is unimportant. The three components are:

Table 5.1 Typical ranges of hydraulic conductivity (k), porosity and hydraulic gradient appropriate to each rock type are presented, based on compilations of data from around the world (*e.g.* Brace (1980), Tellam and Lloyd (1981), Clauser (1992)). The gradients are typical of those found in the UK and are probably equally applicable to the less mountainous parts of Europe. The approximate maximum and minimum fluxes and real (non-Darcian) velocities are calculated from these data. Evaporites have generally similar low hydraulic conductivities to those of plastic clays. The division between above and below 100 m represents the difference between that part of the rock mass which is likely to have undergone some form of stress relief, thereby possessing an enhanced hydraulic conductivity and porosity, and that part of the rock mass which is more likely to represent the host rock environment in which a repository might be located.

Rock type	k ($m\ s^{-1}$)	Porosity	Gradient	Flux ($l\ y^{-1}m^{-2}$)	Velocity ($m\ y^{-1}$)
Plastic clay•					
Above 100 m (max)	10^{-7}	0.5	0.2	640	1.3
(min)	10^{-10}	0.3	0.05	1.6	0.0005
Below 100 m (max)	10^{-8}	0.5	0.2	64	0.13
(min)	10^{-12}	0.3	0.05	0.0016	5×10^{-6}
Shale/mudstone•					
Above 100 m (max)	10^{-6}	0.3	0.2	6400	21
(min)	10^{-9}	0.2	0.05	1.6	0.008
Below 100 m (max)	10^{-7}	0.25	0.2	640	2.6
(min)	10^{-10}	0.05	0.05	0.16	0.003
Crystallines					
Above 100 m (max)	10^{-7}	0.05	0.1	320	6.4
(min)	10^{-9}	0.01	0.001	0.032	0.0032
Below 100 m (max)	10^{-8}	0.01	0.1	32	3.2
(min)	10^{-11}	0.001	0.001	3×10^{-4}	3×10^{-5}
Aquifer					
Maximum	10^{-4}	0.1	0.01	3×10^{4}	320
Minimum	10^{-7}	0.05	0.0005	1.6	0.03

• In formations considered for the disposal of radioactive waste hydraulic conductivities towards the lower value are most appropriate. In such circumstances fluxes may be so low that diffusion is the dominant transport process in some argillaceous rocks, especially at depth. In general, for $k < 10^{-10}\ ms^{-1}$ diffusional transport is likely to be dominant.

Table 5.2 Approximate maximum and minimum values of groundwater travel times for the same rocks and environments as presented in Table 5.1, which can be applied to a potential repository at a depth of 500 m or more.

Rock type	Velocity (m y $^{-1}$)	Path length (m)	Travel time approx.(y)
Plastic clay•			
Above 100 m (min)	1.3	50	40
(max)	0.0005	100	2×10^5
Below 100 m (min)	0.13	50	390
(max)	5×10^{-6}	100	2×10^7
Shale/mudstone•			
Above 100 m (min)	21	50	2
(max)	0.01	200	2×10^4
Below 100 m (min)	2.5	50	20
(max)	0.003	200	6×10^4
Crystallines			
Above 100 m (min)	6	500	80
(max)	0.003	2000	6×10^5
Below 100 m (min)	3	500	160
(max)	0.0003	2000	6×10^6
Aquifer			
Minimum	320	2×10^3	6
Minimum	0.03	2×10^4	6×10^5

• In formations considered for the disposal of radioactive waste hydraulic conductivities towards the lower value are most appropriate. In such circumstances fluxes may be so low that diffusion is the dominant transport process in some argillaceous rocks, especially at depth. In general, for $k < 10^{-10}$ ms^{-1} diffusional transport is likely to be dominant.

- The porosity of that component of the fracture network along which groundwater flow takes place.
- The porosity of the *dead end pores* that are connected to the flowing fractures. Included in this are the components of the fracture network (probably the majority of the network, (*i.e.* as much as 90 %) that plays an insignificant part in the groundwater flow, but is important in the transport of solutes.
- The porosity that is made up of closed pores within the rock, *e.g.* intercrystalline microcracks. This component of the total porosity is unlikely to play any significant part in solute transport.

Sedimentary rocks and evaporites

Sedimentary rocks, and in particularly clays, have very much greater porosities, with the result that they can have very much slower groundwater flow velocities and greater tendencies for matrix diffusion and retardation

Table 5.3 Hierarchy of discontinuities within the Swedish basement rocks used by SKB in their modelling activities. Similar hierarchies of fracture zones within other rock types have not been developed, in part because the situation is more complex in sedimentary rocks and partly because there has been less interest in this subject in non-crystalline environments. Based on data presented in Birgersson *et al.* (1992).

Type of zone	**Spacing** (m)	**Mean k** ($m\ s^{-1}$)
Regional fracture zone	$3.2\text{-}6.9 \times 10^3$	10^{-7}
Local fracture zones	400-800	10^{-8}
Minor fracture zones	50-100	10^{-9}
• Fractures	0.4-0.8	10^{-11}
Subhorizontal fracture zones	700	10^{-7}

• The spacing between fractures lies within this range, but hydraulic measurements indicate that the ratio between the conductive and the total number of fractures is approximately 1:10, giving a distance between conductive fractures of the order of 4-8 m.

mechanisms (these subjects are discussed later in this Chapter and also, in relation to the near-field, in Chapter 4). The more indurated sedimentary rocks, *i.e.* well cemented mudstones and marls, those which tend to be more brittle, can support open fracture networks and will also tend towards having dual porosities (*e.g.* the Opalinus Clay of northern Switzerland (Nagra, 1988), Gera *et al.*, 1992).

Groundwater flow in argillaceous rocks is a very difficult subject to study, since there are complex coupled processes that are believed to take place between the clay minerals and the pore fluids. The subject is currently the subject of a state-of-the-art review funded by the NEA (Horseman *et al.*, 1994) and a recent review for ENRESA (Horseman, 1993) also discusses the subject in detail, but from a different standpoint.

In addition to this problem is the fact that most, if not all, of these rocks are likely to be in a non-equilibrium state, with lower than expected *measured* hydraulic heads being the norm. It is, however, doubtful whether the values of hydraulic head measured are actually what they purport to be. It seems more likely that these values represent the combined effects of both the hydraulic head and the *head* due to chemical effects (*i.e.* the head generated by the effect of the chemistry of the pore fluid on the stress state within the clay). What are described as being *abnormal* head distributions have been found during Swiss site investigations at the potential repository site of Wellenberg, and the complexities of groundwater flow in these types of rock have been reviewed by Neuzil (1986, 1993, 1994) and by Horseman (1993) and Horseman *et al.* (1994). In tectonically more active systems, such as are present in Switzerland, tectonically driven groundwater flow may also be the norm (Ge and Garven, 1992). The potential problems associated with the use of such rocks for disposal purposes is discussed in Section 5.1.5

Evaporites vary very considerably in their reponse to stress, with the extreme end-members being represented by anhydrite, which is very brittle, and several different types of K-bearing salts which exhibit very low yield strengths (Sandia, 1992). Anhydrites are, therefore, often fractured and possess a dual porosity, whereas many other evaporites are effectively self-sealing and have very low porosities (Sandia, 1992). Halite (rock salt) lies in the latter category (PAGIS, 1984).

5.1.3 The complexity of geological systems

Even potentially suitable rock types are almost certain to have faster geosphere pathways and this will be true of all types of geological environment. The contrast in transport rates between such zones and the surrounding rock mass can be several orders of magnitude and these pathways can, therefore, by-pass significant parts of the geological barrier. Their possible presence and their effect on the performance of a repository need to be taken into account probabilistically, and their detection and the analysis of their distributions is often an important and difficult part of a site characterisation programme. The problem of these fast pathways can be described under the heading of the *spatial variability* of the rock mass, a factor which needs to be considered at all scales (Hodgkinson and McEwen, 1991; other papers included within NEA (1991)).

The problems are related to a variety of factors which vary between the rock types considered for disposal, but can be categorised into five general areas:

1. Discontinuities, normally fractures, fracture zones and faults. There will be a continuous spectrum in the size of such discontinuities, the properties of which may also represent a similar spectrum. For example, it is possible that the larger the fracture zone the more likely it is to have a higher transmissivity (as is generally assumed in performance assessment of crystalline rocks (*e.g.* SKB, 1992a; Nagra, 1994)), though this relationship cannot be assumed for all rock types. Faults in some geological environments do not appear to be associated with enhanced transmissivities (*e.g.* faults in the Borrowdale Volcanic Group at Sellafield (Nirex, 1993)).
2. Changes in lithology. This is not significant for many hard rocks, but in sedimentary rocks lithological changes can be very pronounced and occur

gradationally or abruptly (Andrews *et al.*, 1991). Lithological changes in evaporite environments can be particularly significant, *e.g.* the deposition of anhydrite and K-bearing salts within an halite-dominated sequence, especially since the potential host rock, which is most likely to be halite, has excellent transport properties (PAGIS, 1984; Langer *et al.*, 1991).

3. Effects due to the depositional environment of a sediment or volcaniclastic that can impart a heterogeneity to the rock mass at all scales (Andrews *et al.*, 1991). The scale of this heterogeneity is likely to be larger in volcaniclastic rocks. Included in this area are diagenetic effects, those due to sediment dewatering and general syn-depositional changes (an example of this in the Culebra Dolomite at the WIPP site (Beauheim, 1991)).
4. The emplacement of an igneous mass or a salt dome often introduces some structural complexity due to deformation and/or cooling, and such structures are likely to provide some form of heterogeneity to the rock mass (Langer *et al.*, 1991). The emplacement of large volumes of volcanic material, especially if it occurs rapidly, as in the case of tuffs for example, can be associated with syn-depositional deformation on a large scale due to collapse processes (*e.g.* the deposition of the Borrowdale Volcanic Group at Sellafield (Nirex, 1993)). The geometry of these syn-depositional deformation structures is likely to be complex.
5. Deformation of the rock mass that imparts a heterogeneity, often of a variable nature, to the rock. Such heterogeneity can occur within rocks that have suffered intense deformation at high metamorphic grade, *e.g.* mylonites and high-grade gneisses, all the way down to rocks that have suffered from weak deformation with possibly none, or limited, associated metamorphism (*e.g.* the Borrowdale Volcanic Group at Sellafield (Nirex, 1993)). This area is obviously directly linked with item1.

Spatial variability in fractured hard rocks

The spatial variability of a rock mass cannot necessarily be determined solely by determining the location and orientation of faster pathways, because the pathways themselves have internal variability and their potential intersections need also to be examined (*e.g.* see Olsson *et al.*, 1992 for a discussion of work within the Site Characterisation and Validation (SCV) block at Stripa). A single fracture plane will, for example, have an associated tortuosity and an *equivalent wetted surface area* (*i.e.* the area which can be assumed to be in contact with migration solutes) (Hodgkinson and Cooper, 1992).

The intersection of this particular fracture plane, which will possess a limited area, with other fracture planes and the properties of these intersections will also determine the transport properties of the rock. What is invariably found is that only a limited number of fractures transmit significant volumes of groundwater and, therefore in terms of analysing groundwater flow, it is only necessary to consider a limited percentage of such fractures (Olsson *et al.*, 1992). Unfortunately, a proper understanding of solute transport requires that all fractures are considered, since diffusion away from the flowing fractures is an important component in transport calculations (Hodgkinson and Cooper, 1992; Hodgkinson and McEwen, 1990). It is obviously not possible to follow a purely deterministic route and treat fractures individually. A more promising approach is to construct a numerical model of the sheetwork of intersecting fracture planes, whose characteristics are specified to conform to statistical distributions derived from field measurements (Hooper *et al.*, 1993). Such discrete fracture models provide a useful tool for assimilating and analysing fracture characteristics and hydraulic tests, as has been demonstrated at Stripa (*e.g.* Dverstorp and Andersson, 1990; Dverstorp, *et al.*, 1992). They can be used to extrapolate information from single fractures to fracture systems, for example to derive large scale permeabilities, but are too computationally intensive to perform detailed regional studies.

Nirex have used such a modelling approach, firstly over several km^3 of the Borrowdale Volcanic Group at Sellafield, where the codes NAPSAC and FracMan have been used to generate equivalent transmissivity fields and model hydraulic head distributions, and at smaller scales where REVs (Representative Elementary Volumes) for groundwater flow and for solute transport have been developed using a fracture network approach (Hooper *et al.*, 1993). As part of this work they are also examining scale effects in modelling the permeability of the rock mass and in the determination of effective transport parameters. Scale effects are being examined over scales up to 50 m cubes of the rock mass to examine the usefulness of continuum approximations for modelling flow and transport. Ideally, it would be preferable to use the fracture network models directly where sufficient data exist to characterise them, However, there will be occasions where there are insufficient data to describe the network. In such cases it may be possible to use information about the site to constrain the effective properties of formations and, in these circumstances, it is useful to know how fracture network systems behave as a function of the investigation scale (Hooper *et al.*, 1993).

More importantly, however, there are current significant limitations on the size of the region and the complexity of the physical processes that can be addressed directly. Approximations have to be made, and whether these involve simplified networks or continuum approximations, the validity of the chosen approach needs to be demonstrated, by comparison with the results of a detailed model over a smaller volume of the rock (Hooper *et al.*, 1993).

Spatial variability in sedimentary rocks

Spatial variability in sedimentary environments can be due to a combination of the variability introduced by the depositional environment combined with

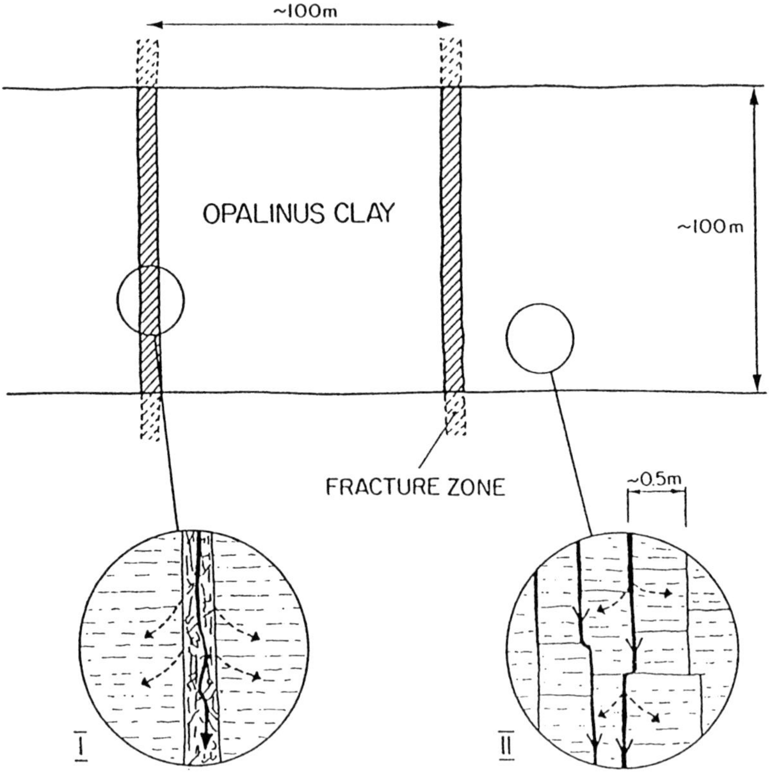

Figure 5.2 Heterogeneity within a sediment illustrated by a schematic model of the flow system within the Opalinus Clay (Switzerland). Flow system 1 represents fracture zones as observed in a large clay pit, with a separation of these zones of approximately 100 m. Flow system 2 represents a closely spaced system of joints and fractures (spacing 0.5 m approx.) as observed in outcrops (from Nagra, 1988). Reproduced by permission of Nagra.

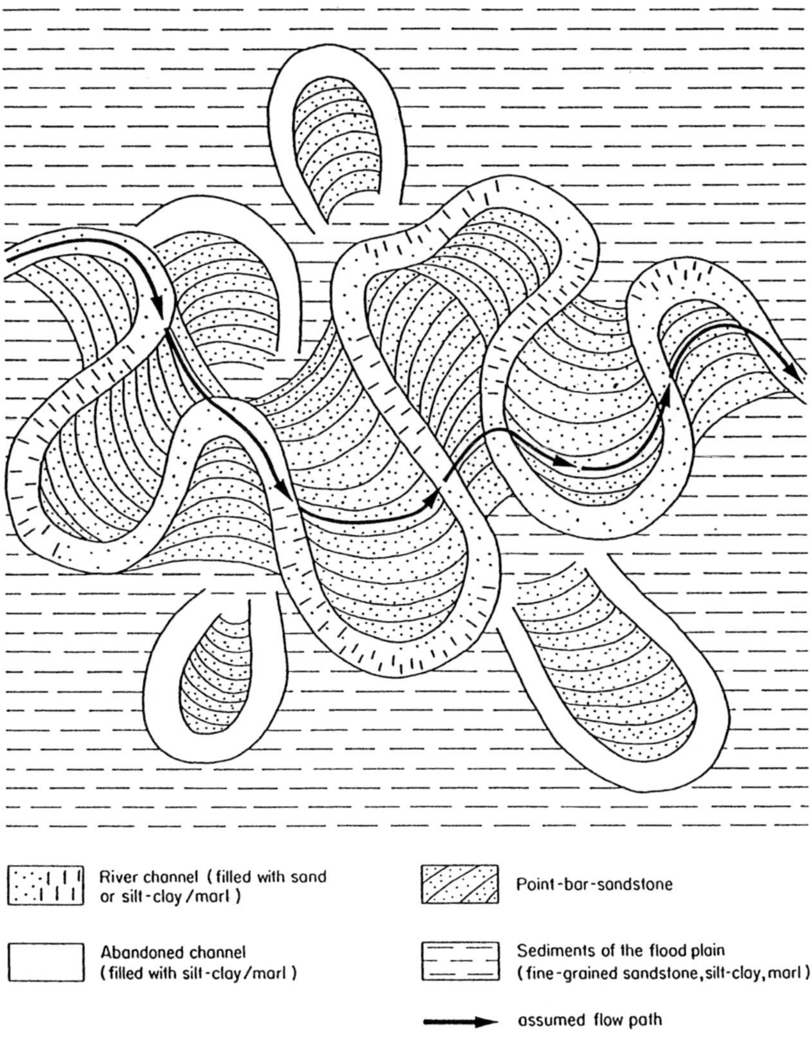

Figure 5.3 Schematic representation of heterogeneity within a sedimentary environment considered for radioactive waste disposal (the Lower Freshwater Molasse, Switzerland). The flow system is composed of sand bodies in a marl/shale matrix. Only a small number of sand bodies are connected and flow occurs through the higher permeability zones of the connected sand channels and sand beds (from Nagra, 1988). Reproduced by permission of Nagra.

later compaction and diagenetic changes and perhaps deformation. Different types of spatial variability in sediments are illustrated with reference to waste disposal in Figures 5.2 and 5.3. A large amount of work has been carried out in the oil industry examining the spatial variability of reservoirs. In radioactive waste disposal it is the low permeability rocks that are of more interest and far less work has been carried out examining the variability within shales and clays (Mackay and O'Connell, 1991). In the more indurated argillaceous rocks it may well be the fracture system that is of most importance in solute transport (*e.g.* the conceptual model of the Opalinus Clay in northern Switzerland (Nagra, 1988)). Spatial variability within a bedded evaporite sequence has been examined in extreme detail as part of the assessment of the WIPP repository site in New Mexico (*e.g.* LaVenue and RamaRao, 1992; Sandia, 1992; Beauheim, 1991).

5.1.4 Hard rock environments

Hard rocks provide some of the easiest constructional environments at depth and the unit cost of disposal can be the lowest of any of the rock types. However, this environment suffers from several problems:

- There often are frequent fast, or faster, pathways within the rock mass and the detection of and the measurement of representative properties of such pathways is inherently difficult. It is only by linking hydrogeological measurements and modelling with geochemical and structural geological studies that sufficient understanding is likely to be obtained (*e.g.* SKB, 1992a; Mazurek *et al.*, 1993; Nagra, 1994; Mazurek, 1994).
- Limited retardation is often provided as the effective porosity is often very low, the minerals lining and filling the fracture network, through which the majority of the radionuclides pass, do not necessarily possess high sorption capacities and the possibilities for retardation by matrix diffusion can be limited (Barker, 1991). In addition, in contrast to the potentially long pathways that are possible in sedimentary environments, pathways in crystalline rocks are invariably short by comparison (SKB, 1992a; Mazurek *et al.*, 1993).
- The groundwater flow and radionuclide transport paths are frequently complex with further complexities introduced by the extent of tortuosity within the fracture planes (*e.g.* Mazurek, 1994).
- The EDZ around the repository can be the most marked of any of the rock types with the possibility of more rapid transport in the zone between the near- and far-fields (Barton *et al.*, 1992; NEA, 1988).

The critical factors within hard rock environments are:

- the properties of the *averagely fractured rock*, *i.e.* that rock between the repository and the closest fast pathway (normally a fracture zone);

- developing an understanding of the geometry of the fracture network;
- determining the extent to which retardation mechanisms, such as matrix diffusion, can be relied on to operate;
- the extent to which fractures created during construction of the repository (the EDZ) and the fractures and fracture zones that intersect the repository can be adequately sealed.

5.1.5 Argillaceous rocks

Argillaceous rocks are potentially suitable for deep disposal because they tend to have low hydraulic conductivities, are associated with low solute transport rates (due to their generally high sorption capacities) and can have discontinuities which are self-sealing. Groundwater modelling has shown that transit times for water from within a relatively deep clay to the surface can, at least in theory, be extremely long (Billington *et al.*, 1989; Sumerling, 1992). There is ample evidence of such rocks providing seals to gas and oil over very extended periods of time and natural analogues have illustrated that plastic clays provide very efficient chemical barriers (Miller *et al.*, 1994).

There are, however, certain problems associated with argillaceous environments which can jeopardise their role as suitable host rocks for long-lived waste (Gera, 1992 *et al.*, 1992; Horseman, 1993; Horseman *et al.*, 1994):

- Although argillaceous rocks do, in general, possess low hydraulic conductivities and can have high sorption capacities, they can contain sedimentary or tectonic structures which can have considerably higher conductivities. Such potential fast transport pathways can be on a small scale due to diagenetic effects (*e.g.* dewatering structures) or can be on a larger scale and due to changes in the depositional environment (*e.g.* sand lenses) or could be due to tectonic effects (*e.g.* faults and associated fracturing).
- There is a general lack of understanding of groundwater flow and solute transport in these types of rocks.
- It is difficult to measure properties *in situ.*
- There are likely to be non-equilibrium conditions within such rocks, partly due to their intrinsic low hydraulic conductivities and partly because of the complex coupling that takes place between the various processes.
- There are often large hydraulic anisotropies within argillaceous rock environments, thereby allowing the propagation of hydraulic heads over large distances.
- There are constructional constraints which can, in the extreme, be very severe, making disposal considerably more difficult and expensive than in other rock types.
- There are thermal problems, due to the lower thermal diffusivity of clays.

Temperature rises within the clay induce increases in pore pressures which are difficult to dissipate due to the low hydraulic conductivity.

- The transport of gas through many clays is, on average, very slow and this could result in excess gas pressures being developed in the near-field (due to gas production within the repository), with the possibility of the rapid, and possibly pulsed transport of gas through the rock mass (Horseman and Harrington, 1994) which could increase radionuclide transport rates.

5.1.6 Evaporites

Pure rock salt (halite) appears to provide one of the most suitable environments for deep disposal since the transport rates of radionuclides within such an environment are very slow for most situations that can be envisaged. Rock salt has a higher thermal diffusivity than crystalline or argillaceous rocks and therefore temperature rises can be lower, it is also essentially self-sealing. Suitable salt deposits are those that possess low contents of impurities of certain other evaporitic minerals, some of which could cause significant problems, and low clay contents. It tends to be easier to find suitable environments within salt domes than within bedded salt deposits and some salt domes, because of their size, can offer very large, potentially suitable disposal volumes (PAGIS, 1984; Langer *et al.*, 1991). The potential problems with rock salt are (Chapman and McKinley, 1987; PAGIS, 1984; Schmitt and Klarr, 1991):

- Evaporites, especially those in the form of salt domes, are more often associated with hydrocarbons than other rock types of interest for deep disposal.
- Salt domes and salt pillows can be in structurally complex environments, in particular the margins to the salt can be very complex and the country rocks into which the salt has been injected can also be structurally complicated. A salt dome in such a geological environment could result in an overly complex site characterisation programme, but would normally indicate that the geological environment was unsuitable.
- Even though the rock salt may have a simple structure, any performance assessment would almost certainly need to consider the surrounding country rocks and their complexity could eliminate the salt from consideration.
- Salt domes, in particular, are frequently composed of multiple injections of a variety of salt types, only some of which may be suitable for disposal. A large salt dome may only have a relatively small volume of salt within a suitable depth range. (There will have been more mixing of different salt horizons within a dome than within a salt pillow).
- Associated with halite are often other evaporite minerals, some of which are unstable and some, such as anhydrite, are much more brittle than

halite and cannot accommodate the strain rates imposed by the halite flow. Fracture of the anhydrite results and the larger ruptured anhydrite horizons are frequently accompanied by brine pockets which can be at lithostatic pressure.

- Faults can cross-cut salt formations and remain as potentially fast transport pathways, even if faulting salt against salt, due to the amounts of insoluble residue within the fault zone.
- Salt domes can have relatively large rates of uplift and be subject to long-term dissolution effects. Such effects are made more pronounced in areas subject to glaciation.
- Site investigations associated with salt domes may need to be extensive and protracted, as the structures and distribution of lithofacies in the region of the dome can be complex. This is the situation even where the releases possible from a repository within the dome are likely to be minimal.
- If retrievability is required of the disposal system then disposal within rock salt is substantially more difficult than in other rock types.
- Bedded evaporites, such as at the WIPP site, can, under most scenarios, provide good long-term containment of radionuclides, but inhomogeneities within the evaporite sequence pose problems for modelling purposes. In general, bedded salt is more heterogeneous and spatially more variable than domal salt. In addition, the WIPP site is associated with high pressure brines below the disposal horizon and with the presence of local deposits of both oil and gas. High pressure fluids within the rock mass are more likely to be found within evaporites than in other rock types.

5.1.7 The potential problems of ILW disposal as compared with the disposal of HLW

The geological structure of rock masses imposes constraints on the disposal of long-lived ILW, where the volumes of waste for disposal are often considerably in excess of the volumes of HLW, and the development of a safety case for an ILW repository may well be more difficult than that for HLW (McCombie *et al.*, 1991). This problem tends to be exacerbated for those countries that carry out reprocessing, as the volume of long-lived ILW is increased.

In hard rocks the separation of fractures and fracture zones within the rock mass is likely to be of the order of the size of the underground structures necessary for waste disposal, and it will not be possible to separate the waste from these potentially faster transport pathways in the same way that is possible for the disposal of HLW. The source term and the evolution of the near-field used in the performance assessment for the site are likely to be more complex than is the case for HLW spent fuel disposal. In argillaceous rocks the size of the underground openings is likely to be restricted for geotechnical reasons and the use of tunnels with circular cross-sectional linings, a requirement in the weaker sediments, limits the waste packing density and the size of objects possible for disposal (Nirex, 1989; Billington *et al.*, 1989).

5.2 Geochemical aspects

This Section discusses the chemical and mineralogical properties of the far-field relevant to the performance assessment of a deep geological repository for radioactive wastes. Geochemical information for the far-field is important for two fundamental reasons:

- chemical compositional data for groundwaters can be useful to establish residence times, directions and rates of groundwater flow and may be used as an indirect means of validating hydrogeological flow concepts for a specific repository site,
- to establish radionuclide transport behaviour, *e.g.* solubility, sorption properties.

In order to understand the concepts concerned with radionuclide migration and retardation it is necessary to have some basic knowledge of geochemical features of rocks and groundwaters considered suitable for waste disposal.

5.2.1 Chemical and mineralogical properties of the far-field

There have been many different rock types with widely different properties considered suitable for the disposal of radioactive wastes (Section 5.1). The main rock types are: granite; basalt; tuff; metamorphic rocks such as gneisses or schists; shales; mudstones; clays; and evaporites. Not only are there wide variations in physical properties (Section 5.1), they differ in chemical and mineralogical properties also.

Table 5.4 illustrates the major element composition of these rock types. The salient features of these chemical compositions are that:

- silica is the main component of all rocks considered and may constitute 70 % of granite and tuff,
- aluminium is the next most abundant oxide, emphasising the often amphoteric behaviour of aluminosilicate minerals,
- iron is the most abundant redox-sensitive element, and is considerably greater in rocks such as basalts and amphibolites in comparison with other rock types. The relatively small amounts of reduced iron in most rocks means that there is a finite amount of natural redox buffer capacity,
- the volatile content of rocks such as shale and clay is relatively high and may pose problems for the disposal of heat-emitting waste.

Table 5.4 Typical chemical compositions (wt % oxides) of rocks considered as hosts for radioactive wastes. Data from Smith (1982). Reproduced by permission of Sceptre Books.

Oxide	Granite	Basalt	Tuff	Amphibolite	Schist	Shale
SiO_2	70.8	49.0	72.8	49.3	63.3	62.4
TiO_2	0.4	1.0	0.3	1.2	1.4	1.1
Al_2O_3	14.5	18.2	13.3	16.9	17.9	16.6
Fe_2O_3	1.6	3.2	1.5	3.6	3.6	3.2
FeO	1.8	6.0	1.1	6.8	2.6	2.1
MgO	0.9	7.6	0.4	7.0	1.6	2.5
CaO	1.8	11.0	1.1	9.5	1.9	1.7
Na_2O	3.3	2.5	3.6	2.9	1.3	0.9
K_2O	4.0	0.9	4.3	1.1	3.1	3.0
H_2O	0.8	0.4	1.1	1.5	2.6	5.3
CO_2			0.1			1.0
Total	99.9	99.8	99.6	99.8	99.3	99.8

Table 5.5 Typical mineralogical compositions (volume %) of rocks considered as hosts for radioactive wastes. Data from Smith (1982). Reproduced by permission of Sceptre Books.

Mineral	Granite	Basalt	Tuff	Amphibolite	Schist	Shale
quartz	30		4		32	17
alkali feldspar	60	5	6			
plagioclase	5	45		42	18	
pyroxene		40				
amphibole				50		
olivine		5				
biotite	4			5	7	
muscovite					38	1
magnetite	1	5		3	3	1
staurolite					2	
clays						80
calcite						1
glass + pumice			90			
Total	100	100	100	100	100	100

Table 5.5 shows the mineralogical composition of the same rock types. Features of relevance to waste isolation are:

- Most rock types (except for shales, clays, mudstones *etc.*) are made up of minerals stable at high temperature (and high pressure in some instances) and are intrinsically unstable at earth surface or shallow burial conditions. As cold/warm groundwater circulates through these rocks, chemical reactions will occur, dissolving minerals and precipitating others. This implies that these rock types will have different mineral assemblages lining groundwater flow paths. These 'veins' will consist of minerals stable at low temperature in the presence of water, such as clays, hydroxides, zeolites and micas. The veins may provide a physical seal against the rock matrix, thus limiting diffusion of solutes into the matrix.
- There is a considerable range in the rates of mineral-water reactions (Table 5.6) so that some rock types containing unstable phases such as volcanic glass, calcic plagioclase, pyroxenes, amphiboles and olivine (basalt, gabbros *etc.*) weather appreciably more quickly than others. This may lead in some instances to greater solute loads in groundwaters contained in these rock types, but more usually, to thicker alteration zones along groundwater flowpaths. This may increase sorptive properties in some instances.
- Quartz is essentially inert in low temperature groundwater systems so that water-rock reactions in granitic basement rocks are dominated by feldspar (especially calcic plagioclase feldspar) and mafic minerals such as biotite or amphibole.

Evaporites are highly variable mineralogically, but consist principally of a variety of chlorides and sulphates such as halite (NaCl) gypsum ($CaSO_4.2H_2O$), anhydrite ($CaSO_4$), carnallite ($KMgCl_3.6H_2O$), sylvite (KCl) and carbonates, *e.g.* dolomite ($CaMg[CO_3]_2$) *etc.* Evaporites are important sources of chloride and sulphate ions to groundwaters circulating in sedimentary basin sequences.

5.2.2 Groundwater and water-rock reactions

Fluids occur in all rock types, although in some arid regions the depth to the water table may be considerable. Brine inclusions will even be present in evaporites. The amount and mobility of groundwater will depend upon the physical properties of the rock (porosity, permeability) and its regional setting (Section 5.1). The chemical composition of groundwaters can be complex and will reflect the history of water presence in the rock. Groundwaters may be dilute enough to be potable, or be as saline as brines capable of precipitating halite. First and foremost, groundwater compositions reflect geological history (*e.g.* sea-level changes, proximity to evaporite deposits *etc.*), and secondly, the composition of the host rock. For example, a sandstone aquifer in the

Table 5.6 Relative dissolution rates of principal rock-forming minerals at 25 °C and neutral pH. Reprinted from *Geochimica et Cosmochimica Acta* Volume 53, Brady, P.V. and Walther, J.V., 'Controls on silicate dissolution rates in neutral and basic solutions at 25 °C', pages 2823-2830, Copyright 1989, with kind permission from Elsevier Science Ltd., The Boulevard, Langford Lane, Kidlington OX5 1GB, U.K.

Mineral	-log rate mol cm^{-2} s^{-1}
Nepheline, $NaAlSiO_4$	13.3
Olivine (forsterite), Mg_2SiO_4	14.2
Clinopyroxene (diopside), $CaMgSi_2O_6$	15.0
Andalusite, Al_2SiO_5	15.1
Orthopyroxene (enstatite), $MgSiO_3$	15.2
Ca-plagioclase (anorthite), $CaAl_2Si_2O_8$	15.5
Analcime, $NaAlSi_2O_6.H_2O$	15.7
Na-plagioclase (albite), $NaAlSi_3O_8$	15.9
Quartz, SiO_2	15.9
Chrysotile, $Mg_3Si_2O_5(OH)_4$	16.3
Pyrophyllite, $Al_2Si_4O_{10}(OH)_2$	16.3
Kaolinite, $Al_2Si_2O_5(OH)_4$	17.4

centre of a sedimentary basin may have a groundwater with a salinity several times that of seawater, whereas the same sandstone at the basin margin may have experienced meteoric recharge and may contain a groundwater dilute enough for potable water supplies. Water-rock reactions may have affected the composition of both groundwaters, but clearly, in this instance, geological history and location are more important factors. It is outwith the scope of this section to provide a complete survey of all possible groundwater types, but a brief summary of important features is relevant.

Although all groundwaters will tend to evolve towards thermodynamic equilibrium with the minerals of the host rock, in practise, true equilibrium is rarely achieved. This is a result of extremely slow silicate dissolution and precipitation reactions and perturbations to equilibrium induced by groundwater flow. Often, groundwater compositions will reflect a 'steady-state' composition which is not a true reversible equilibrium, but is essentially 'rock-buffered' (see Hanor, 1994 for further discussion of this concept).

The two most important parameters governing water chemistry are temperature and chloride concentration. Chloride is usually the dominant anion in any groundwater. Chloride acts 'conservatively', *i.e.* it is not in

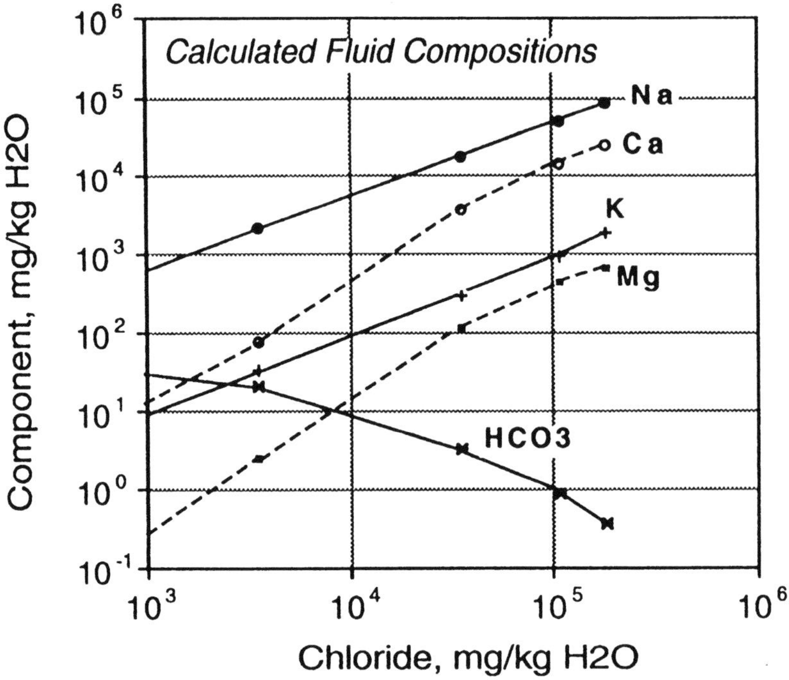

Figure 5.4 Calculated fluid compositions in equilibrium with a mineral assemblage of quartz + muscovite + albite + K-feldspar + calcite + dolomite at 100 °C as a function of dissolved chloride. Reprinted from *Marine and Petroleum Geology,* Volume 11, Hanor, J.S., 'Physical and chemical controls on the compositions of waters in sedimentary basins', pages 31-45, Copyright 1994 by permission of the publishers, Butterworth-Heinemann Ltd.

general incorporated into minerals (other than evaporites), and thus can be treated as a 'mobile' ion (*e.g.* Michard, 1987, 1990). Often, definition of temperature and chloride ion content can fix all other ion concentrations, with some simplifying assumptions concerning minerals in equilibrium with the groundwater (Figure 5.4). Minerals which are often in equilibrium with groundwaters include carbonates, sulphates, silicates such as clays, and a silica polymorph, usually chalcedony.

A table of 'typical' compositions of groundwaters in rocks considered for waste disposal are presented in Table 5.7. However, it should be noted that there is no such concept as a groundwater characteristic of a certain rock type. Saline groundwaters can be just as common in fractured hard rocks such as granites, as in clays. Often the 'salinity' may be externally derived or

Table 5.7 Groundwater compositions in fractured hard rocks, tuff and clay utilised in radioactive waste performance assessment studies (mg/l).

Rock type Location Country	Granite[1] Finnsjön Sweden	Granite[2] Böttstein Switzerland	Basalt[3] Hanford USA	Tuff[4] Yucca Mtn USA	Clay[5] Boom Clay Belgium
Temp (°C)	25	55	60	25	25
pH	6.9	6.8	9.4	7.6	9.2
Eh, mV	-200	-230	-450		-280
Na	23.0	4038	250	43.9	120
K	3.2	45	1.0	5.1	8.0
Ca	76.0	870	1.3	12.5	1.2
Mg	6.3	2.6	0.4	1.9	0.5
Fe(tot)	9.0	0.45	0.016		
SiO_2	12.0	17.0	121	57.8	
HCO_3	220.0	117	46	125.3	250.0
Cl	61.0	6620.0	148	6.9	5.1
SO_4	8.9	1560	108	18.7	2.0
F	0.6	3.6	37	2.2	3.0
TDS, mg/l	420	13274	712	275	390

[1]Bruno and Sellin (1991); [2]Nagra (1985); [3]Smith *et al.* (1980); [4]Delany (1985); [5]Baeyens *et al.* (1985).

allochthonous, *i.e.* from adjacent or overlying sediments (*e.g.* Canadian Shield brines [Bottomley *et al.*, 1994]; groundwaters in the the Borrowdale Volcanic Group, Sellafield, UK [Savage *et al.*, 1993]), or from seawater (*e.g.* saline groundwaters in the Fennoscandian Shield [Lahermo and Lampen, 1987]), but in some instances appears to be derived 'internally' (*autochthonous*) either through water-rock reaction (*e.g.* the Stripa granite, Sweden [Nordstrom *et al.*, 1989]; the Carnmenellis granite, UK [Edmunds *et al.*, 1985]) or as evidence of a much older source of solutes. In general, the salinity of groundwaters increases with depth and age, and often the correlation between depth and age is intuitively applied in groundwater systems.

Information on sources and 'ages' of groundwater solutes can provide an indirect confirmation of hydrogeological conceptual models for a geological site of interest. Isotopic information on groundwater composition can be extremely important in characterising solute and solvent sources, temperature of recharge and degree of equilibration with the host rock. For a detailed discussion of isotope hydrogeology the reader is referred to IAEA (1987, 1992). A summary of isotopes and trace constituents of use to groundwater studies is presented in Table 5.8.

Table 5.8 Isotopes and trace constituents of use to groundwater characterisation studies.

Application	Isotope/Component
Temperature of recharge (palaeoclimate)	Rare gases: Ar; Ne; Kr; Xe; He Stable isotopes: $^{2}H/^{1}H$; $^{18}O/^{16}O$; $^{13}C/^{12}C$
Groundwater age, mixing	^{14}C; ^{3}H (tritium)
Solute sources	$^{11}B/^{10}B$ $^{87}Sr/^{86}Sr$ ^{34}S $^{143}Nd/^{144}Nd$ ^{36}Cl
Water-rock equilibration	$^{234}U/^{238}U$
Hydrogeological parameters, *e.g.* fracture widths, porosity	^{222}Rn; He loss
Deep crustal processes	$^{3}He/^{4}He$ ratio

5.2.3 Groundwater characteristics relevant to radioactive waste disposal

In terms of considerations for the disposal of radioactive wastes, the most important features of the chemical compositions of groundwaters are:

- redox, defined either as Eh, pe or oxygen fugacity,
- pH, or the negative logarithm of hydrogen ion concentration,
- the concentrations of potential radionuclide-complexing ligands, such as chloride, carbonate/bicarbonate, phosphate, sulphate, sulphide, fluoride,
- the absolute abundances and type of naturally-occurring organic substances,
- the abundance and nature of colloids.

Eh and pH are the two most important parameters (other than temperature) governing radionuclide solubility and sorption behaviour. Eh is usually controlled by water-rock reactions, often microbially catalysed. Iron minerals are the most important controls/buffers, although manganese, sulphur and organics may be important. Reactions controlling redox may not be true 'buffers' in the sense that the reactions are reversible. For example, redox may be dominated by the irreversible dissolution of ferrous iron silicates in many igneous or metamorphic rock types.

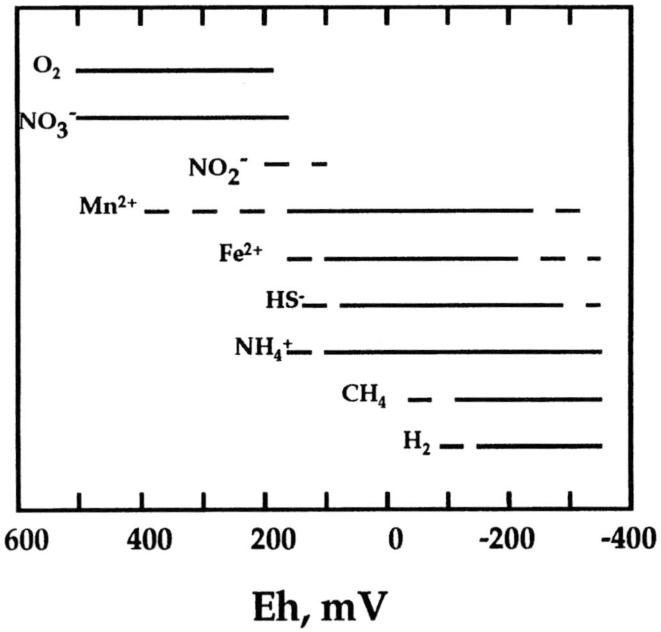

Figure 5.5 Summary of downgradient stability ranges of various redox-sensitive species encountered in three UK aquifers (the Chalk, Lincolnshire Limestone and Sherwood Sandstone). Changes occur over a horizontal distance of 20-30 km and depths of several hundred metres. From Edmunds *et al.* (1984). Reproduced by permission of IAHS.

Such reactions are often slow and redox disequilibrium between different redox 'couples' is a common feature in groundwaters (see Hostettler, 1984; Lindberg and Runnells, 1984 for further discussion). It is then a moot point whether measurements of redox in groundwaters using conventional platinum electrodes are physically meaningful. Rather than use a global redox value for radionuclide speciation and solubility calculations in performance assessment, it may be more appropriate to use redox values as defined by a range of redox couples in order to 'bracket' likely redox conditions at depth (*e.g.* Pearson and Scholtis, 1993). In general, groundwaters become more reducing with depth and residence time (Figure 5.5).

The pH of most groundwaters lies in the range 5-9, although values less than and greater than this may occur. Reactions involving carbonate/bicarbonate ions and calcite are relatively fast and are thus the most common controls on

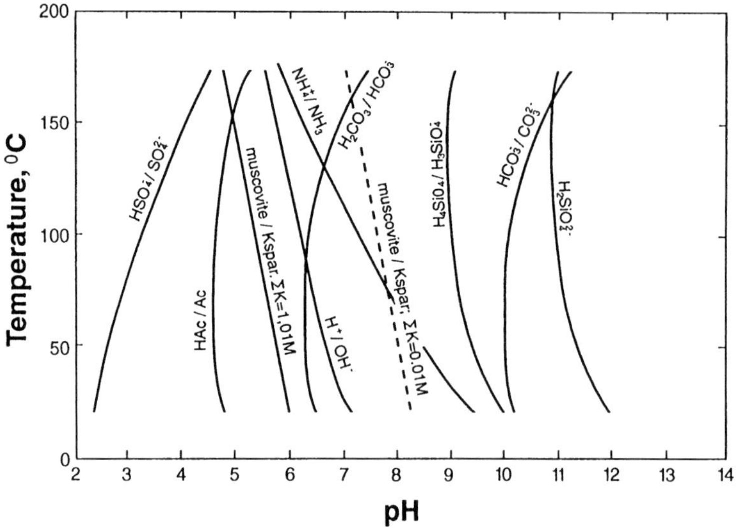

Figure 5.6 pH buffers operative in groundwater systems. Adapted from Barnes, H.L., 'Measuring thermodynamically interpretable solubilities at high pressures and temperatures', pages 321-338 in *Chemistry and Geochemistry of Solutions at High Temperatures and Pressures*, (eds. D.T. Rickard and F.E. Wickman), Copyright 1981, with kind permission from Elsevier Science Ltd., The Boulevard, Langford Lane, Kidlington OX5 1GB, U.K.

pH:

$$CaCO_3 + H^+ \rightarrow Ca^{2+} + HCO_3^- \quad (5.3)$$

Aluminosilicates such as feldspars, clays and micas may also be important. The following illustrates the buffering between potassium feldspar and muscovite mica:

$$3KAlSi_3O_8 + 2H^+ + 12H_2O \rightarrow KAl_3Si_3O_{10}(OH)_2 + 2K^+ + 6H_4SiO_4 \quad (5.4)$$

Similarly, that between potassium feldspar and kaolinite:

$$2KAlSi_3O_8 + 2H^+ + 9H_2O \rightarrow Al_2Si_2O_5(OH)_4 + 2K^+ + 4H_4SiO_4 \quad (5.5)$$

Such reactions are the most important buffers of far-field pH (Figure 5.6). Occasionally, the concentrations of organic acids in sedimentary rocks may be great enough to also act as pH buffers.

Organic substances in deep groundwaters tend to be molecules such as carboxylic acids or methane, whereas shallower systems may contain amounts of complex molecules such as humic or fulvic acids. These latter compounds are more associated with the breakdown of biological material such as peat (Aiken *et al.*, 1985). Humic and fulvic acids form strong complexes with heavy metal cations and are significant in the assessment of the mobility of radionuclides. Falck (1989) has reviewed the properties of these materials and methods of modelling their behaviour in aqueous systems.

Colloidal material may exist in groundwaters and consists of fine particulate mineral material (silica, clays) or organics (Tipping, 1988). Radionuclides may sorb preferentially onto such materials and thus the migration of colloids becomes an issue for performance assessment calculations (*e.g.* Smith, 1993). A number of interlaboratory exercises for the sampling and characterisation of colloids with regard to waste disposal issues has now been conducted (*e.g.* Longworth *et al.*, 1990; Degueldre, 1990). Modelling the migration of colloids is addressed later in this Chapter (Section 5.3.5).

5.2.4 Brines, argillaceous rocks and membrane phenomena

Groundwaters with a total dissolved solids content greater than 100,000 mg/l may be considered to be 'brines'. Brines are common features of both sedimentary basins (Hanor, 1983) and hard rock basement areas such as the Canadian Shield (Bottomley *et al.*, 1994). The occurrence of brines poses special problems for waste disposal considerations in that geochemical modelling techniques are currently inadequate to model silicate-water (and heavy metal-water) equilibria in such saline solutions. The evaluation of chemical equilibria in brines requires the adoption of the 'ion interaction' approach developed by K.S. Pitzer and co-workers at the University of California (Pitzer, 1980). However, relevant coefficients are currently unavailable for all elements of interest. Consequently, predictive modelling of solubilities and sorption properties is greatly hindered in such groundwaters.

Brines are thought to have originated by one or more of the following mechanisms (Hanor, 1983):

- subsurface dissolution of evaporites,
- infiltration and burial of brines produced by sub-aerial evaporation of surface waters,
- membrane filtration, in which water molecules are preferentially forced through a mudrock 'membrane' and solutes are concentrated in a residual brine.

The latter mechanism is especially relevant to understanding the evolution of the chemical composition of groundwaters in mudrocks. Because mudrocks (*i.e.* clays, shales, mudstones) consist of densely packed charged clay particles (particularly smectite clays) they may act as osmotic membranes where

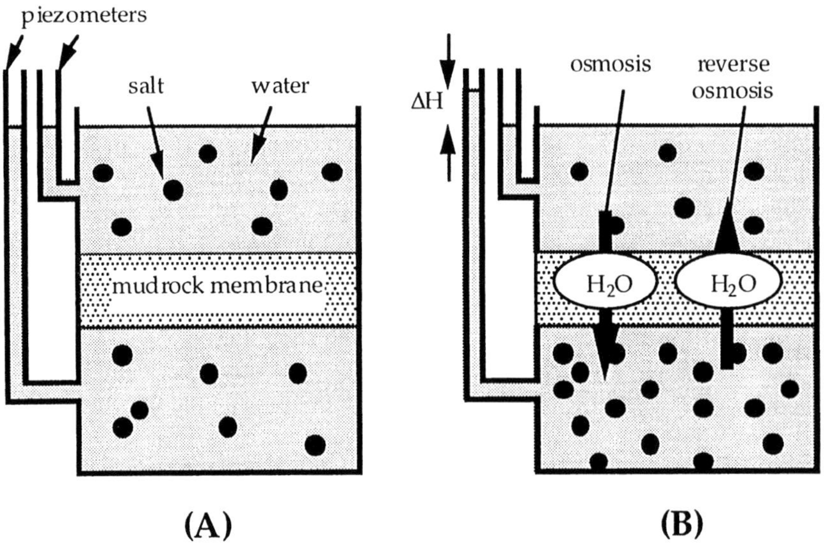

Figure 5.7 (A) Two sands containing salty water are separated by a semi-permeable mudrock membrane. Piezometers measure the hydraulic head of the pore fluids in each sand. (B) The pore fluid in the lower sand is saltier than the pore fluid in the upper sand, and water molecules migrate downwards across the mudrock as a result of *osmosis*. If the differences in hydraulic head are sufficient, however, water molecules can be made to migrate in the reverse direction, a process known as *reverse osmosis* or *membrane filtration*. After Hanor, 1983. Reproduced by permission of Kendall/Hunt.

water molecules may pass through, but dissolved ions are retarded. If aquifers containing groundwaters of different salinity are separated by a semi-permeable mudrock formation, diffusion of water molecules from the less saline to the more saline groundwaters by osmosis will occur (Figure 5.7). Such behaviour can lead to 'overpressuring' of mudrock formations. If, however, hydraulic heads are such that water molecules are forced to flow in the reverse direction, groundwaters in the saline aquifer become saltier. This mechanism is known as 'reverse osmosis' or 'hyperfiltration' and may lead to the development of brines.

The relevance of these phenomena to waste disposal considerations is that there may be a strong coupling of physical and chemical behaviour in mudrock formations. Consequently, osmotic behaviour should be considered in the analysis of hydraulic, mechanical and chemical behaviour in mudrocks (or clay backfills) contemplated as hosts for radioactive waste.

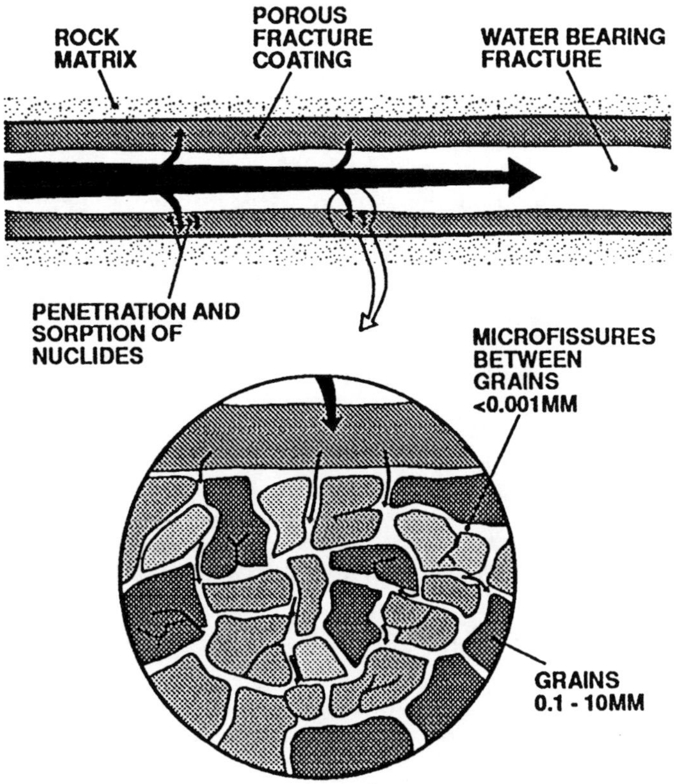

Figure 5.8 Schematic illustration of advective flow and matrix diffusion in fractured rock. Reproduced by permission of SKi.

5.2.5 Retardation processes

Retardation processes for radionuclides migrating from a radioactive waste repository have been discussed in Chapter 4 (near-field). These are:

- exchange of cations from the aqueous phase with minerals,
- repulsion of similarly charged ions by ion exclusion,
- sorption of ions on mineral surfaces,
- direct incorporation of ions into mineral structures, 'mineralisation',
- precipitation of ions as discrete solids.

These mechanisms are equally applicable to the far-field and constitute the principal means by which radionuclides may be retarded relative to the rate of groundwater flow.

Also of importance to retardation in the far-field is the concept of 'matrix diffusion'. Essentially we can consider matrix diffusion to be the diffusion of radionuclides out of principal flow paths into 'dead-end' pore space. This concept is shown schematically in Figure 5.8. The availability of rock volume adjacent to groundwater flow paths in fractured hard rocks may delay radionuclide migration by several thousand years (see Section 5.3). The availability of matrix pores for diffusion will be strongly dependent upon the style of rock alteration along groundwater flow paths. For example, the development of thick clay-hydroxide veins due to water-rock interaction which will effectively 'seal' the rock matrix from groundwater may rule out matrix diffusion.

5.2.6 Disturbed zone behaviour

Repository excavation will lead to the development of a zone in the repository rock with different characteristics from the host rock (Section 5.1 above). In addition to physical changes, other effects will be:

- the oxidation of ferrous iron and reduced sulphur bearing minerals, leading to the development of hydrous iron oxides and partial acidification of groundwaters, and
- the introduction of additional microbial activity into the far-field.

In cementitious repositories for LLW/ILW, the elevated pH of near-field pore fluids may have considerable effects upon far-field behaviour. After repository closure, pore fluids equilibrated with cement minerals will migrate into the far-field. The hyperalkaline nature of these pore fluids (pH 12-13) will be chemically reactive with the far-field, because of the enhanced solubility and rate of reaction of most silicates and aluminosilicates at alkaline pH. 'Early' cement pore fluids, *i.e.* those dominated by Na-K hydroxides (see Chapter 4), may lead to the development of increased porosity adjacent to the engineered barriers of the repository (up to 1 m distant), and the replacement of feldspars and micas by zeolites and calcium silicate hydrate (CSH) minerals (Rochelle *et al.*, 1992; Lichtner and Eikenberg, 1993; Savage and Rochelle, 1993). However, rock matrices may be sealed from groundwater flow paths by mineral precipitation, removing the possibility of radionuclide retardation by matrix diffusion (Steefel and Lichtner, 1994). With time, this alteration will lead to the filling of porosity and sealing of groundwater flow paths. This preferential sealing of certain flow paths may lead to channelling of groundwater (and gas) in flow paths which remain open (Figure 5.9). The size of this chemically-altered zone is sensitively related to the rates of groundwater flow and chemical reaction.

These interactions are complex and because there are large solid volume changes accompanying the precipitation of CSH minerals and zeolites, there is a strong coupling effect between hydraulic and chemical behaviour. These problems are best addressed by the use of coupled numerical models which

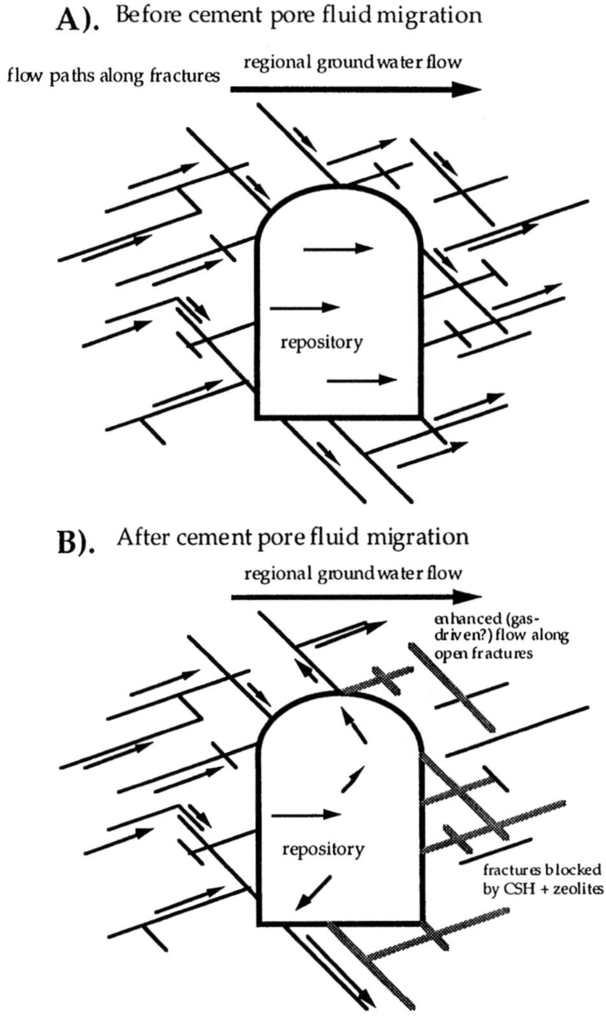

Figure 5.9 Schematic diagram of possible changes in direction and magnitude of groundwater flow in a repository with a cementitious near-field.

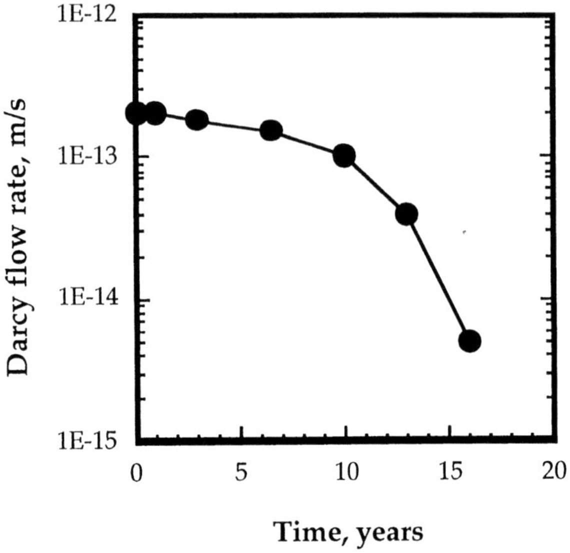

Figure 5.10 Variation of flow rate with time as predicted by the coupled flow and chemistry computer code, 'PRECIP' for a cement pore fluid entering a 20 μm fracture in granitic rock. From Rochelle *et al.* (1992). Reprinted from Kharaka, Y.K. and A.S. Maest (eds.), *Water-Rock Interaction - Proceedings of the 7th International Symposium*, WRI-7, Park City, Utah, 13-18 July 1992. 1992. 1730 pp., 2 vols., Hfl.285/£106.00. A.A. Balkema, P.O. Box 1675, Rotterdam, Netherlands.

explicitly account for the effects of changing porosity on the flow field (Figure 5.10).

5.3 Far-field transport modelling

5.3.1 Preliminaries

The performance of the far-field or geosphere is subject to a number of uncertainties. Almost all arise from limited site characterisation, or the consequent choice of conceptual and mathematical modelling approach. The fundamental processes governing solute transport and dispersal within groundwaters are well understood. They are:

- *advection* by the groundwater flow;
- molecular *diffusion*;
- retardation due to interaction with available surfaces; the role of *chemical sorption* (reaction with minerals in the rock surface), and surface forces.

For each of these processes modellers are forced to adopt simplifications owing to the fact that the flow path geometry is unknown. The actual fluid flow regime is unavailable and must be approximated. Consequently the dispersal process must be augmented by a:

- hydrodynamic *dispersion* process which can account for microscopic fluctuations within the flow geometry which are ignored by the flow model.

It must be stressed that the need for such a factor is governed solely by the rock heterogeneity on scales beneath that modelled, or else by the limited description of the local flow profiles. Similarly, retardation must be extended to include:

- *physical retardation* such as diffusion into the rock matrix, or into dead end pores: again processes which are not prescribed by the geometrical model.

Thus the lack of knowledge of the site coupled with the necessarily limited model description of the rock yields five key sources of model uncertainty. Of these the characterisation of the groundwater flow can be, perhaps, the most well understood, since validation only requires that the major flow conduits are approximated (the fluxes being largely unaffected by smaller scale structures). The others (diffusion, dispersion, chemical and physical retardation) are to some extent functions of the description of the *flow path geometry* adopted. Tortuosities, surface areas, sorption isotherms, Peclet numbers, dispersivities, scaling laws, all provide ways of extrapolating observed data to account for transport and dispersal at large far-field length scales. Their use and abuse is an evolving story and it would be misleading to imply that any such concepts possess any authority other than on a site by site validated basis.

In this section we begin by sketching some of the broad issues and choices before concentrating on more specific examples and issues in later sections.

5.3.2 Fractured and porous media

In modelling the behaviour of fluid flow and transport within natural rock, a distinction is usually made between porous and fractured media. The difference is one of scale, yet these are treated in fundamentally different manners.

Scale and scaling is of fundamental importance in natural systems: it is no coincidence that geologists place a hammer or compass as a scale bar in photographs of rocks. Thus geological media may be considered as fractured,

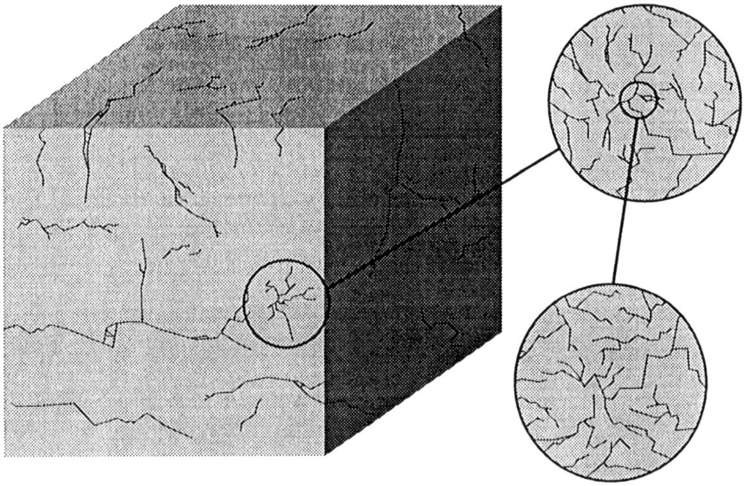

Figure 5.11 Generic rock types may naturally give rise to non-linear, statistical scaling behaviour.

highly fractured, or porous depending upon the scale at which it is to be modelled (Figure 5.11).

In fractured rock the movement of fluid is assumed to be confined to a few well defined fractures or fracture zones. The intervening blocks of rock are usually considered to be impermeable to groundwater, or else to contain only static water within pore spaces, and hence are not conductive to flow. Thus the flow path geometry is restricted to (relatively) sparse sub-volumes of the geosphere; fracture intersections and connectivity being crucial for both fluid flow and solute transport.

For porous media, flow paths are considered to exist on relatively small (microscopic) scales. Thus rock properties will be averaged over some intermediate scale so that the whole rock is considered to be conductive to flow (to some possibly variable extent) at the (macroscopic) scales associated with the calculation to be made. In this case the rock properties must be characterised statistically. For example, consider the porosity (fraction of the rock volume available to fluid flow) of a sandstone block. This may be measured locally so that information concerning the spatial variation of the porosity field, θ, can be considered as a function of the position, $\mathbf{x}$, say. This variation can be observed by plotting the quality

$$\gamma(h) = \frac{1}{2}E[(\theta(\mathbf{x}+) - \theta(\mathbf{x}))^2]$$

where E denotes the expected value over all pairs of points separated by a

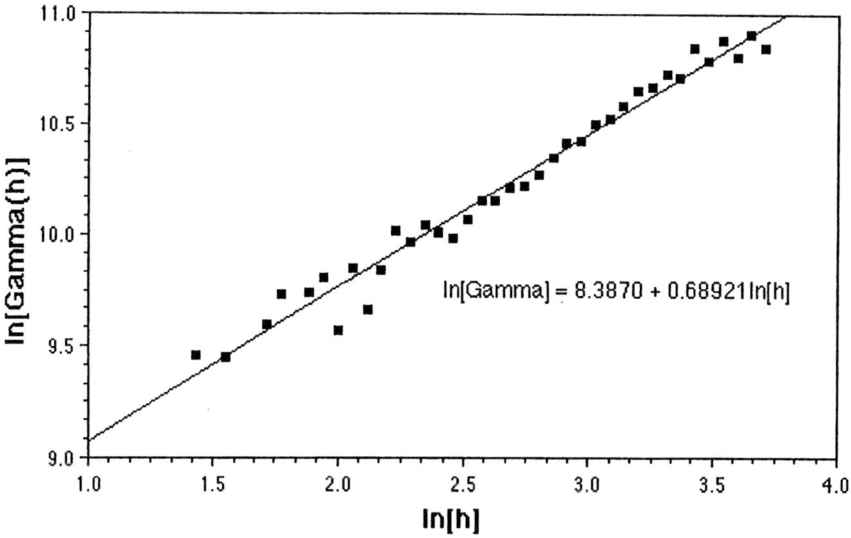

Figure 5.12 Spatial variability of 300 porosity measurements from a sandstone block, and a fitted power law: $\gamma \approx 4400h^{0.69}$.

distance $h = ||$. If θ is modelled conceptually as a realisation of a stationary stochastic (random) function, then γ is usually referred to as the observed variogram. In practice, expected values must be measured by averaging over some binned observations. In Figure 5.12 we show an experimentally observed γ for a sandstone block calculated from 300 individual measurements. For large h, γ is constant, whereas it approaches 0 as $h \to 0$. There is a number of ways of classifying the behaviour of γ, and thereby expressing the statistical scaling behaviour of the medium. If $\gamma(h)$ obeys a power law rule ($\approx h^{\alpha}$) in the limit as $h \to 0$, then the variability of θ may be modelled by generating fractal fields. This ties in nicely with our experience of scaling properties of rock structures: the larger the scales, the greater the possibility of having a large flow path, see Figure 5.11.

Sometimes geosphere models for flow path geometry invoke multiple channel, or multiple porosity ideas. The aim of those is to allow for alternative conduits for flow through rock volumes without being too specific as to their geometrical layout. Thus in dual porosity models, for example, a 'mainline' flow path is complemented by solute access to the surrounding porous rock volume where no flow takes place (see Figure 5.13). Similarly channelling models superimpose many flow paths from one point to another (possibly allowing for interaction) without specifying any detail as to their fundamental geometry.

For fractured and porous rock, flow and transport calculations can be made over a range of scales: for example, in modelling laboratory experiments; well characterised aquifers, or fracture zones; or large, relatively unknown rock masses.

5.3.3 Groundwater flow modelling

There is an immense literature on groundwater flow and hydrogeology and much of this is applicable to the consideration of the performance of deep geological repositories. However a common fault is that site investigation programmes treat the hydrogeology in extreme detail (see Chapter 8), with large effort invested in flow predictions; whereas the subsequent transport and dispersal problems for radionuclides are considered by the adoption of a few simplistic assumptions. Recall that relatively simple flow networks may capture the dominant fluid flux behaviour (Section 5.1.3). There is a need to change this emphasis in order to focus the flow modelling towards the needs of the transport and dispersal calculations. This bias results from the historical ties between the geosciences and hydrogeology; thus permeability values and mean travel times are often quoted, see Table 5.2. However, in considering solute migration, the *primary* role of the far-field may be in providing a *dispersal mechanism.*

Within a saturated rock, groundwater moves through the available volume as a consequence of spatial differences in pressure, temperature, chemical potential, and so on. Of these, the pressure, or head, is usually considered as the dominant factor: it is also easily measurable in the field, and controllable in the laboratory.

If the flow path geometry were known, the task of describing the consequent flow involves the complete specification of the pressure field and the consequent discharge of water. These are clearly linked conceptually, although alternative relations may be defined mathematically. Sources and sinks must also be described (recharge, discharge, boreholes, and so on). For known geometries, the most common is a slow flow, or Stokes flow, approximation which relates the steady fluid velocity $\mathbf{v}$ to the pressure field P via

$$\Delta \mathbf{v} = \frac{1}{\mu} \nabla P$$

where μ denotes the fluid viscosity.

At water rock surfaces, a no-slip condition $\mathbf{v} = \mathbf{0}$ is imposed. In addition, water is usually assumed to be incompressible; mathematically this implies

$$\nabla . \mathbf{v} = 0$$

Fortunately, for narrow fissures of aperture $2h$ say, the slow flow equation may be averaged to obtain a more direct model

$$\overline{\mathbf{v}} = -\frac{h^2}{3\mu}\nabla P \tag{5.6}$$

where $\overline{\mathbf{v}}$ is the cross fracture average fluid velocity. The flow rate through the fissure (per unit width) is given by $\Phi = 2\overline{\mathbf{v}}\mathbf{h}$. Hence, the smaller the aperture, the less conductive is the fissure (the same pressure gradient drives less fluid flux). A similar expression is available even when the aperture h is variable, and similar considerations apply to flow paths through variable capillaries. Time dependence may also be considered.

In contrast, when a medium is highly fractured, or consists of packed grains (for example, sands, clay, shales, conglomerates), an explicit representation of the microscopic geometry is undesirable. Darcy was first to investigate empirical relationships between applied pressure gradients and the consequent flux of fluid through such porous media. Darcy's law, usually written

$$\mathbf{q} = -k\nabla P, \tag{5.7}$$

introduces the notion of a permeability k relating the specific discharge of fluid $\mathbf{q}$ to the pressure gradient. For any direction, the corresponding component of $\mathbf{q}$ denotes the volume of water passing through an orthogonal unit surface area within a medium. It is usually called simply the Darcy velocity (or specific discharge) and is related to the mean fluid velocity $\overline{\mathbf{v}}$ via

$$\overline{\mathbf{v}} = \frac{\mathbf{q}}{\theta}$$

where θ denotes the effective porosity (the fraction of rock volume available to the flowing water). $\mathbf{q}$ is sometimes described as the equivalent velocity of the water assuming the total volume of rock were available to the flow. Thus it is always true that

$$\overline{\mathbf{v}} > \mathbf{q}$$

Permeability is a non-physical parameter in the sense that it can only be calibrated from experiment. Its utility comes from the implicit averaging of the geometric rock properties controlling the conduction of the fluid.

The comparison between (5.6) and (5.7) is useful. Both apertures and porous media are conductive. If (5.6) is used for a distribution of fissures (per unit volume), then an equivalent porosity and permeability can be calculated and the use of Darcy's law motivated. However (5.7) remains fundamentally empirical and its use in 'averaging' flow path geometry should be validated for particular media under consideration. Darcy's law is also sometimes extended by incorporation of higher order, non-linear terms. In addition, time-dependent responses to changing head conditions may be considered in non steady-state calculations which introduce the notion of storativity.

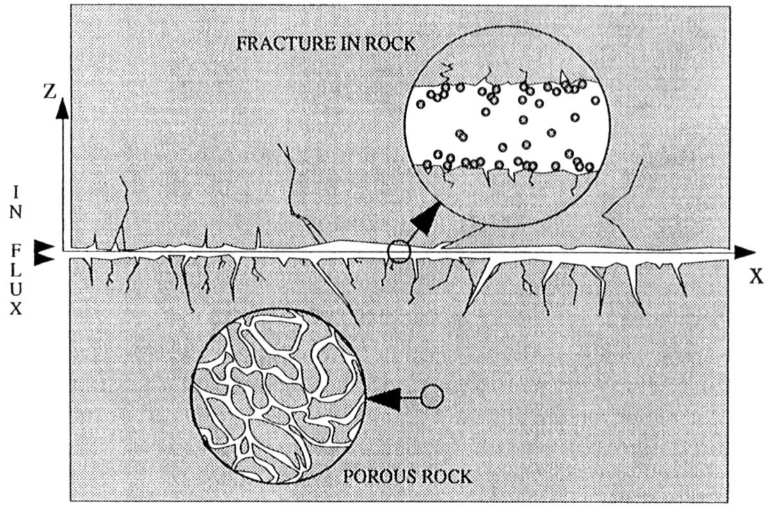

Figure 5.13 A single fracture, surrounded by porous saturated rock.

The use of Darcy's law is widespread owing to the large amount of site specific and experimental validation, and its linearity, which allows for numerical and analytical solution. It may also be incorporated into models which allow for the porosity of the medium, θ, to be time-dependent. For example, if the groundwater is incompressible, then conservation of fluid mass implies

$$\frac{\partial \theta}{\partial t} + \nabla . \mathbf{q} = 0 \Leftrightarrow \frac{\partial \theta}{\partial t} = \nabla .[k \nabla P]$$

For large rock domains a mixture of porous media and explicit fissures may be suggested. This is really a way of imposing spatial structure (at some model scale). However, whichever concepts are employed there will always be some submodel scale at which the flow path geometry is unrealistic, too smooth, or else unrepresented. The impact of this on various flow models is of importance in flow calculations although it may often have an even greater impact upon tracer migration and dispersal. For example flow through a fractured rock may be conceptualised as a single transmissive fracture (representing a fracture zone) through a porous rock matrix where the water is static, see Figure 5.13.

For both fractured and porous media two distinct modelling approaches should be distinguished:

- deterministic, exact calculations from explicitly defined flow path geometry, or medium properties;
- ensemble simulation, where many realisations of plausible flow fields are generated according to certain rules or constraints.

In both cases, flow may be conditioned by known data (head measurements, discharges, mean permeabilities).

Two forms of ensemble simulation are particularly important and correspond to the extremes of a sparsely fractured impermeable rock, and a permeable (though variable) rock matrix: discrete and continuum modelling.

Discrete simulation

The first ensemble simulation is embodied in discrete fracture network codes. Fractures are generated and distributed according to input distributions controlling density, strike, dip, size, aperture and so on. Then a flow calculation is made assuming some prescribed flow or head boundary conditions. This can be carried out many times: each realisation generated as a representative from some statistical ensemble. A number of attempts to validate such approaches have been made, and it is generally accepted that such an approach is successful in generating highly channelled flow, where discharge is concentrated to a few zones within fracture networks. The variability within ensembles provides a statistical problem in their interpretation; and also their comparison with observed hydrogeological data. Each representation provides a specific geometrical interpretation of the conditioning information (heads, locations of major zones, ...), yet the reality is simply one of such realisations.

Fluid flow may be dominated by one or two fractures, thus the performance of realisations in matching observed flows may be 'all or nothing'; an indication that a more intense distribution of fractures, or more realisations are required. For crystalline rock, the assumption of an impermeable rock mass may be satisfactory as far as the flow calculation goes, since the water in fine pores or fissures is likely to be relatively stagnant. However difficulties may ensue for this approach where a transport calculation is to be made, since access to the rock pore water is a strong physical retardation factor in tracer dispersal. Hence such simulations may be biased towards transport through 'fast paths' unless 'equivalent' (substitute) retardation models are invoked.

From a positive perspective, discrete network simulation allows the modeller to have a wide variety of control: fracture density, fracture intersection rates, fracture size distributions, dips, strikes, and so on. Thus the main conduits for flow can be controlled and conditioned to a variety of data. The connectivity of the network can also be quantified and it is straightforward to model and locate known fracture zones. Results also tend to exhibit a large degree of chanelling with highly heterogeneous flow; an effect often observed in the field.

Continuum simulation

For continuum modelling of porous media there are a number of ensemble simulation approaches. These are usually based on Darcy's law (or its generalisations). The idea is that permeability fields (and also porosity fields) should be generated at some spatial resolution over a model domain. Given flow or head boundary conditions the steady or time dependent Darcy velocity may be obtained numerically. Conceptual choices arise in a number of areas:

- at what resolution should the variable permeabilities be specified?
- at what resolution should the head and velocity fields be solved for?
- how should different realisations be developed within an ensemble?
- how can these be conditioned by known data or head values?

The classical approach to this problem uses a technique known as kriging. This takes some known data values at specific locations and gives a maximum likelihood estimation of the associated field at any other location, based upon both the given values and the hypothesis that the real field is one realisation of a particular type of random function. The theory of random functions is highly developed, but it is worth pointing out here the difference between the spatial variation of a particular realisation of a random function, and the variation of a random function evaluated at a single point within the domain. One should not be confused with the other, although they can be directly related (the assumption of 'ergodicity' often being invoked to make one equivalent to the other).

In general, although kriging gives the statistically best estimate of the unknown field at each point, it does not generate typical features or structures. To get around this, so-called 'conditional kriging' is employed in which the difference between the actual field and a particular, or seed, representative random field is kriged. The resulting smooth estimate is then added to the seed random field to obtain a final realisation which both interpolates the observed data and also possesses the structure of the seed field at small scales. The consistency between the statistics of the random field (at fixed points) used in the kriging calculation and the spatial characteristics of the seed field can be ensured (assuming ergodicity). However the final conditioned realisation may not display the desired scaling behaviour when sampled *a posteriori*.

Similar realisations can also be generated with a more direct approach. A class of rough, heterogeneous fields can be generated or sampled, and known data interpolated by using a linear combination of such fields and minimizing some specific objective function. A particularly interesting class of fields are 'statistically self-affine fractals'. These possess a specific statistical scaling property and are highly suited to numerical generation. Their adoption is equivalent to the choice of a power law variogram in describing for the spatial variation (and/or the behaviour of the random function used in kriging). In Figures 5.14 and 5.15 we show two such fractal permeability fields defined over a two dimensional layer of rock, together with the resultant Darcy velocity fields. In each case the same pressure boundary conditions have been applied. Both fractals have the same fractal dimension (2.7 in this case, corresponding to $\gamma \approx h^{0.6}$ as $h \to 0$).

Note that the high resolution of these calculations coupled with the heterogeneity of the permeability fields have yielded flow regimes with strongly channelled subdomains. It is the channelling phenomenon which is typical, or generic, not the location or size of any particular feature. This technique is conditioned by the observed data directly since, if sampled, the two point spatial correlation function for each fractal will be statistically indistinguishable from the one fitted to the observed data. This structural variability is ensured for each realisation, not the statistical variability of an ensemble as in kriging.

5.3.4 Tracer migration

Consider a solute, subject to diffusive processes, within saturated porous or fractured medium. For the purposes of transport calculations we may assume that the fluid flow is already known. For definiteness we begin by assuming a non-sorbing tracer, the extension being obvious for those cases where sorption may be modelled via a linear partition between the free and sorbed species, or where a more sophisticated approach is adopted. The classical approach for solutes in porous media is based on the assumed existence of representative elementary volumes (REVs) over which microscopic fluctuations in the rock geometry and flow field may be averaged. Typically one obtains a model for the tracer concentration in groundwater, c say:

$$\frac{\partial(\theta c)}{\partial t} = \nabla.(D\nabla c - c\mathbf{q}) \tag{5.8}$$

Here θ is the porosity, $\mathbf{q}$ is the Darcy velocity for the fluid flow, and D is the dispersion tensor usually written as:

$$D = D_T I + (D_L - D_T)\frac{[\mathbf{q}\mathbf{q}^T]}{|\mathbf{q}|^2}$$

where I is the identity, the constant D_L is the longitudinal dispersion coefficient, effective in the direction parallel to $\mathbf{q}$, and the constant D_T is the transverse dispersion coefficient, effective in all directions perpendicular to the direction of $\mathbf{q}$. The dispersivities include the effect of molecular diffusion, which is also usually represented via a Fickian diffusion process (the associated flux being proportional to the gradient in tracer concentration).

For dispersal and transport through porous media, we usually have D of the form:

$$D_L = \theta D_{\text{mol}} + \alpha_L|\mathbf{q}|$$

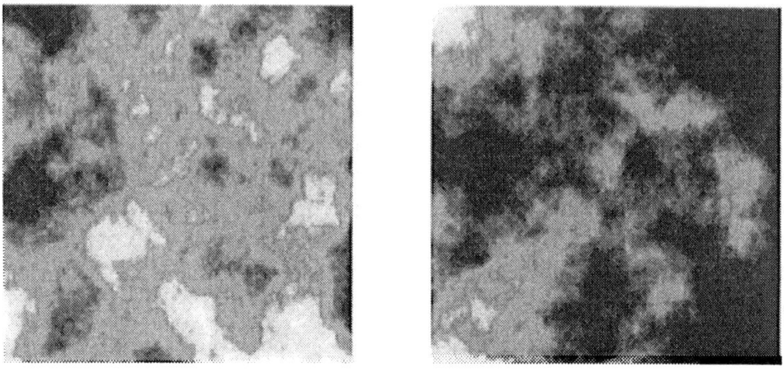

Figure 5.14 Two permeability fields represented by 2.7 dimensional fractals defined at a resolution of 384 × 512.

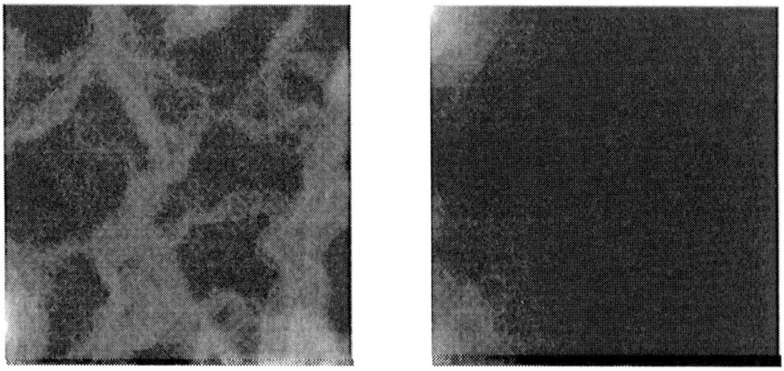

Figure 5.15 Flow fields corresponding to the permeability fields in Figure 5.14 calculated with similar head boundary conditions using Darcy's law.

$$D_T = \theta D_{\text{mol}} + \alpha_T |\mathbf{q}|.$$

Here D_{mol} is the molecular diffusion coefficient of the tracer in groundwater and α_L and α_T are constants having units of length. Hence they are often known as dispersion lengths or dispersivities. In the laboratory, α_L is typically of the order of centimetres. α_T is smaller, taking values of between 1-20 % of that of α_L. In the field it may be up to 100 metres or so. In Figure 5.16 we show various measured values for dispersivities, α_L say, as a function model length.

The problem with this approach is that a fixed scale for a REV is not necessarily available. The larger the scale of the experiment, the larger the scale of the fractures and heterogeneities which may be incorporated whilst other smaller heterogeneities are 'averaged' away. Indeed, there is a natural self-similarity which might be exploited when using a fractal approach within a continuum model.

In field tests injected tracer distributions are known and breakthrough distributions can be observed at distances from the injection site. Such measurements are used to calculate dispersion lengths, since by assuming (5.8), the constants α_T and α_L may be related in the variance of the breakthrough curves. In Figure 5.16 for example it is found that α_L tends to increase with the distance over which the tracer experiment is carried out. This is sometimes used to motivate the choice of constant Peclet number (see later on) in specifying dispersivities. However the problem is really related to the fact that the observed dispersal is not necessarily Fickian-like at the breakthrough scale.

For Fickian processes similar to the above macroscopic description of dispersion, the spatial variance of any tracer distribution increases linearly with time. Consider a one dimensional version of (5.8)

$$\frac{\partial c}{\partial t} = \frac{D}{\theta}\frac{\partial^2 c}{\partial x^2} - \frac{q}{\theta}\frac{\partial c}{\partial x} \qquad x \in (-\infty, \infty), \quad t > 0$$

Considering the first two spatial moments we obtain the average position of the tracer plume, as a function of time, given by

$$\overline{x}(t) = \frac{\int xc\,dx}{\int c\,dx} = \frac{q}{\theta}t + \overline{x}(0),$$

whilst the variance is given by

$$\sigma^2(t) = \frac{\int (x-\overline{x})^2 c\,dx}{\int c\,dx} = \frac{2D}{\theta}t + \sigma^2(0).$$

Thus if $\sigma^2(t)$ is known at times t_1 and t_2, the term D/θ can be found. Similarly, if $c(x,t)$ were known for fixed x, then the quantities

$$\tilde{t} = \int tc(x,t)\,dt \qquad \tilde{\sigma}^2 = \frac{\int (t-\overline{t})^2 c(x,t)\,dt}{\int c\,dt}$$

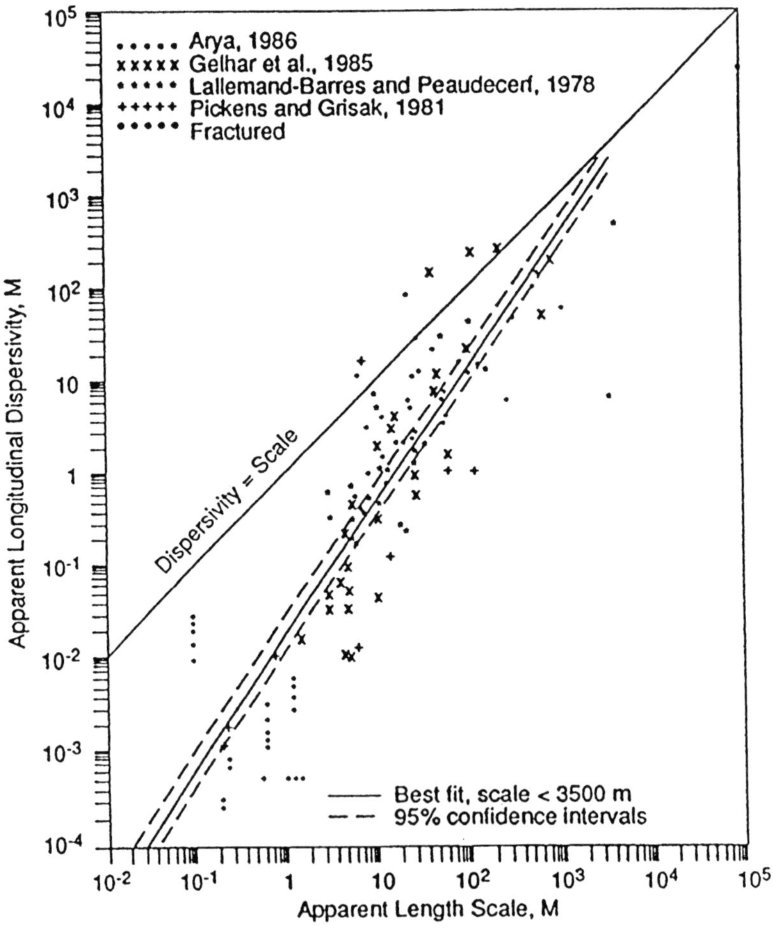

Figure 5.16 Scaling dependence of dispersivities. From Neuman, S.P. *Water Resources Research*, Volume 26, pages 1749-1758, 1990, copyright by the American Geophysical Union.

may also be used to determine v and D, via

$$\frac{\tilde{t}}{\theta} = \frac{2D}{q^2} + \frac{x}{q}$$

$$\frac{\tilde{\sigma}^2}{\theta^2} = \frac{8D^2 + 2Dqx}{q^4}$$

If D and hence α_L is found to depend on t when computed from the spatial variation, or x when computed via $\tilde{\sigma}^2$ and $\tilde{t}$, then the standard model incorporating diffusion and dispersion as Fickian processes is inappropriate, and alternative models should be sought.

Given the level of effort and expenditure invested into models based on REVs and their solutions, there is some natural inclination to avoid overhauling the modelling perspective. However unless the phenomenon of dispersion is better understood there can be little hope of extrapolating results based on the laboratory scale to the scale of the field.

In special cases where the geometry is well defined (5.8) is simplified further. For example it can be shown that for slow flow along a uniform capillary, or through a near uniform planar fracture (5.8) may be approximated by a one-dimensional equation

$$\overline{c}_t = D^*\overline{c}_{xx} - q^*\overline{c}_x - s^*\overline{c} \tag{5.9}$$

for the cross capillary (or fracture) average concentration $\overline{c}$. Here D^*, q^* and s^* are determined from the geometry of the fracture and the outflow or sorption allowed at the capillary, or fracture, boundaries. If there is no flux through, or sorption at, such faces, then $s^* \equiv 0$. For quite general cases D^* behaves like $|q^*|^2$, where q^* is the average velocity. Compare this to the use of a dispersion length, where D was proportional to $|\mathbf{q}|$ (ignoring diffusion). Such results generalise classical Taylor dispersion theory (Taylor, 1953). The one-dimensional equation (5.9) is asymptotically valid where the solution is dominated by terms having small wave numbers (that is; is relatively smooth), and hence over large times. The contribution to the dispersion of order $|v|^2$ is due to the diffusion of the tracer across the non-uniform flow profile. The resulting time dependent increase in spatial variance is known as Taylor dispersion. However the dispersivity itself remains constant and dispersal process is still Fickian in character. Attempts to use non Fickian dispersal models have been made and some of these models will be discussed later in a later chapter.

Of course, for any medium, fractured or porous, the transport modelling involves conceptual decisions which cannot be justified via the theory alone. However as pointed out in Chapter 10, modelling may be possible only at scales inappropriate to the geosphere. Thus care should be taken when assumptions are adopted for tracer transport which have implications for the upscaled behaviour. There is no substitute for field injection tests.

5.3.5 Colloids

In any consideration of the transport and dispersal of radionuclides through saturated fractured and porous rock it is important to assess the possible effect of colloids. To date, this is poorly understood (although both laboratory and field data regarding colloid and groundwater chemistry are available), and there have been few attempts to incorporate such information into a dynamic colloid migration model able to quantify the impact of colloids upon radionuclide breakthrough.

Some radionuclides present in groundwater sorb strongly on to rock surfaces. Not all such surfaces are immobile though, and it is well known that radionuclides may become sorbed to colloidal particles present within the groundwater. Despite their small size, the colloids present a large surface area to the radionuclides, and volume for volume, may sorb the solute species more efficiently than the surfaces of the surrounding rock mass. Once sorbed to colloids, the transportation characteristics of the radionuclides become altered radically to those of the colloids. For instance, colloids may have an opposite net charge to the radionuclide species sorbed to them, such as when a cationic metal becomes sorbed to an organic particle yet the net charge remains negative. Hence the net retardation of the colloid-bound species within the fracture is dependent upon the host colloid population, and is distinct from that of the solute species.

Within fractures, mobile colloids tend to stay centre stream where the flow rate is highest (as in hydrodynamic chromatography experiments), and are also subject to surface forces causing their motion relative to the fracture surfaces. In addition, since colloid-bound radionuclides are excluded from the rock pores, the effectiveness of matrix diffusion and sorption as a retardation mechanism will be lessened.

Let us consider colloid dispersal within an idealised fracture. This must present the colloids with their best opportunity to migrate over *long* distances, at average flow rates 20-30 % in excess of that of any solutes present. Since the real flow paths are irregular and tortuous, this would both perturb the assumed flow rate, and increase the rock surface area available for colloid adsorption. Radionuclides are present in one of the following forms:

- free, in solute form within the fracture;
- sorbed to free colloids;
- sorbed to stationary colloids;
- in solute within the water-filled pore space;
- sorbed to the rock mass, in equilibrium with the concentration within the pore space.

This is summarised schematically in Figure 5.17.

There are three major effects of radionuclide colloid interaction upon the transport and breakthrough of radionuclides:

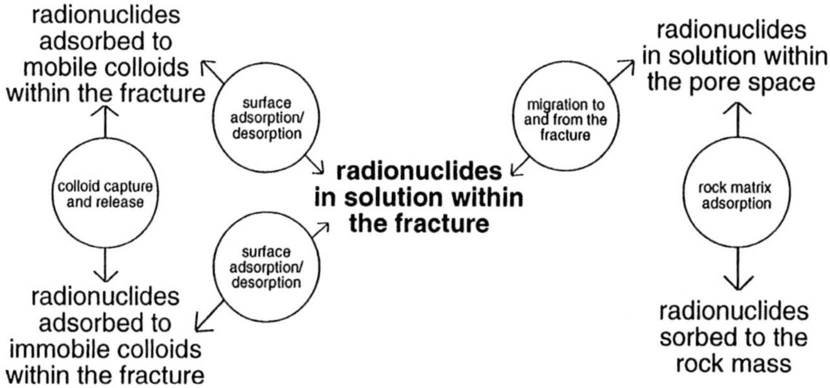

Figure 5.17 The partition of the radionuclide population, and exchange between the classes.

1. Colloid-bound radionuclides proceed at a faster flow rate (on average) than the solute species. This may be shown to be around a 30 % increase, for a variety of situations. Indeed, this is also seen in evolution rates of colloids through both packed columns and capillaries.
2. The exclusion of colloid bound radionuclides from the rock matrix means that their retardation behaviour is radically altered. Colloids may become immobile, thus also entrapping currently adsorbed radionuclides. The retardation of radionuclides due to colloid-rock sorption is dependent upon both the colloid radionuclide sorption behaviour as well as the colloid-rock sorption/desorption rates. However, it is clear that the more colloidal material that becomes immobile, the stronger will be the effective retardation. For this reason natural colloids would be advantageous within porous media since they present a large surface area for adsorption of radionuclides, and yet are filtered out efficiently in such circumstances (as described by standard theory of filtration).
3. The dispersion behaviour of colloids is distinct from that of solutes. Accordingly, the dispersion of free colloid-bound radionuclides must be then taken into account. On a small scale, the dispersion of colloids within a single fracture can be analyzed via a generalisation of Taylor dispersion theory. Such dispersion is a result of the colloids diffusing or being advected (due to surface forces) across the flow profile of the groundwater.

It is possible to analyse these effects and to make scoping calculations for radionuclide dispersal in the presence of colloids, extending the cases calculated for the geosphere in the absence of colloids. It can be shown that breakthrough curves could be both enhanced and retarded by the presence of colloids, according to the levels assumed for sorption to free and immobile colloid material.

Typically, flow path geometry is ill-characterized in models of natural systems and as a consequence, macroscopic (mean) flow rates and dispersion tensors are utilised in order to account for the microscopic fluctuations in media structure (and the consequent hydrodynamic profile) which are unaccounted for within the model. Even for tracer migration and dispersal this issue is far from settled. Given such uncertainty, we suggest that within any flow system (actual or theoretical), colloid and tracer migration phenomena should be treated consistently.

One important area which has been largely overlooked in modelling solute and colloid transport problems involves non-linear retardation effects, which appear naturally in a consideration of colloids in terms of the formation of self-limiting monolayers of adsorbed colloids. This can be due to either a limited distribution of active sorption sites or steric interaction between mobile and adsorbed colloids.

5.3.6 Examples and illustrations

In this section various examples are discussed to illustrate some issues in far-field modelling and to give some simple results which help in understanding the behaviour of the far-field. The topics covered are:

- one-dimensional transport modelling;
- fracture network modelling;
- diffusion against flow;
- flow in tunnels.

Example 1: transport modelling in one-dimension

In practice, it is often possible to reduce the dimensionality of a transport calculation in various ways. This can be justified in cases where the flow follows a restricted path (*e.g.* a channel through a fracture system), or when a stream-tube can be followed from a localised release, or, more generally, when averages can be taken perpendicular to the main transport direction. This final case occurs in many systems, where transport is from a (large) repository to a (large) biosphere receptor or aquifer. The details of the spatial location of the transported material are not of primary interest – only total arrival rates at the biosphere.

As examples of the type of modelling that can be performed for such a system we consider the following:

- a one-dimensional advection-dispersion model incorporating matrix diffu-

sion;
- a pure channelling model, with tracer distributed across a collection of independent channels;
- a channel network model with transfer of tracer between the various channels.

Each of these models is a particular idealisation of the real system and requires lumped parameters which are not directly measurable.

1-D advection-dispersion. The first model is widely used. For example, SKI Project-90 utilised a model of this type, using the CRYSTAL code. The model incorporates the following processes:

- advection of tracer in flowing groundwater;
- equilibrium sorption between dissolved tracer and rock surfaces;
- hydrodynamic dispersion;
- diffusion into rock-matrix pore water; and
- radioactive decay and in-growth.

Advection of tracer occurs as the groundwater moves through the interconnected pore space of the rock. The local rate of transport, *i.e.* the local flow velocity, v, is related to the local flow velocity, q, by $q = \phi v$, where ϕ is the flow porosity. Thus, even rock with low Darcy flow rates can still have rapid advective transport if they have low porosities.

Sorption is really a collection of processes involving interactions between dissolved tracer and tracer attached to surfaces. The equilibrium assumption is usually made for simplicity. The ratio between concentration and amount of sorbed tracer depends on the tendency of the tracer to sorb (as measured by a K_D) and on the available surface area for sorption. Both of these parameters are very difficult to estimate with any precision.

Hydrodynamic dispersion accounts for the variability in flow rates from point to point and the spreading out of tracer which results. A Fickian (diffusion-like) model is used for simplicity although this is not well-founded theoretically or experimentally. The observed scale-dependence of dispersion shows that the Fickian model is not adequate – and that there is not a rock property that can be used to fully model dispersion in this way. When calculations have to be performed, the best approach would seem to be to assume a fixed Peclet number (typically between 4 and 10), determined as vL/D, with v the flow velocity, L the transport distance and D the effective dispersion coefficient. The inclusion of L here makes this procedure hard to justify theoretically – since it implies that the dispersion seen depends on the experimental scale.

When it occurs, diffusion in the rock matrix is a very important mechanism for slowing the transport of tracers. The process is simply diffusive transfer

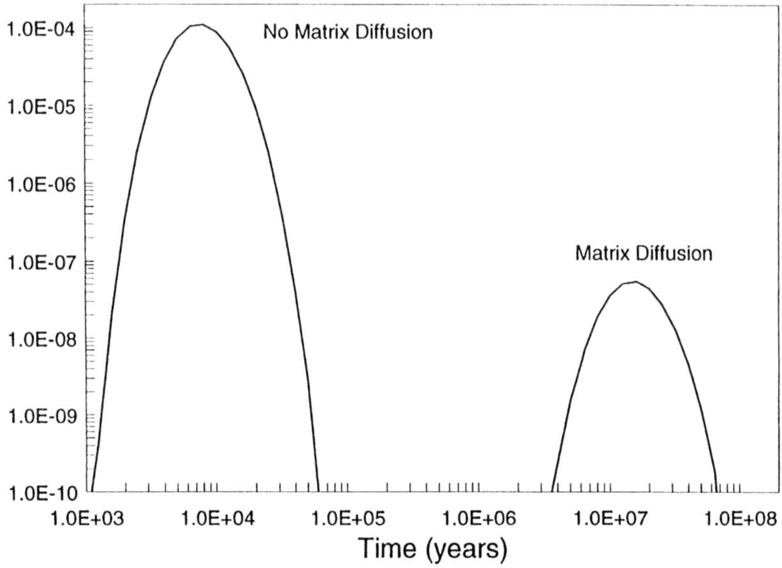

Figure 5.18 The potential significance of matrix diffusion.

between the flowing water and stagnant water in bulk rock. Convincing demonstrations of the effect are hard to make, particular on field scales and therefore appropriate parameter values are very uncertain. The major uncertainties are over the relevant contact area per unit volume between the flowing and stagnant water and the sorption capacity of the rock matrix. The importance of matrix diffusion is shown in Figure 5.18 which shows the breakthrough of a pulse with and without matrix diffusion.

Radioactive decay and in-growth can be added simply to this type of linear model. The modelling of decay chains can add some interesting structure to the breakthrough curve obtained. It is also worth noting that a daughter nuclide's concentration will increase as the K_D of its parent increases and that consequently great care must be taken in assigning conservative K_D values.

Another observation that is worth making is that by omitting the matrix diffusion term from this model a porous model is recovered – with a small re-interpretation of the terms. This is useful for coding and verification. This model is linear and can be very successfully solved using transform methods (Laplace transform and numerical inversion using, *e.g.* Talbot's algorithm). In fact, numerical methods using spatial and temporal discretisation can find this problem very challenging.

The fundamental characteristic of this model is the very long delay of radionuclides that can occur if matrix diffusion is present. In experimental work, this is hard to distinguish from other reasons for non-recovery of tracer and hence it is hard to give definitive parameter values. In cases where matrix diffusion is not dominant, the model exhibits a spreading which increases as the square root of time (*i.e.* the central second moment increases linearly with time).

Pure channelling model. In this case, a number of separate paths is imagined. Each has its own velocity and no mixing occurs between the channels. Such a model could be appropriate for a fractured rock system where the source is spatially distributed so that channels might be well separated. Channelling models can incorporate the effects of equilibrium sorption and radioactive decay and in-growth just as in the conventional advection-dispersion model. Matrix diffusion from each channel can also be modelled. In fact, the channelling model is really a set of advection-dispersion models with distributed flow velocities. If the velocity distribution is narrow there will be little difference between the channelling model and a single advection-dispersion model. For a wide velocity distribution however the velocity differences will determine the degree of spreading and this will be proportional to time. In fact for large enough times velocity differences will dominate in all cases.

Channel model with mixing. In a channel model with mixing, there is assumed to be some intercommunication between channels. Mixing between them occurs at some rate and the longer time linear growth of the spread of the plume (seen in a pure channelling model) is removed. In the short term, before much mixing can occur, a linear growth is still seen. Thus, this model tends to give a pure channel behaviour at early time and a Fickian behaviour at later times. This is demonstrated in Figure 5.19.

Example 2: fracture network modelling

In fracture network modelling the aim is to reproduce the statistical characteristics of a fracture system, in terms of number density, size and physical properties. By generating many realisations of statistically equivalent systems a measure of variability can be obtained as can an estimate of appropriate scales for continuum models. Fracture network codes are often used to model flow and transport in three dimensional systems, such as at Stripa. It is however worthwhile looking at a simpler two dimensional system in order to gain insight into the fundamentals of this type of model.

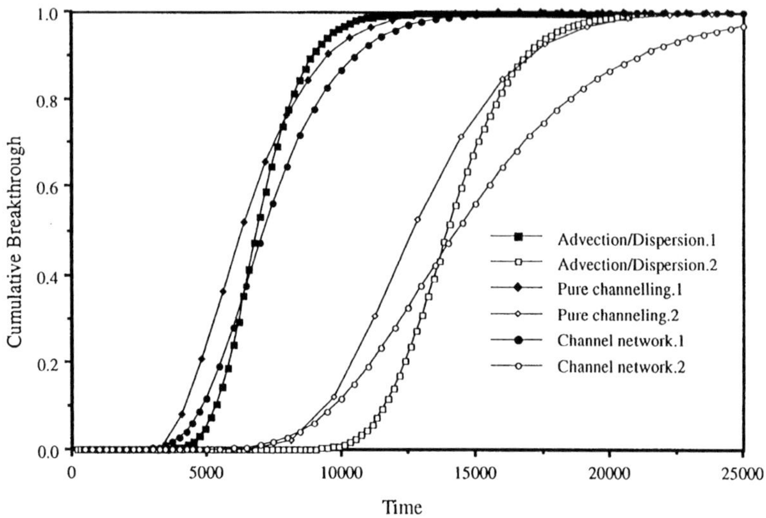

Figure 5.19 Comparison of various 1-D models.

Connectivity. It is interesting to look at the issue of connectivity before going on to consider flow and transport. That is, given a particular density of fractures of given size, is there a connected flow path through the system? It turns out that, for a given distribution of fracture sizes, there is a critical density. Below this density large scale connectivity is not expected; above it large scale flow paths are common. To see this, consider a similar system on a regular square lattice (Figure 5.20). Bonds are present with a given probability, p, (corresponding to the density of fractures). By considering the dual lattice, whose bonds are present with complementary probability $1 - p$ it is easy to see that either there are large scale clusters in the original lattice or in its dual. Since the two lattices are identical, the critical probability is easily seen to be $\frac{1}{2}$. In systems of fractures, represented as finite lines in two dimensions, the probability is replaced by a number density. It is found that the average number of intersections per fracture is the most convenient indicator of connectivity. Numerical calculations for systems of randomly oriented fixed length fractures give the critical density to be so as to give an average of 3.1 intersections per fracture. Other orientation distributions give similar results. In three dimensions the result is substantially lower; an average of 2.0 intersections per fracture plane is sufficient to give a connected system.

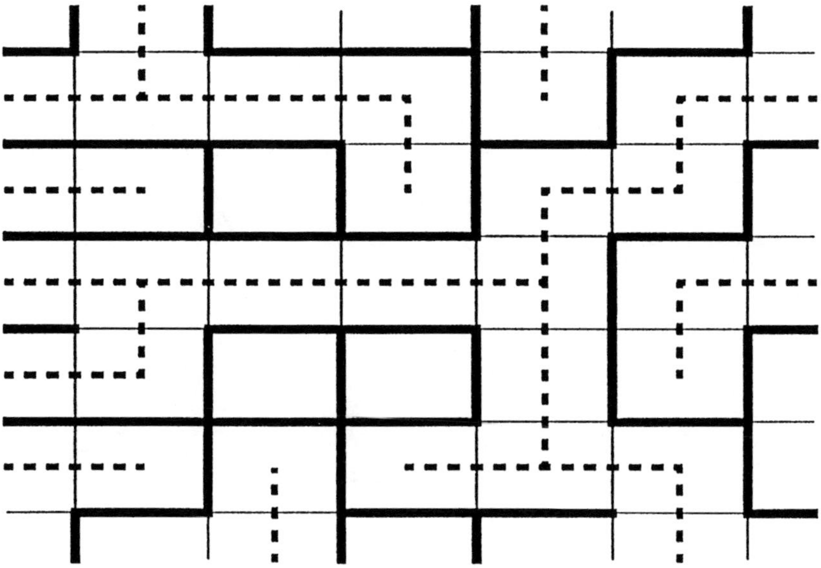

Figure 5.20 Lattice connectivity for original (solid) and dual (dotted) lattices.

Flow. Given a 2-D fracture system it is relatively easy to calculate a flow field through it and find an effective permeability. This can be related back to the density. Below the critical density, the permeability is zero and it climbs approximately linearly with density above this, as can be seen from Figure 5.21. By performing many numerical experiments on systems of different sizes an idea of representative elementary volumes can be obtained, as shown in Figure 5.22.

Transport. Having calculated a flow field in the system, advective transport with full mixing at intersections can be simulated using a particle following method. By calculating breakthrough curves, effective transport properties can be found and again representative elementary volumes considered.

An important lesson from such work is that even for simple, fully characterised systems, the dispersive behaviour is not truly Fickian up to very large scales. Figure 5.23 shows a dispersion length calculated in two different ways from the breakthrough curve. There is no sign of convergence between the two methods with increasing region size.

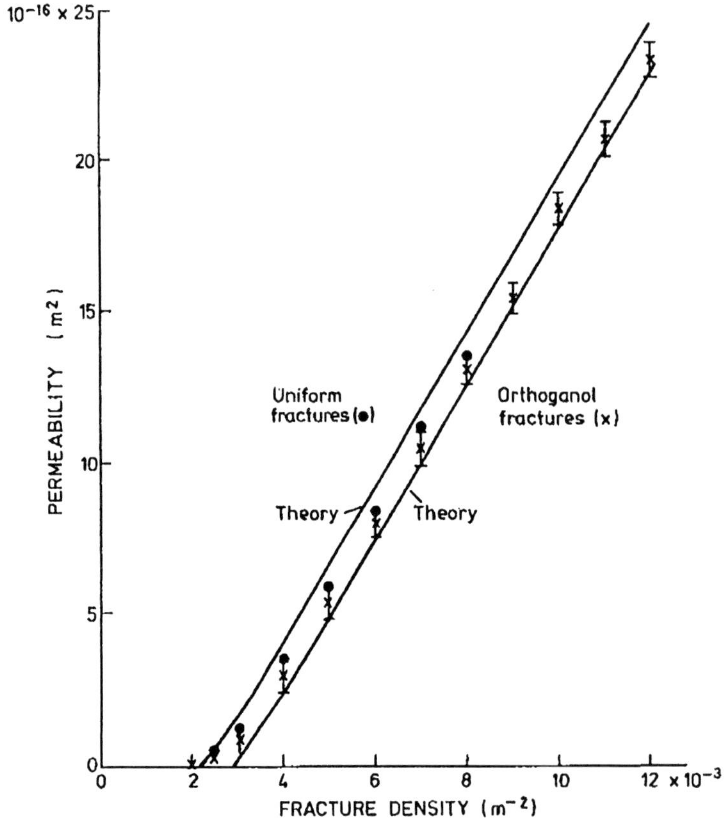

Figure 5.21 Network permeability versus density.

Example 3: an observation on diffusion against the prevailing flow direction

Consider the following situation. A repository imposes a fixed concentration at the edge of a low-permeability layer. At the far side of this layer, a high permeability zone keeps the concentration effects fixed at zero. Thus a typical concentration profile is as shown in Figure 5.24.

How does the rate at which radionuclides cross into the high permeability zone (at $x = L$) depend on the flow velocity in the low permeability layer, for a fixed diffusion coefficient (and no dispersion)?

The resulting steady state rate is dependent on the Peclet number. In fact

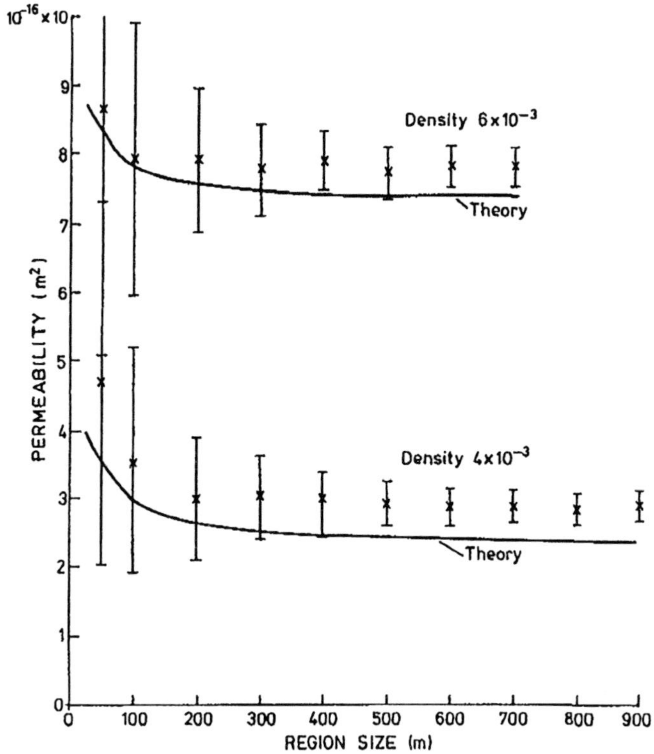

Figure 5.22 Network permeability versus region size for fractures 50 m long.

the rate per unit area is easily shown to be

$$\frac{DC_0}{L}\frac{Pe}{1-e^{-Pe}} \qquad (Pe = \frac{vL}{D})$$

Thus, as the velocity increases the rate becomes proportional to it and tends to vC_0. The formula is still valid for negative Peclet numbers (diffusion against the flow). Although the rate eventually falls off exponentially for negative Peclet numbers, there is a significant release rate for small negative velocities. Overall the relationship is as shown in Figure 5.25.

Perhaps more intriguingly, the shapes of the breakthrough curves for positive or negative Peclet number are identical – so that the flow direction cannot be determined even from the full breakthrough information. The full curves are shown in Figure 5.26.

This example illustrates just how difficult it is to be sure of the details of the true system even with accurate experimental measurements. It shows

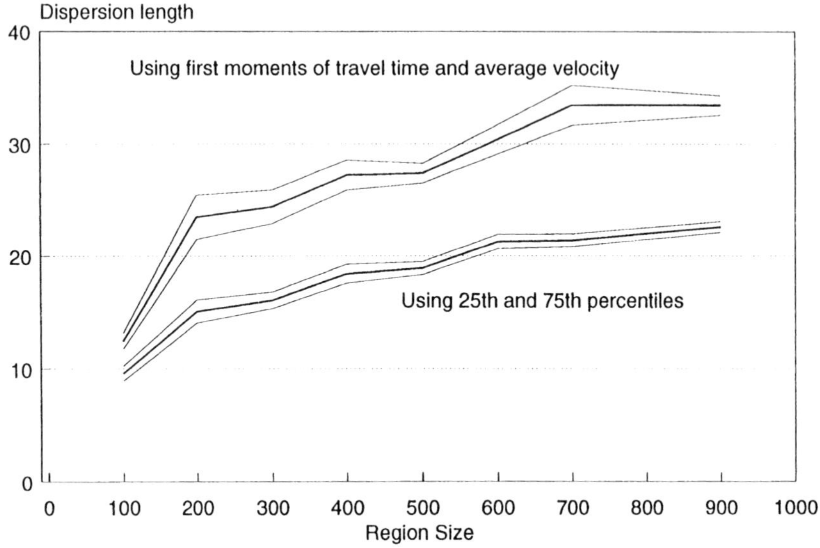

Figure 5.23 Comparison of dispersion length calculated by two different methods from the breakthrough curve.

that collecting tracer breakthrough data is not in itself sufficient to determine something as fundamental as flow direction.

Example 4: flow through tunnels or holes

A useful result to be aware of occurs in the following system. Consider a cylindrical region of hydraulic conductivity k_c embedded in a region of hydraulic conductivity k_r. The system might be an open borehole or tunnel, or a backfilled emplacement hole or tunnel. There is a general flow across the outer region. How does the flow in the inner region depend on the ratio of conductivities?

It is easy to show that the Darcy velocity at $r = 0$ (the centre of the inner region) is given by:

$$q_0 = q_\infty \frac{2k_c}{k_r + k_c}$$

where q_∞ is the Darcy velocity at infinity. This formula shows that the velocity reduction for a low permeability region is proportional to the permeability, where the increase for a highly permeable region is limited to a factor of 2.

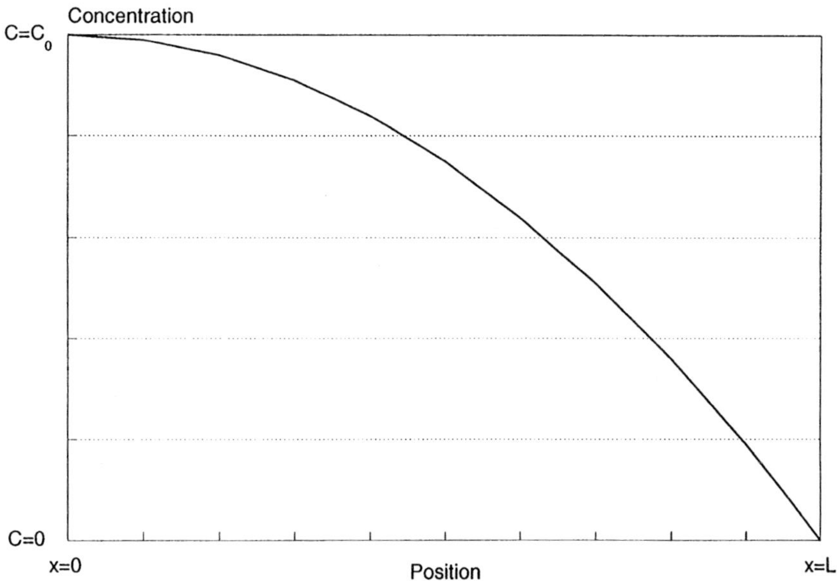

Figure 5.24 Typical concentration profile for diffusion of a species through a low-permeability layer.

The former result gives confidence in the performance of, for example, bentonite buffers. The latter indicates that open tunnel or boreholes are not massive perturbations on the overall system.

A slightly more general form of this result can be obtained for elliptical openings. If the ratio of the axis aligned with the flow to that orthogonal to the flow is denoted by ζ, then the result is:

$$q_0 = \frac{(1+\zeta)k_c}{k_r + \zeta k_c}$$

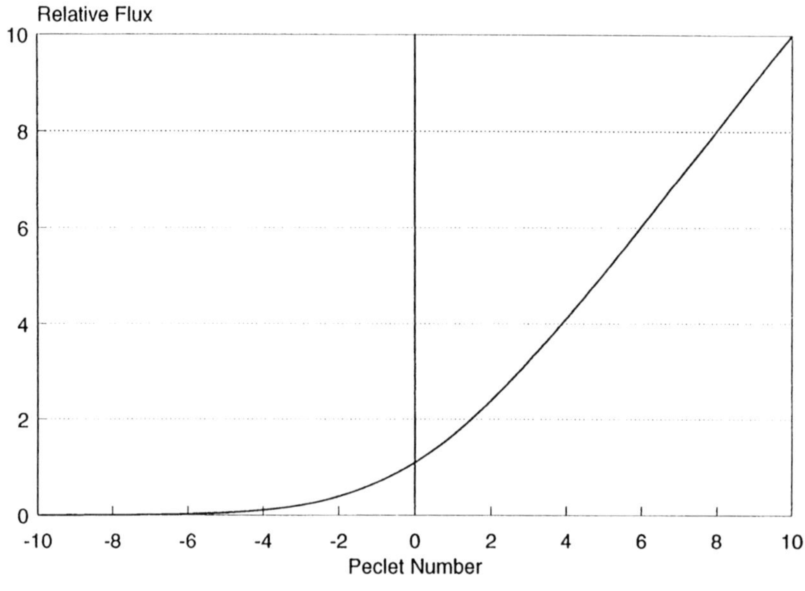

Figure 5.25 Steady state flux versus Peclet number.

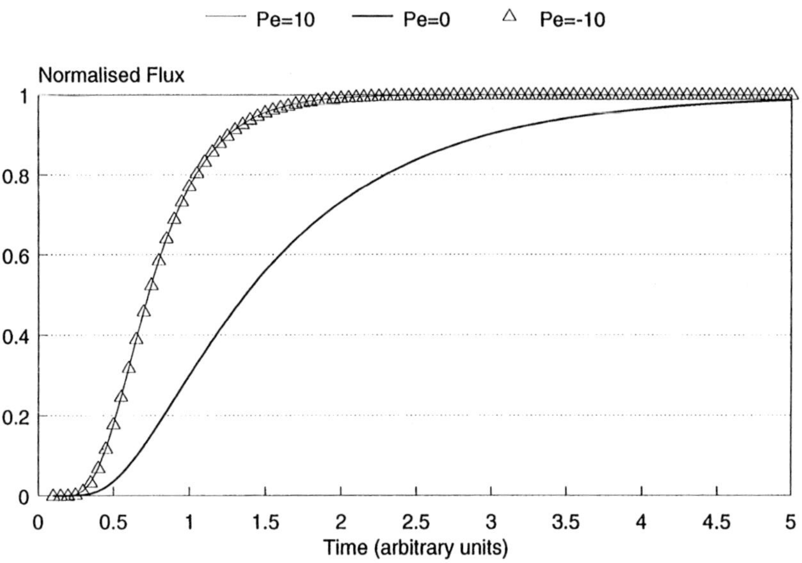

Figure 5.26 Normalised breakthrough curves for various Peclet numbers.

5.4 References

Ahlbom, K., Andersson, J-E., Andersson, P., Ittner, T., Ljunggren, C. and Tíren, S. (1992) Finnsjön study site. Scope of activities and main results. *SKB Technical Report* 92-33. Swedish Nuclear Fuel and Waste Management Company, Stockholm, Sweden.

Aiken, G.R., McKnight, D.M., Weshaw, R.L. and MacCarthy, P. (eds.) (1985) *Humic Substances in Soil, Sediment and Water: Geochemistry, Isolation and Characterization.* John Wiley and Sons, 691pp.

Andersson, P. (*ed.*) (1993) The Fracture Zone Project – Final Report. *SKB Technical Report* 93-20. Swedish Nuclear Fuel and Waste Management Company, Stockholm, Sweden.

Andrews, R.W., Dershowitz, W., Hufschmied, P. and Black, J. (1991) Evaluation strategies for hydrogeological characterisation of a repository area in marine and fluvioterrestrial sedimentary strata. In *NEA/CEC Workshop on Heterogeneities in Hydrogeological Systems*, Paris, October, 1990. NEA/OECD, Paris, France, p 152-170.

Baeyens, B., Maes, A., Cremers, A. (1985) In situ physico-chemical characterization of Boom Clay. *Radioactive Waste Management and the Nuclear Fuel Cycle* **6:** 391-408.

Barker, J. (1991) Transport in fractured rock. In *Applied Groundwater Hydrology – A British Perspective*, (eds. R.A. Downing and W.B. Wilkinson). Clarendon Press, Oxford, p 199-216.

Barnes, H.L. (1981) Measuring thermodynamically interpretable solubilities at high pressures and temperatures. In *Chemistry and Geochemistry of Solutions at High Temperatures and Pressures* (eds. D.T Rickard and F.E. Wickman), *Nobel Symposium on the Physics and Chemistry of the Earth*, Stockholm, Sweden, Vols 13 and 14, 321-338.

Barton, N., Makurat, A., Monson, K., Vik, G. and Tunbridge, L. (1992) Rock mechanics characterisation and modelling of the Disturbed Zone Phenomena at Stripa. *Stripa Project Technical Report* 92-12. Swedish Nuclear Fuel and Waste Management Company, Stockholm, Sweden.

Beauheim, R.L. (1991) Identification of spatial variability and heterogeneity of the Culebra Dolomite at the Waste Isolation Pilot Plant site. In *NEA/CEC Workshop on Heterogeneities in Hydrogeological Systems*, Paris, October, 1990. NEA/OECD, Paris, France, p 131-142.

Billington, D.E., Lever, D.A. and Wisbey, S.J. (1989) Comparative assessment of concepts and areas for deep disposal (Phase 2): Summary. *UK Nirex Report* No. NSS/A301, UK Nirex Ltd, Harwell, U.K.

Birgersson, L., Skagius, K., Wiborgh, M. and Widen, H. (1992) Project Alternative Study – PASS. Analysis of performance and long-term safety of repository concepts. *SKB Technical Report* 92-43. Swedish Nuclear Fuel and Waste Management Company, Stockholm, Sweden.

Bonne, A. (1991) Geoscientific investigations in the Belgian R&D programme concerning the disposal of radioactive waste in clay. In *Geological Problems in Radioactive Waste Isolation – A World-Wide Review*, (ed. P.A. Witherspoon).

Lawrence Berkeley Laboratory Report LBL-29703. Lawrence Berkeley Laboratory, Berkeley, California, U.S.A.

Bottomley, D.J., Gregoire, D.C. and Raven, K.G. (1994) Saline groundwaters and brines in the Canadian Shield: geochemical and isotopic evidence for a residual evaporite brine component. *Geochimica et Cosmochimica Acta* **58:** 1483-1498.

Brace, W.F. (1980) Permeability of crystalline and argillaceous rocks. *International Journal of Rock Mechanics* **17:** 241-251.

Brady, P.V. and Walther, J.V. (1989) Controls on silicate dissolution rates in neutral and basic solutions at 25 °C. *Geochimica et Cosmochmica Acta* **53:** 2823-2830.

Bruno, J. and Sellin, P. (1991) Radionuclide solubilities to be used in SKB 91. *SKB Technical Report* 91-02. Swedish Nuclear Fuel and Waste Management Company, Stockholm, Sweden.

Chapman, N.A., Apted, M., Hodgkinson, D. and Robinson, P. (1993) A review of SKB 91 and SKB R & D Programme 92. *SKI Technical Report* 93:7. Swedish Nuclear Inspectorate, Stockholm, Sweden.

Chapman, N.A. and McEwen, T.J. (1991) Geological aspects of the British programme for deep disposal of nuclear wastes. In *Applied Groundwater Hydrology – A British Perspective*, (eds. R.A Downing and W. B. Wilkinson). Clarendon Press, Oxford, p 143-164.

Chapman, N.A., McEwen, T.J. and Beale, H. (1986) Geological environments for deep disposal of intermediate level wastes in the UK. *Proc. of IAEA Symposium on Siting, Design and Construction of Underground Repositories for Radioactive Wastes*, Hannover, March 1986. IAEA, Vienna, Austria, pp 311-328.

Chapman, N.A. and McKinley, I.G. (1987) *The Geological Disposal of Nuclear Waste*. John Wiley and Sons, New York, 280 pp.

Clauser, C. (1992) Permeability of crystalline rocks. *EOS Transactions, American Geophysical Union* **73:** 233-238.

Degueldre, C. (1990) Grimsel colloid exercise. *CEC Report* EUR 12660. Commission of the European Communities, Luxembourg.

Delany, J.M. (1985) Reaction of Topopah Spring Tuff with J-13 water: a geochemical modeling approach using the EQ3/6 reaction path code. *Lawrence Livermore National Laboratory Report* UCRL-53631. Lawrence Livermore National Laboratory, California, USA.

Dverstorp, B. and Andersson, J. (1990) Analysing transport in low permeability fractured rock using the Discrete Fracture Network Concept. *Proceedings of GEOVAL 1990 Symposium*. Swedish Nuclear Power Inspectorate (SKI), Stockholm, Sweden.

Dverstorp, B., Andersson, J. and Nordqvist, W. (1992) Discrete fracture network interpretation of field tracer migration in sparsely fractured rock. *Water Resources Research* **28:** 2327-2344.

Edmunds, W.M., Andrews, J.N., Burgess, W.B., Kay, R.L.F. and Lee, D.J. (1985) The evolution of saline and thermal groundwaters in the Carnmenellis granite. *Mineralogical Magazine* **48:** 407-424.

Edmunds, W.M., Miles, D.L. and Cook, J.M. (1984) A comparative study of sequential redox processes in three British aquifers. In *Hydrochemical Balances of Freshwater Systems*, Proceedings of IAHS Symposium (ed E. Eriksson), Uppsala,

Sweden, 55-70.

Falck, W.E. (1989) A review of modelling the interaction between natural organic matter and metal cations. *CEC Report* EUR12531. Commission of the European Communities, Luxembourg.

Ge, S. and Garven, G. (1992) Hydromechanical modelling of tectonically driven groundwater flow with applications to the Arkoma Foreland Basin. *Journal of Geophysical Research* **97, B6:** 9119-9144.

Gera, F., Andretta, D., Bocala, W., Chiantore, V. and Schneider, A. (1992) State of the art report: Disposal of radioactive waste in deep argillaceous formations. *ENRESA Technical Report* 01/92. ENRESA, Madrid, Spain.

Hanor, J.S. (1983) Fifty years of development of thought on the origin and evolution of subsurface sedimentary brines. In *Revolution in Earth Sciences* (ed. S.J. Boardman), 99-111.

Hanor, J.S. (1994) Physical and chemical controls on the composition of waters in sedimentary basins. *Marine and Petroleum Geology* **11:** 31-45.

Hodgkinson, D.P. (1991) A comparison of predictions and measurements for the Stripa Simulated Drift Experiment. *Stripa Project Technical Report* TR 91-10. Swedish Nuclear Fuel and Waste Management Company, Stockholm, Sweden.

Hodgkinson, D.P. and Cooper, N. (1992) A comparison of measurements and calculations for the Stripa Tracer Experiments. *Stripa Project Technical Report* 92-20. Swedish Nuclear Fuel and Waste Management Company, Stockholm, Sweden.

Hodgkinson, D.P. and McEwen, T.J. (1991) Spatial variability: A repository performance perspective. In *NEA/CEC Workshop on Heterogeneities in Hydrogeological Systems*, Paris, October, 1990. NEA/OECD, Paris, France, p 27-35.

Hooper, A.J., Billington, D.E. and Herbert, A.W. (1993) The use of conceptual models in the assessment of the proposed UK Nirex deep geological repository for intermediate and low level wastes. *NEA SEDE/PAAG Workshop on the role of conceptual models in demonstrating post-closure safety*, November 1993, Saclay, France. NEA/OECD, Paris, France

Horseman, S.T. (1993) Disposal of high-level waste in argillaceous rocks. *Intera Report* IG2506-4 to ENRESA, to be published as ENRESA Report.

Horseman, S.T. *et al.* (1994) State of the art report on basic concepts and mechanisms of water, gas and solute movement in argillaceous media. *OECD/NEA Report* (in prep.). NEA/OECD, Paris, France.

Horseman, S.T. and Harrington, J. (1994) Migration of repository gases through an overconsolidated clay. *British Geological Survey Technical Report* WE/94/7. British Geological Suvey, Keyworth, UK.

Hostettler, J.D. (1984) Electrode electrons, aqueous electrons, and redox potentials in natural waters. *American Journal of Science* **284:** 734-759.

IAEA (1987) *Isotope Techniques in Water Resources Development.* International Atomic Energy Agency, Vienna, Austria.

IAEA (1992) *Isotope Techniques in Water Resources Development 1991.* International Atomic Energy Agency, Vienna, Austria.

Lahermo, P.W. and Lampen, P.H. (1987) Brackish and saline groundwaters in Finland. In *Saline Waters and Gases in Crystalline Rocks* (eds. P. Fritz and S.K. Frape), *Geological Association of Canada Special Paper* **33:** 103-109.

Langer, M., Schneider, H. and Kühn, K. (1991) The salt dome of Gorleben –

Target site for the German radioactive waste repository. In *Geological problems in radioactive waste isolation – A world wide review*, (ed. P.A. Witherspoon). *Lawrence Berkeley Laboratory Report* LBL-29703. Lawrence Berkeley Laboratory, Berkeley, California, USA.

LaVenue, A.M. and RamaRao, B.S. (1992) A modelling approach to address spatial variability within the Culebra Dolomite transmissivity field. *Sandia Report* SAND92-7306. Sandia National Laboratory, New Mexico, USA.

Lichtner, P.C. and Eikenberg, J. (1993) Propagation of a hyperalkaline plume into the geological barrier surrounding a radioactive waste repository. *Nagra Technical Report* NTB 93-16. Nagra, Wettingen, Switzerland.

Lindberg, R.D. and Runnells, D.D. (1984) Groundwater redox reactions: an analysis of equilibrium state applied to Eh measurements and geochemical modeling. *Science* **225:** 925-927.

Longworth, G., Ross, C.A.M., Degueldre, C. and Ivanovich, M. (1990) Interlaboratory study of sampling and characterization techniques for groundwater colloids: the Markham Clinton exercise. *CEC Report* EUR 12843. Commission of the European Communities, Luxembourg.

Mackay, R. and O'Connell, P.E. (1991) Statistical methods of characterising hydrogeological parameters. In *Applied Groundwater Hydrology – A British Perspective.* (eds. R.A. Downing and W.B. Wilkinson), Clarendon Press, Oxford, p 217-242.

Mazurek, M. (1994) Geology of the crystalline basement of northern Switzerland and derivation of geological input data for hydrodynamic modelling and safety performance analysis. *Nagra Technical Report* NTB 93-12. Nagra, Wettingen, Switzerland.

Mazurek, M., Gautschi, A. and Vomvoris, S. (1993) Deriving input data for solute transport models from deep borehole investigations – An approach for crystalline rocks. *International IAEA Symposium on Geological Disposal of Spent fuel, High-level and Alpha-bearing Wastes*, Antwerp, Belgium, October 1992. International Atomic Energy Agency, Vienna, Austria.

McCombie, C., McKinley, I.G. and Zuidema, P. (1991) Sufficient validation: The value of robustness in performance assessment and system design. *GEOVAL-90 Symposium*, Stockholm, May 1990. NEA/OECD, Paris, France.

Michard, G. (1987) Controls of the chemical composition of geothermal waters. In *Chemical Transport in Metasomatic Processes*, (ed. H.C. Helgeson). *NATO Advanced Study Institutes Series. Series C Mathematics and Physical Sciences* **218:** 323-353.

Michard, G. (1990) La composition chimique des eaux thermominérales et géothermales. *Hydrogéologie* **4:** 253-266.

Miller, W., Alexander, R., Chapman, N., McKinley, I. and Smellie, J. (1994) *Natural Analogue Studies in the Geological Disposal of Nuclear Wastes.* Studies in Environmental Sciences 57, Elsevier, Amsterdam, 395pp.

Nagra (1985) Project Gewähr 1985: Nuclear Waste Management in Switzerland: Feasibility Studies and Safety Analyses. *Nagra Gewähr Report Series* NGB 85-01/09. Nagra, Baden, Switzerland.

Nagra (1988) Sediment Study-Intermediate Report. Disposal options for long-lived radioactive waste in Swiss sedimentary formations (Executive Summary). *Nagra Technical Report* 88-25E. Nagra, Wettingen, Switzerland.

Nagra (1994) Conclusions from the regional investigation programme for siting a HLW repository in the crystalline basement of northern Switzerland. *Nagra Technical Report* NTB 93-09E. Nagra, Wettingen, Switzerland.

NEA (1988) Proceedings of an NEA Workshop on *Excavation Response in Geological Repositories for Radioactive Waste,* Winnipeg, Canada, April 1988. NEA/OECD, Paris, France.

NEA (1991) Proceedings of an NEA/CEC Workshop on *Heterogeneities in Hydrogeological Systems,* Paris, October, 1990.

Neuman, S.P. (1990) Universal scaling of hydraulic conductivities and dispersivities in geologic media. *Water Resources Research* **26:** 1749-1758.

Neuzil, C.E. (1986) Groundwater flow in low permeability environments. *Water Resources Research* **22 (8):** 1163-1195.

Neuzil, C.E. (1993) Low fluid pressure within the Pierre Shale: A transient response to erosion. *Water Resources Research* **29:** 2007-2020.

Neuzil, C.E. (1994) How permeable are clays and shales? *Water Resources Research* **30:** 145-150.

Nirex (1989) Deep Repository Project. Preliminary Environmental and Radiological Assessment and Preliminary Safety Report (PERA/PSR). *UK Nirex Report* No. 71. UK Nirex Ltd, Harwell, U.K.

Nirex (1993) The geology and hydrogeology of the Sellafield area – Interim Assessment. *UK Nirex Report* 524, Vols. 1 to 4. UK Nirex Ltd, Harwell, U.K.

Nordstrom, D.K., Ball, J.W., Donahoe, R.J. and Whittemore, D. (1989) Groundwater chemistry and water-rock interactions at Stripa. *Geochimica et Cosmochimica Acta* **53:** 1727-1740.

Olsson, O. *et al. (ed.)* (1992) The Stripa Project. Site Characterisation and Validation – Final Report. *Stripa Project Technical Report* TR 92-22. Swedish Nuclear Fuel and Waste Management Company, Stockholm, Sweden.

PAGIS (1984) Performance Assessment of Geological Isolation Systems – Phase 1 report for the Salt Option. *CEC R&D Programme on Radioactive Waste Management and Storage.* Commission of the European Communities, Luxembourg.

Pearson, F.J. and Scholtis, A. (1993) Chemistry of reference waters of the crystalline basement of Northern Switzerland. *Nagra Technical Report* NTB 93-07. Nagra, Wettingen, Switzerland.

K.S. Pitzer (1980) Theory: ion interaction approach. In *Activity Coefficients in Electrolyte Solutions* (ed. R.M. Pytkowicz). CRC Press, Boca Raton, U.S.A, 157-208.

Rochelle, C.A., Bateman, K., Milodowski, A.E., Noy, D.J., Pearce, J. and Savage, D. (1992) Reactions of cement pore fluids with rock: implications for the migration of radionuclides. In *Water-Rock Interaction* (eds. Y.K. Kharaka and A.S. Maest). Balkema, Rotterdam, 423-426.

Sandia (1992) Preliminary performance assessment for the Waste Isolation Pilot Plant, December 1992. Volumes 1 to 3. *Sandia Report* SAND92 0700/1. Sandia National Laboratory, New Mexico, USA.

Savage, D., Bath, A.H., Crawford, M.B., Gillespie, M.R., Hawkins, M.P., Metcalfe, R., Milodowski, A.E. and Naden, J. (1993) Groundwater relationships in cover and basement at the margins of the East Irish Sea Basin, West Cumbria, UK. *Terra Abstracts* **5:** 643-644.

Savage, D. and Rochelle, C.A. (1993) Modelling reactions between cement pore fluids and rock: implications for porosity change. *Journal of Contaminant Hydrology* **13:** 365-378.

Schmitt, R. and Klarr, K. (1991) Parameters, methodologies and priorities of site selection for radioactive waste disposal in rock salt formations. *ENRESA Technical Report* 06/91. ENRESA, Madrid, Spain.

SKB (1983) Final storage of spent nuclear fuel – KBS-3, Vols. 1-5. Swedish Nuclear Fuel and Waste Management Company, Stockholm, Sweden.

SKB (1992a) SKB-91. Final disposal of spent nuclear fuel. Importance of the bedrock for safety. SKB Technical Report 92-20. Swedish Nuclear Fuel and Waste Management Company, Stockholm, Sweden.

SKB (1992b) Treatment and final disposal of nuclear waste. Detailed R&D programme 1993-1998. SKB, September 1992. Swedish Nuclear Fuel and Waste Management Company, Stockholm, Sweden.

SKI (1991) SKI Project-90, Volumes 1 and 2. *SKI Technical Report* 91:23. Swedish Nuclear Inspectorate, Stockholm, Sweden.

Smith, D.G. (ed.) (1982) *The Cambridge Encyclopedia of Earth Sciences.* Cambridge University Press, 496 pp.

Smith, M.J., Anttonen, G.J., Barney, G.S., Coons, W.E., Hodges, F.N., Johnston, R.G., Kaser, J.D., Manabe, R.M., McCarel, S.C., Moore, E.L., Noonan, A.F., O'Rourke, J.E., Schulz, W.W., Taylor, C.L., Wood, B.J. and Wood, M.I. (1980) Engineered barrier development for a nuclear waste repository located in basalt - an integration of current knowledge. *Rockwell Hanford Operations Report* RHO-BWI-ST-7. Rockwell Hanford Operations, Richland, Washington State, USA.

Smith, P.A. (1993) A model for colloid-facilitated transport through fractured media. *Nagra Technical Report* NTB 93-32. Nagra, Wettingen, Switzerland.

Steefel, C.I. and Lichtner, P.C. (1994) Diffusion and reaction in rock matrix bordering a hyperalkaline fluid-filled fracture. *Geochimica et Cosmochimica Acta* **58:** 3595-3612.

Sumerling, T.J. (*ed.*) (1992) Dry Run 3 – A trial assesment of underground disposal of radioactive wastes based on probabilistic risk analysis, (Summary). *UKDOE Report* No. DoE/HMIP/RR/92. XXX. UK Department of the Environment, London, UK.

Taylor, Sir Geoffrey (1953) Dispersion of soluble matter in solvent flowing slowly through a tube. *Proceedings of the Royal Society* **A219:** 186.

Tellam, J.H. and Lloyd, J.W. (1981) A review of the hydrogeology of British onshore non-carbonate mudrocks. *Quarterly Journal of Engineering Geology London* **14:** 347-355.

Tipping, E. (1988) Colloids in the aquatic environment. *Chemistry and Industry* 485-490.

6

The biosphere and radiological effects

6.1 The biosphere in the context of PA

Development of disposal systems for radioactive waste has a considerable history, as does the corresponding development of assessment methodologies to determine the associated risks. The whole process has been supported by major research programmes around the world to investigate the performance of alternative and combinations of engineered barriers and the isolating effect of the natural geological barrier. These barriers to radionuclide migration are suitable and convenient subjects for investigation using predictive modelling techniques.

In the biosphere, which we take to mean the surface environment in which humankind carries on its normal business, detailed evaluation is more problematic. The reasons for this are many-fold but are largely concerned with two basic factors.

Firstly, the surface environment is naturally subject to variation over, geologically speaking, comparatively short timescales. Conditions in the surface environment will change considerably, affecting corresponding estimates of human radiation exposure. Such changes can be anticipated both before any geosphere releases to the surface environment and during the period of time typically predicted for the maximum rate of release. Results from the European Commission study PAGIS (Performance Assessment of Geological Isolation Systems) show the annual discharge of the most significant radionuclides to the biosphere extending over many thousands of years within a factor of a few of the maximum value (Cadelli *et al.*, 1988). The same is indicated in the Swiss Project Gewähr (Nagra, 1985) and Project-90 from Sweden (SKI, 1991). Within such long periods of release many different surface environment conditions will arise.

Secondly, one of the major contributors to changes in the surface environment will be the various influences of many kinds of human activities. Such activities are very difficult to predict over comparatively short timescales, let alone over thousands of years. Furthermore, any evaluation of radiation doses to man has to include assumptions for human behaviour, which will have been influenced by social, technical and other developments. The issues have been widely discussed, for example, in Land and Schneider (1987). Not only is human behaviour very complicated (which could equally be said of other aspects of the repository safety assessment), it has associated with it an inherent lack of predictability which can be related to free will or, some would say, the perversity of human nature.

It follows that the unpredictability of human behaviour is different in nature from the unpredictability associated, say, with rocks and groundwater flow, which behave according to a set of physical laws. Our understanding of these geophysical and geochemical laws may not be complete in every detail. However, our understanding of the relevant phenomena in the geosphere operates over longer timescales than for the corresponding biosphere considerations. Noting also man's influence on the biosphere, long term predictions for the biosphere are seen by some as impracticable.

Nevertheless, it has to be recognised that there is a need to say something about the consequences of release of radionuclides to the surface environment. This need arises because of the requirement to demonstrate compliance with the protection objectives of waste disposal which are defined in terms of the impacts occurring in the biosphere. Perhaps, more importantly, there is also a need to satisfy public curiosity in this matter. The most commonly considered radiological impacts are doses and risks to man, but these are not the only potentially relevant quantities. Modelling of the biosphere is required to convert the predictions of release from the geosphere into these corresponding radiological impacts.

Some forms of regulatory criteria apparently avoid the problems of long term dose assessments by specifying limits on release of radionuclides from particular engineered barriers or pre-defined regions of the geosphere. Alternatively, prescriptions can be set on the physical lifetime or other performance of these barriers. These forms of criteria are useful and relevant to the specification of engineering design requirements. However, if the objective is to ensure (among other things) that risks or doses to man are acceptable according to some radiological standard (*e.g.* as currently proposed by the International Commission on Radiological Protection (ICRP, 1985; see discussion in Chapter 1), then these prescriptive criteria can only be justified by consideration of the consequences of any leakage through and beyond the barriers to the surface.

A very important aspect of the biosphere methodology is the requirement to assess the radiological impacts associated with the different types of geosphere release which are anticipated. Typically, such release is assumed to occur in groundwater, but release of radionuclides in radioactive gases, or as the ground surface is eroded, may also be important. It is important to develop consistent boundary conditions between modelling of radionuclide transport in the geosphere and biosphere, at the so-called geosphere-biosphere interface. Further kinds of release may be associated with intrusion by man, either into the contaminated geosphere or into the repository itself. Such intrusion may directly affect the intruders or may modify the effectiveness of engineered and geosphere barriers and, hence, affect the subsequent releases to the biosphere.

The treatment of different postulated geosphere releases, the level of detailed evaluation of radionuclide migration and accumulation through the biosphere and the evaluation of associated radiation doses will all depend in part on the specific objectives of the assessment. These, in turn, will be different according to different regulatory requirements and according to the stage of development of the disposal project. Two examples serve to illustrate the differences which may arise. Firstly, the criteria involved may include consideration of individual dose (the dose to a single person) as well as collective doses (the sum of doses to a particular population). The area over which radionuclide dispersion has to be followed will be different in each case. Secondly, the level of detail required in biosphere modelling at the stage of site selection might be less than at the stage of licensing actual disposals.

The radiological consequences of potential interest include not only effects on human health but also the effects on non-human biota. Both types of effects may be put into perspective by comparisons with natural background. The implications for non-human biota of measures to limit exposure of humans have recently been reviewed by the International Atomic Energy Agency (IAEA, 1992). It is concluded that, in general, if releases of radionuclides are kept small enough so that annual individual dose limits for humans are not exceeded then the consequences for populations of other biota will be small. However, site-specific studies may be necessary in some circumstances. The calculational end points for biosphere models may include not only doses to man and other biota, but also concentrations in and fluxes through various biosphere media.

A list of potentially relevant end points for biosphere models is suggested in Smith (1989), as follows:

- annual individual doses to man, *i.e.* to members of (hypothetical) critical groups,
- estimates of where and when these doses are expected to arise,
- annual individual risks to members of (hypothetical) critical groups, noting the probability of a deleterious health effect following a given exposure, and also the probability that the exposure arises,
- temporal and spatial development of collective dose rate and truncated collective dose commitment to man,
- doses and dose rates to biota,
- distribution of waste radionuclides released from the geosphere into the

biosphere for comparison with naturally occurring concentrations and fluxes,
- non-radiological environmental impacts of radioactive waste disposal,
- expressions of confidence in any of the estimates for the quantities referred to above.

It still appears to be the case that few, if any, assessments have dealt with all these points, although any of them might be raised in the debate of safety of geological disposal. In addition, each end point could be further refined. For example, in many solid waste disposal assessments, only doses to adults are evaluated. Doses and risks to infants and children could be different, as could dose to the foetus.

It is clear that the results of any assessment must be relevant to and understood by a wide range of audiences, each with its own special interests and requirements. In the broadest sense, all the scientific information available can have some relevance and it should be part of the assessment programme to determine how all such information can be applied and presented. It need not be assumed that prevailing national criteria or international recommendations provide all the background for what is required. A watching brief should be kept on international developments such as the workshop on HLW criteria hosted by the Nuclear Energy Agency in Paris (NEA, 1990). Such workshops are often attended primarily by professionally interested parties. In addition, advice should be sought about the relevance of standards and the presentation of results on a wider basis, but particularly where the impacts affect man directly, *i.e.* in his own environment. It is because the surface environment is where the impact arises that assessment of the surface environment is most likely to be affected by criteria development. The biosphere assessment strategy should therefore include procedures for continually updating the assessment objectives according to developments in radiological and other protection objectives.

6.2 Migration and accumulation of radionuclides

Estimating the transport of radionuclides through the biosphere in the context of releases from deep geological repositories, *i.e.* in the long-term, is not a new subject. The modelling capability has been available for some time, see for example Bergstrom *et al.* (1982), Lawson and Smith (1985) and Grogan (1985). Some of the models were originally relatively simple extensions of models developed for other purposes, such as estimating the consequence of routine discharges from power plants. Projects such as BIOMOVS (BIOMOVS 1991, 1994) have contributed to a greater understanding of relevant long-term processes and how to model them. Also important has been the development of

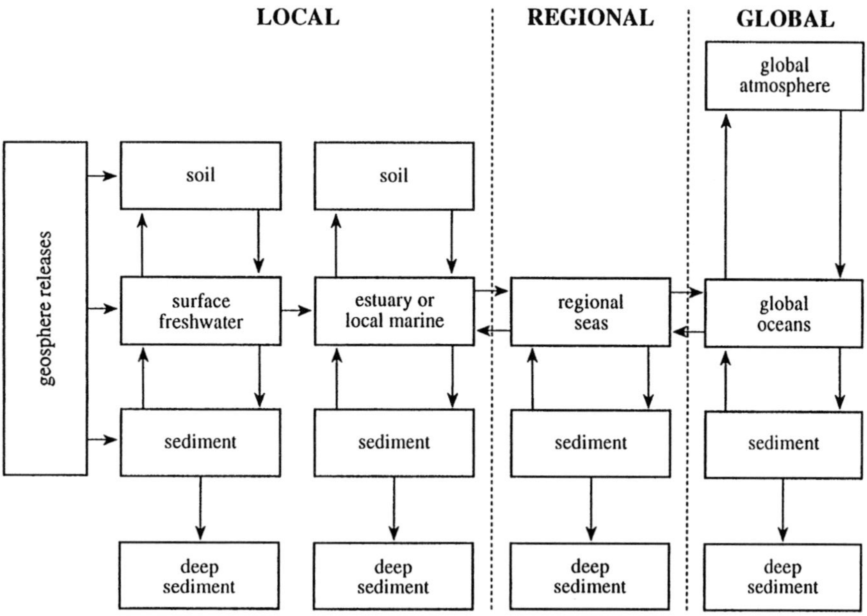

Figure 6.1 Schematic of biospheric transport model.

appropriate background assumptions for biosphere calculations, as illustrated in a report issued by the Swedish Nuclear Power Inspectorate (SKI, 1989) as part of a scenario development project organised with the Swedish Nuclear Fuel and Waste Management Company (SKB).

All of the models which have been used in biosphere assessment for post-disposal impacts are comparatively simple in mathematical terms. Nearly all are based on linear donor-controlled compartment models. Figure 6.1 shows the typical environmental compartments among which radionuclide migration is considered. Dispersion among local compartments where geosphere release occurs are important for estimating the highest radionuclide concentrations and fluxes. Regional and global compartments are generally only necessary if it is intended to calculate collective doses over large populations. Special models have been developed for this purpose specifically for very long-lived and mobile radionuclides such as I-129 (Smith and White, 1983).

Figure 6.2 (from Lawson and Smith, 1985), shows the greater detail which can be included in the number and type of compartments. For such a model, it is assumed that, as soon as a radionuclide enters a compartment, instantaneous mixing occurs so that there is a uniform concentration of the radionuclide over the whole compartment. Each compartment must be chosen to represent a region of the environment for which this assumption

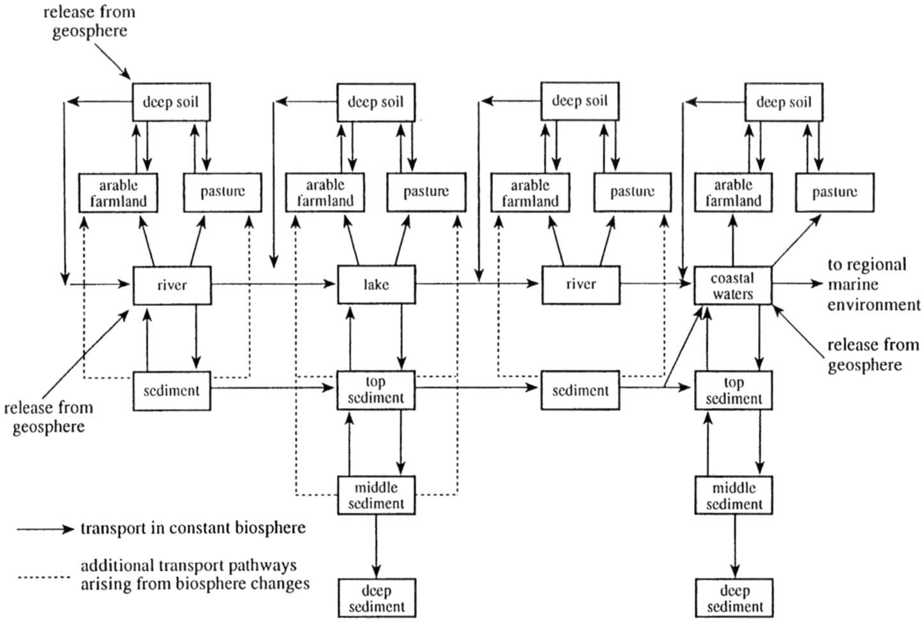

Figure 6.2 Schematic of local biospheric transport.

is reasonable. In general, the tendency should be to reduce the number of compartments to a minimum in order to simplify the model but the limit of such simplification is set by the above requirement. Transfer between compartments is described by 'transfer coefficients', which represent the fraction of the activity in a particular compartment transferred from that compartment to another one in unit time. Thus, the rate of change of a radionuclide in any compartment is given by:

$$\frac{dN_{\dot{A}}}{dt} = S - K_{AB}N_{\dot{A}} + K_{BA}N_{\dot{B}} - \dot{\lambda}N_{\dot{A}} + \ddot{\lambda}N_{\ddot{A}}$$

where S = input of radionuclide to compartment A from the geosphere (atoms per unit time)

$N_{\dot{A}}$ = number of atoms of radionuclide in compartment A

$N_{\dot{B}}$ = number of atoms of radionuclide in compartment B

$N_{\ddot{A}}$ = number of atoms of parent radionuclide in compartment A

K_{AB}, K_{BA} = transfer coefficients (per unit time) from compartments A to B, and from B to A

$\dot{\lambda}$ = radioactive decay constant (per unit time) of radionuclide

$\ddot{\lambda}$ = radioactive decay constant (per unit time) of parent radionuclide
and t = time

The solution of the matrix of equations given above provides the time-dependent inventory of each compartment. Assumptions for compartment sizes then result in estimates of concentrations in the corresponding media. Compartment sizes and transfer coefficients can be made time-dependent (Lawson and Smith, 1985), but some solution methods applied do not allow for this. Conversion from atoms to units of radioactivity (Bq) is trivial.

The transfer coefficients are commonly determined by the major physical characteristics of the compartments involved. For example, the transfer coefficient from one river section to another would be given by the volumetric flow in the donor section of the river (m^3y^{-1}) divided by its volume (m^3). Account also has to be taken of the chemical and biological processes which could be important for particular radionuclides. For example, transfer from soil to a river compartment will be affected by sorption of the radionuclide on the soil. This sorption may result in a relatively long-term accumulation of a radionuclide in soil, given continuing release from the geosphere or continuing irrigation with contaminated river or groundwater. High sorption in the geosphere tends to reduce impacts, but the opposite can be true in the biosphere.

It is beyond the scope of this book to consider all the potentially relevant radionuclide migration and accumulation processes. In summary, they include advection and dispersion in water bodies, leaching from and absorption and adsorption in soils and sediments, wind and water erosion and other geomorphological processes, bioturbation of soils and sediments and the effects of human activities. How these processes need to be treated depends on the objectives of the assessment and the calculational end points of interest. Uptake of radionuclides into biota may also result in significant radionuclide transfer. However, these processes are even more important when it comes to calculating radiological impacts.

6.3 Calculation of radiation doses

Having determined activity concentrations in various environmental media, calculation of the resulting radiation doses relies substantially on the application of simple formulae derived from models and monitoring of current day releases into the environment. For example, fixed relationships between activity concentrations in soil and in animal and vegetable foodstuffs are commonly applied. Apart from ingestion and exposure via the food-chain, account also has to be taken of external irradiation and inhalation exposure pathways. Assumptions also have to be made concerning the particular behaviour of the individuals or populations for whom a dose estimate is required. Of special interest are the assumptions for the behaviour of critical groups whose exposure is representative of the highest that may be expected.

A typical calculation for the dose due to root crop consumption has the following components:

$$\begin{bmatrix} \text{concentration} \\ \text{in soil} \\ (\text{Bq m}^{-3}) \end{bmatrix} \times \begin{bmatrix} \text{root} \\ \text{uptake} \\ \text{factor} \\ (\text{Bq kg}^{-1} \\ \text{per} \\ \text{Bq m}^{-3}) \end{bmatrix} \times \begin{bmatrix} \text{individual} \\ \text{consumption} \\ \text{rate} \\ (\text{kg y}^{-1}) \end{bmatrix} \times \begin{bmatrix} \text{dose per} \\ \text{unit intake} \\ (\text{Sv Bq}^{-1}) \end{bmatrix}$$

the product of which is Sv y^{-1} or annual dose. In addition, the root crop could become contaminated due to irrigation and translocation of foliarly intercepted deposition to the roots. The data required are common to radio-ecology and radiological assessments generally and are not uniquely a problem for solid waste management. However, there is generally less data available for some of the long-lived radionuclides relevant to solid waste management. A significant range of exposure pathways and the corresponding data requirements are described in Smith *et al.* (1989) for relevant individual dose and collective dose calculations. Ranges in many of the parameter values are also presented.

The large uncertainties in some of the parameters reflect variability in farming practices and soil types and inter-species variation. For a specific location today, such uncertainty can be reduced. However, it is not possible to make the relevant measurements for releases occurring in the far future at locations which geosphere modelling is inadequate to identify precisely.

Factors to convert intake of activity into a level of dose (values of dose per unit intake) are provided by the ICRP, *e.g.* in Publication 30 (ICRP, various dates). However, these data are continuously being updated according to new recommendations on metabolic models and data (Clarke, 1994), as well as due to modifications in the definitions of radiation dose (ICRP, 1991). Simple multiplying factors for converting a level of dose to a level of risk of fatal cancer and other deleterious effects are also provided by the ICRP (ICRP, 1991). These factors are population dependent.

6.4 Timescales

The timescales before radionuclides are predicted to enter the biosphere from a deep repository are long, typically more than 10^4 years after disposal. A time cut-off may be thought appropriate for the calculations, bearing in mind the difficulty of interpreting the significance of such long-term results. Resolution of this issue is largely a matter for criteria development. However, if a cut-off is adopted, then, in presentation of the overall results, the justification for the cut-off should not be presumed. It should be demonstrated in the context of the particular proposals being considered.

The timescale for a system of the type illustrated in Figure 6.2 to come into equilibrium, given a continuous release from the geosphere, is also typically of the order of 10^4 y. That is, by or within 10^4 y the rate of loss of activity from the area local to the geosphere release will be equivalent to the release into it from the geosphere. Beyond this time, dose rates do not increase. In fact, for many radionuclides, the peak dose rate is reached much sooner than 10^4 y.

Provided that the maximum rate of release from the geosphere continues for 10^4 or so, there is no need to couple the geosphere and biosphere models directly in order to calculate peak annual dose rates. These can be obtained by multiplying peak geosphere release rates (Bq y^{-1}) by peak dose rates per unit geosphere release rates (Sv y^{-1} per Bq y^{-1}). The latter can be determined independently from the geosphere model, providing the types of geosphere releases are adequately described. This procedure was adopted within SKI's Project 90 (Charles and Smith, 1991).

In general, uncertainties will increase the further into the future predictions are made. Billington *et al.* (1990) have suggested that assessments should be considered in the context of four time frames: 0-10^2 y, 10^2-10^4 y, 10^4-10^6 y and 10^6-10^8 y, with an implicit cut-off at 10^8 y. These time frames were suggested as an aid to interpretation and presentation of results. The UK National Radiological Protection Board (NRPB) has adopted a similar idea in proposing limits on how far into the future detailed assumptions about the human environment and behaviour should be made (NRPB, 1992). Since gradual changes in assumptions are said to be difficult to implement in assessments, the NRPB recommends, for simplicity, that general assumptions only should be applied after about ten thousand years and that risk calculations should not be continued beyond one million years. Such a suggestion may be unhelpfully narrow. On the one hand, it would seem hard to justify particular assumptions for human behaviour for anything like as long as ten thousand years whereas some other aspects of the environment may be more amenable to longer-term evaluation.

However long the delay might be before release to the biosphere commences, it is likely that such a release will continue at near its peak rate for a considerable period relative to the rate of surface environment changes. Assuming release is predicted to occur before any specified time cut-offs, then the assessment ought take account of such environmental changes during the period of peak release, particularly if the change can result in higher radiation exposures. Such changes could involve conversion of lake-bed sediments into farmland, as indicated in Figure 6.2 and BIOMOVS (1989).

6.5 Environmental change and human activities

To omit consideration of human activities is to ignore a problem, not a mechanism for its solution. The assumptions made concerning human behaviour will affect the exposure pathways requiring consideration and will also influence the definition of critical groups. However, human activities may also influence the sub-surface environment and geosphere/biosphere interface. The assessment strategy should include consideration of such questions, or risk criticism through lack of completeness. Other important changes which will occur with or without modification by man are associated with climate and geomorphology.

The Radioactive Waste Management Committee of the Nuclear Energy Agency (NEA) and corresponding advisers from the International Atomic Energy Agency (IAEA) and the Commission of the European Communities (CEC) have said in a collective opinion (NEA/IAEA/CEC, 1991) that any

> 'estimate of far future living conditions would be largely speculative. Such calculations are, therefore, generally viewed as an illustration of what the doses would be if the release occurred today, rather than as a prediction of the actual dose to some human living in the far future'.

This is a pragmatic compromise arising out of the need to provide an estimate of radiological impact and the difficulty with predicting human behaviour. It is reasonable, however, to expand on what is meant by *'if the release occurred today'*. The results of long-term assessments should provide an indication of the general levels of dose (or risk) given an assumed set of biosphere conditions, rather than be based on predictions, taking account of all the possibilities for evolution of the surface environment. More than one set of biosphere conditions may be specified. Even if the release were actually occurring today, the range of possible conditions in the area of release is large. In identifying such sets of biosphere conditions, or reference biospheres, it would not be the intention to avoid the implications of the full range of possible human behaviour and other factors affecting the evolution of the surface environment. Rather, it is intended to simplify their analysis and provide for a clearer presentation of results and help distinguish the performance of alternative disposal concepts.

The concept of reference biospheres as a standard approach to biosphere modelling for long-term safety assessments has been discussed at both an international and a national level (SKI, 1989) and the potential value of such an approach was recognised during BIOMOVS (BIOMOVS, 1991).

Thus a Working Group has been established within the second phase of BIOMOVS with the objective of reaching a consensus on the development and application of the reference biosphere approach in the evaluation of long-term consequences of waste disposal systems. Details are given in BIOMOVS II (1994).

The reference biosphere could take the form of descriptions of the general situations to be modelled, or may even include numerical recommendations on how such situations could be modelled or may also include results for relevant end points. International consensus on the latter may be difficult because there is no firm agreement on what the end points are. It matters less how far down this route one goes than that the first step should be taken, *i.e.* of agreeing guidelines for the general situations to be modelled. Where appropriate, such guidelines could be interpreted in a site specific way. Examples of reference biosphere characteristics are suggested in SKI (1989), based on categories of climatic conditions and the state of local development and exploitation in the region of the release. There is no presumption that conditions are static within each reference biosphere, so that on-going changes during a period of release can be taken into account, such as the accumulation of radionuclides in sediments and soils.

All the above discussion relates to how to model the radiological consequences of releases from the geosphere. In fact, biosphere changes may influence the geosphere release; for example, a reduction in rainfall may result in modified groundwater flow. Glaciation and permafrost may have similarly important consequences (McEwen and de Marsily, 1991). Integrating biosphere and geophere changes into a single system model is described in Sumerling and Thompson (1992).

6.6 Uncertainty and variability

Considerable effort has been extended towards development of technical methods for determining the uncertainty in model results due to uncertainty in the model parameter values, and in the variability to be expected in real outcomes. In addition, especially for the near surface, major uncertainty arises in the conceptual description of the environments on which the mathematical models are based. The adoption of reference biospheres can go some way to reducing this. The uncertainties and variability will have been given boundaries by the limitations inherent in the individual reference biosphere descriptions. However, this makes it imperative that a sufficiently representative set of reference biospheres should be developed in the first place.

The assessment strategy must include evaluation of all aspects of uncertainty and, to the extent that some aspects may be intractable, set the limits on the practical interpretation of the results which are then to be presented. In addition, the strategy must include investigation and development of techniques for explaining such analyses to a wide range of audiences with different technical and other backgrounds. There is considerable emphasis given to trying to evaluate adequately the significance of extreme values in parameter distributions and high consequence but low probability events. Bearing in mind the need for credibility in modelling predictions of risk *etc.*, it is relevant that, in some disciplines (and hence in some audiences) *'model building means setting aside the incidental, atypical or accidental'* (Brandel, 1982).

6.7 Surface environment assessment programme

Programmes for geological disposal are being developed and implemented over timescales of the order of a decade or longer. It appears inappropriate to plan the assessment procedure on a shorter timescale, although provisional results may be required as various programme sub-goals have to be met. For assessment of the surface environment, four stages of development are proposed, comprising

1. **Establishment of concepts and methodology**
   - identify the generally relevant issues,
   - identify the specific objectives for the surface assessment within the assessment programme,
   - identify specific regulatory and other requirements,
   - establish an agreed methodology taking account of international developments, such as the reference biosphere concept, as well as the locally identified issues. The methodology should define the radiological basis, the end points to be evaluated, the approach to giving estimates of confidence in results and presentation of results to various audiences.
2. **Development of tools for biosphere model calculations**
   - develop familiarity with available mathematical and calculational methods
   - tailor most appropriate methods according to the end points required, building in new features as necessary
3. **Developments of supporting databases**
   - review of literature and existing databases compiled for other programmes
   - compilation of a database for the particular assessment, taking special account of the relevance of current-day data to the particular timescales in the assessment

- documentation of the database contents.

4. **Testing of the surface environment assessment methodology**
   - noting that assessment and development of the disposal system is an iterative process (NEA/IAEA/CEC, 1991), the methodology should be tested at appropriate sub-goal stages in the overall programme, taking account of regulatory and other developments and proving the effectiveness of the methodology at meeting its objectives.

6.8 References

Bergstrom, U., Edlund, O., Evans, S. and Röjder, B. (1982) BIOPATH. A computer code for calculation of the turnover of nuclides in the biosphere and the resulting doses to man. Basic description. *Studsvik Report* NW-82/261.

Billington, D.E., Lever, D.A. and Wisbey, S.J. (1990) Radiological assessment of deep geological disposal: work for UK Nirex Ltd. In *Proceedings of Paris Symposium on Safety Assessment of Radioactive Waste Repositories, 1989.* Nuclear Energy Agency, International Atomic Energy Agency, Commission of The European Commission.

BIOMOVS (1989) BIOMOVS Technical Report No. 5, Scenario B5, Ageing of a Lake. Swedish Radiation Protection Institute, Stockholm.

BIOMOVS (1991) Proceedings of a Symposium on the Validity of Environmental Transfer Models. Stockholm, October 1990. Swedish Radiation Protection Institute.

BIOMOVS II (1994) An interim report on reference biospheres for radioactive waste disposal. Development by a Working Group of the BIOMOVS II Study. BIOMOVS II Technical Report No. 2, published by the Swedish Radiation Protection Institute on behalf of the BIOMOVS II Steering Committee, Stockholm.

Brandel, F. (1982) *Civilisation and Capitalism, 15th to 18th Century. Vol. 2. The Wheels of Commerce.* Collins, London.

Cadelli, N., Cottone, G., Orlowski, S., Bertozzi, G., Girardi, F. and Saltelli, A. (1988) Performance Assessment of Geological Isolation Systems for Radioactive Waste. Summary. *CEC Report* EUR 11775 En. Commission of the European Communities, Luxembourg.

Charles, D. and Smith, G.M. (1991) Project-90, conversion of releases from the geosphere to estimates of individual doses to man. *SKI Report* TR 91:14. Swedish Nuclear Power Inspectorate, Stockholm, Sweden.

Clarke, R.H. (1994) 1994 Meeting of ICRP. *Radiological Protection Bulletin* No. 156. National Radiological Protection Board, Chilton, U.K.

Grogan, H. (1985) Biosphere modelling for a HLW repository. Scenario and parameter variations. *Nagra Technical Report* NTB 85-48. Nagra, Baden, Switzerland.

IAEA (1992) Effects of ionising radiation on plants and animals at levels implied by current radiation standards. *IAEA Technical Report Series* No. 332. International Atomic Energy Agency, Vienna, Austria.

ICRP (1985) International Commission on Radiological Protection. Radiation

Protection Principles for the Disposal of Solid Radioactive Waste, ICRP Publication 46. Pergamon Press, Oxford, U.K.

ICRP (1991) 1990 Recommendations of the International Commission on Radiological Protection. ICRP Publication 60 ICRP 21, No. 1-3. Pergamon Press, Oxford, U.K.

ICRP (various dates) Limits for intakes of radionuclides by workers. ICRP Publication 30, Parts 1, 2 and 3, Annals of the ICRP, Vol. 2, No 3/4, 1979; Vol. 4, No. 3/4, 1980; Vol. 6, No. 2/3, 1981; Vol. 8, No. 4, 1983; and Supplements to Parts 1, 2 and 3 of ICRP-30, Annals of the ICRP, Vol. 3, No. 1-4, 1979; Vol. 5, No. 1-6, 1981; Vol. 7 No. 1-3, 1982; Vol. 8 No. 1-3, 1982. Pergamon Press, Oxford, U.K.

Land, K.C. and Schneider, S. H. (eds.) (1987) *Forecasting in the Social and Natural Sciences*. Reidel, Dordrecht.

Lawson, G. and Smith G.M. (1985) BIOS: a model to predict radionuclide transfer and doses to man following releases from geological repositories for radioactive wastes. *NRPB Report* NRPB-R169. National Radiological Protection Board, Chilton, U.K.

McEwen, T.J. and de Marsily, G.H. (1991) The potential significance of permafrost to the behaviour of a deep radioactive waste repository. *SKI Report* TR 91:8. Swedish Nuclear Power Inspectorate, Stockholm, Sweden.

Nagra (1985) Project Gewähr 1985. Feasibility and safety studies for the final disposal of radioactive wastes in Switzerland. Nagra, Baden, Switzerland.

NEA (1990) Proceedings of the Joint RWMC/CRPPH Workshop of Radiation Protection and Safety Criteria for the Disposal of High Level Waste. Paris 5-7 November, 1990. Nuclear Energy Agency, Paris, France.

NEA/IAEA/CEC (1991) Disposal of radioactive waste: can long-term safety be evaluated? An international collective opinion. Nuclear Energy Agency, International Atomic Energy Agency, Commission of the European Communities, Paris, France.

NRPB (1992) Radiological protection objectives for solid radioactive waste disposal. Documents of the NRPB, Vol. 3, No. 3, HMSO.

SKI (1989) Biosphere scenario development: an interim report of an SKI/SSI/SKB Working Group. *SKI Report* TR 89:15. Swedish Nuclear Power Inspectorate, Stockholm, Sweden.

SKI (1991) SKI Project-90. *SKI Technical Report* 91:23. Swedish Nuclear Power Inspectorate, Stockholm, Sweden.

Smith, G.M. (1989) An overview of biosphere modelling for the assessment of solid waste disposal. In *Proceedings of Paris Symposium on Safety Assessment of Radioactive Waste Repositories, 1989*. NEA/OECD, Paris, France.

Smith, G.M., Fearn, H.S., Smith, K.R., Davis, J.P. and Klos, R. (1989) Assessment of the radiological impact of solid radioactive waste at Drigg. *NRPB Report* NRPB-M148. National Radiological Protection Board, Chilton, U.K.

Smith, G.M. and White, I.F. (1983) A revised global circulation model for Iodine-129. *NRPB Report* NRPB-M81. National Radiological Protection Board, Chilton, U.K.

Sumerling, T.J. and Thompson, B.G.J. (1992) Application of a probabilistic system-model based methodology for the performance assessment of deep underground disposal of nuclear wastes. *Proceedings of the Third International*

Conference on High-level Waste Radioactive Waste Management, Las Vegas.

7

Selection of waste disposal sites

7.1 Introduction

The concepts of waste disposal, both deep and shallow, and the requirements of the multi-barrier systems have been discussed in previous chapters. In this chapter these technical aspects will be linked with non-technical and social and economic factors in a discussion of the selection of disposal sites.

It is possible to divide the factors relevant in terms of site selection initially into two main types, technical and non-technical. The precise nature of these factors will depend to some extent on the type of disposal system envisaged, as the importance of the geological and hydrogeological environments, for example, are greater for deep than they are for shallow disposal. Nevertheless, it would appear that there is general consensus amongst the countries that already operate repositories, or are in the process of selecting and/or constructing such structures, on the factors that need to be taken into account in their selection.

A distinction needs to be made at this point between repositories that were selected in the first two decades of the nuclear industry and those that have been selected in subsequent years. The earlier repository sites were chosen, in almost all cases, for their convenience to the industry, and no real form of site selection procedure was used. Examples of such sites in the U.K. are Drigg and Dounreay, and there are numerous sites in this category in the U.S.A., many of them associated with the military, the largest being at Hanford, where liquid HLW is still stored in underground tanks. Drigg was sited on a disused munitions factory, immediately south of Sellafield, and the first waste was buried in a railway cutting. At Dounreay, small pits were constructed in Devonian flagstones close to the small sea cliffs. The locations for these sites were not selected using any form of safety assessment, although more recent developments or renewals of authorisations at such sites have required safety assessments. In the former U.S.S.R. and in some of the countries of eastern Europe, largely unevaluated disposal continued until very recently.

Sites that have been selected since about 1970 are generally in a different category, and many of the sites that existed before that date have upgraded their disposal systems to make them more acceptable to modern safety standards. In the last two decades and, in particular, since the late seventies, sites have been selected using a variety of generally acceptable and defensible site selection approaches. It is these site selection programmes that will be discussed in this Chapter.

Advice and information on site selection is available from a variety of sources:

1. The IAEA, who have produced a variety of documents in their Safety Series (Safety Standards, Safety Guidelines and Recommendations) and their Technical Report Series, *e.g.* IAEA (1977, 1981, 1983, 1985, 1989, 1994 and 1995).
2. Reports, some officially published and others more for internal consumption, from a variety of countries interested in the disposal of waste. Some countries, notably the U.S.A., have produced very extensive documents describing the requirements for site selection and, in the case of the U.S.A., these are set within a legal framework. Examples of such reports are: Äikäs (1985), Bataille (1991, 1993), Castaing (1982, 1983, 1984), ENRESA (1990), Goguel (1987), Nagra (1981, 1983, 1988), Nirex (1989), Salmi *et al.* (1985), SPTF (1987), US DOE (1984, 1985, 1986, 1987), SKB (1992a, b) and Rosén and Gustafson (1993). Most of these are reviewed in McEwen and Balch (1993).
3. From published papers, which have been presented mainly at specialised conferences on radioactive waste disposal, arranged by organisations such as the NEA/OECD, IAEA, SKI, *etc.*, *e.g.* Chapman *et al.* (1986), Kowalski *et al.* (1986).

Disposal of radioactive waste in a repository, especially in one at depth, involves a substantial reliance on the geological and hydrogeological environment to provide long-term confinement of the waste. The goal of the selection process is to select a site which is suitable under the following headings:

- long-term safety,
- safety of the repository operating system (short-term safety),
- technical feasibility of the repository system,
- environmental considerations,
- social acceptance (public acceptability),
- considerations of cost.

The selection of a suitable site involves a process of optimisation, with some of the factors having a greater importance than others, in particular those factors concerned with the long-term safety of the repository. A more flexible approach can be taken to other factors, in particular those involving aspects of planning and public acceptance. The goal of the selection process is to identify a site which is acceptable with regard to the defined requirements, and it is important to realise that the *best site* is not being chosen. In fact, the very concept of such a site is misleading, as it would not be practically possible to demonstrate that any site were the best, in terms of its long-term safety, except by examining the whole of a country to a comparable level of detail.

7.2 Suitability of a site

The siting of a repository needs to consider a large number of factors, which can be placed under the headings listed above, and several of these factors can be derived from the more general requirements that will have to be met by the disposal system.

7.2.1 Long-term safety

The necessary requirement for demonstrating the long-term safety of a repository system, particularly for a deep repository, implies that suitable geological and hydrogeological conditions are important. The general requirements for a suitable environment for deep disposal have been presented in Chapter 5.

The distribution of some, or all, of these geological/hydrogeological environments is thus the logical starting point for the search for a deep disposal site. In the case of the sedimentary environments, their boundaries can often be reasonably well defined, using the large amount of seismic and deep borehole data that are available for many countries within sedimentary areas, and by using depth cut-offs, below and above which it might be considered that conditions would be unsuitable for disposal purposes. Cut-offs were defined at 200 m and 1000 m depth within the U.K. programme, for a variety of reasons from geotechnics to groundwater transit times, with the minimum depth being determined by the requirement to prevent erosion by glacial effects, *etc.* Greater depths are possible, for example the former iron ore mine at Konrad (Germany), which is due to become a repository for ILW in the near future, covers a depth range from 800-1300 m. It is unlikely, however, if a depth less than 200 m will be suitable for long-term disposal; for example, the proposed disposal depth at Mol (Belgium) is at 215 m below ground.

Areas of hard rocks in low relief terrain can be less easy to define, as there is no accepted definition of 'low relief' nor, necessarily, of hard rock. In Sweden and Finland, both these factors cause little problem, but in countries such as the U.K. and France potentially suitable hard rocks cover a wide range of rock types. In both cases a large amount of subjective judgement may have to be used, and the area of search defined by this geological environment could undergo several changes during the site selection process. Although low relief is a desirable attribute for any repository environment, if extremes of relief are avoided there are other attributes, such as having a geologically simple environment, which are considerably more important. It is, therefore, important not to overrate the significance of this particular attribute and to restrict unduly the areas of search for this geological environment.

Areas of low relief can be defined by taking the first derivative of digitised topographic data averaged over 1 km squares, which are normally used for correcting gravity data, but this may be an unwarranted sophistication, and subjective judgement may be entirely acceptable. Hydraulic heads are, in any case, not normally transmitted very far in hard rocks, where the groundwater flow system is controlled by the fracture network and where effective hydraulic anisotropies are not large.

In countries with varied geological environments, it may be considered prudent to concentrate effort, at least initially, on those environments which are believed to offer most potential for disposal purposes. For example, in the U.K. site selection programe, it was decided initially that three of the environments should be considered, namely hard rocks in low relief terrain, seaward dipping sediments in areas of simple structure and small islands. However, during the site selection exercise, because it was felt that as many different types of potentially suitable geological environment as possible should be considered in the early stages of the selection process, the basement under sedimentary cover (BUSC) environment was subsequently added. In addition, a constant re-evaluation of the areas of search, particularly in hard rock areas, resulted in there being several modifications to the areas of search, mainly in Scotland. There needs to be flexibility in the approach that is followed at this stage, so that potential geological environments and sites are not eliminated unnecessarily.

7.2.2 Safety of the repository operating system and technical feasibility

The repository will have to fulfill certain minimum requirements in order to make it technically and practically feasible. There will be certain geotechnical constraints on the positioning and possible depth range of any repository in a particular host rock. The design of the repository needs to be compatible with aspects, such as the waste handling system, the type of backfill and its mode of emplacement and, in more mountainous regions, the accessibility of the repository needs to be considered.

7.2.3 Environmental considerations

The siting a repository will have to be carried out so as to minimise its environmental impact. The construction of the repository, its operation and its long-term presence all have different effects on the environment. Some of these effects, *e.g.* additional traffic, disposal of excavated rock spoil, *etc.*, are directly related to the public acceptability of such an operation and can vary very substantially between different sites. These subjects are discussed in Section 7.3, under non-geological factors in site selection.

7.2.4 Costs

The total costs for a repository are large but, particularly for a deep repository, are spread out over many decades and discounted cost calculations can be applied. The costs for waste disposal as a percentage of each kWh generated are in fact very small. There can be, however, very large differences in the total cost between the possible disposal options; for example UK Nirex Ltd. carried out cost comparisons between offshore engineered disposal options and disposal of the same ILW and LLW in hard rock caverns, and the costs varied from £2.5B to £6.5B (Nirex, 1989). The safety of the disposal system, and not the cost, is, however, normally stated as being of prime significance.

7.2.5 Site selection programmes

The geological and hydrogeological factors which influence the choice of sites for the deep disposal of long-lived waste will be considered in four parts, in relation to the type of geological environment which is being considered. The shallow, or surface disposal of short-lived waste will be considered separately.

Crystalline or hard rocks

The best published examples of the principles involved in a comparison of different hard rock areas for assessment are provided by KBS-3 (SKB, 1983), in which a simplified analysis of the geological investigations at four sites in Sweden is presented, in several reports which describe the selection of potential HLW disposal sites in Finland (Vuorela and Hakkarainen, 1982; Äikäs, 1985; Salmi *et al.*, 1985; TVO, 1985; TVO, 1992) and in the reports which describe the selection of the Hard Rock Laboratory (HRL) on the island of Äspö (SKB, 1989a, b; Wikberg *et al.*, 1991; Stanfors *et al.*, 1991; Almén and Zellman, 1991; Gustafson *et al.*, 1991). The most recent documents from Sweden, (SKB, 1992a, b) and Rosén and Gustafson (1993), describe the planned programme and possible processes that could be applied in the selection of a deep disposal site in that country. The current site selection activities in Sweden draw on the extensive experience gained in siting the interim storage facility for spent nuclear fuel (CLAB), the LLW repository at Forsmark and the Äspö HRL site. This experience is summarised in SKB (1992a).

Other documents which provide methodologies for the assessment of granitic areas, such as SERE/87 (Peyrus, 1987) and US DOE (1987), do not report on the application of these methodologies to specific areas, or make the process so complex that it is difficult to follow, and it is, therefore, more difficult to evaluate how they were, or might be, applied.

The following description of the process of selecting areas of hard rock for disposal purposes, although written with reference to the Swedish and Finnish programmes, is applicable to any country and to any type of hard rock (*cf.* Figure 7.1).

The blocks of rock selected in Sweden and Finland for examination were defined in the following manner in terms of geological and hydrogeological suitability. Regional fracture zones were mapped from aerial photographs, and areas were chosen that lay between these regional zones. The selection factors that were considered at this stage were:

1. Topography.
2. Distance between regional and the larger local fracture zones.
3. Frequency of small local fracture zones and fracture densities.
4. The distribution of different rock types.
5. The geological structure present between the fracture zones.
6. The extent of mineralisation.
7. Any hydrogeological data from nearby boreholes of the likely transmissivities of the fracture zones and/or the rock mass between them.

Preference was given to areas with the following attributes:

1. Areas with low relief, although an absolute measure of the relief, such as the mean slope, was not determined.
2. Areas in which the distance between the regional fracture zones was maximised, although, in most cases there were, at this stage of the selection process, no data on the dip direction and dip angle of these zones.
3. Areas in which there was also a large separation of the local fracture zones.
4. Areas of the rock mass which exhibited low fracture densities – an estimate of this parameter was obviously only possible where exposure was sufficiently extensive.
5. Areas of the rock mass that were as homogeneous as possible, with respect to their rock type(s) and their geological structure.

Geological and hydrogeological reconnaissance studies were carried out in areas that emerged as having good potential from this part of the selection process, and the results of these studies served as a basis for decisions on whether to drill reconnaissance boreholes. The results from these boreholes enabled decisions to be made on where to conduct the much larger site investigations that are summarised in KBS-3 (SKB, 1983). A similar operation for the selection of the site for HRL is described in a series of reports (SKB, 1989; Wikberg *et al.*, 1991; Stanfors *et al.*, 1991; Almén and Zellman, 1991; Gustafson *et al.*, 1991). The programme carried out in Finland is described in TVO (1992). The sizes of the areas eventually chosen by this process for detailed investigation were of the order of 5-10 km^2.

These areas chosen for investigation had relatively low densities of local fracture zones, which for the most part were also visible from aerial photographs. The important factor here is that for a repository of known dimensions, SKB and later other waste disposal organisations, have all decided to maintain what has been termed by SKB (SKB, 1983) a *respect distance* of at least 100 m between any fracture zone and the repository. This was believed to be the minimum acceptable distance for transport of radionuclides from a repository through rock of relatively low hydraulic conductivity (termed *averagely fractured rock*) before their arrival at a fracture zone, which would then almost certainly result in much more rapid transport. In KBS-3 the generic design of the repository was tailored for each site, depending on the separation of the fracture zones, so that the repository could have one or several layers, and a similar procedure has been followed by Nagra in Kristallin-1 (Nagra, 1994).

As part of their current site selection activities for a deep disposal site for spent fuel, SKB are examining various strategies for developing a geoscientific classifying procedure for use in the siting process (Rosén and Gustafson, 1993). It was assumed in the development of this procedure that any feasible classification methodology would need to integrate existing data with expert knowledge using the Bayesian approach. The suggested strategy for identifying areas for geoscientific classification is based on simple set theory. Unsuitable areas are discriminated with respect to the costs associated with site selection factors, both geoscientifc and socio-economic *etc.* It is assumed that the method of delimiting areas of search can be performed cost-effectively by making decisions sequentially in increasing order of associated cost. This is believed to form the basis for a straightforward identification process that is both credible and possible to defend. The classification methodology is based on approaches that have been used in areas such as groundwater vulnerability mapping (*e.g.* the DRASTIC system (Aller *et al.*, 1987)) and rock quality classification *e.g.* the RMR system (Barton *et al.*, 1974). The set of critical parameters to be included within the classification system was based on a literature review of the issues thought most important for repository safety. It is not known whether this process is being applied during the current site selection exercise.

Sedimentary rocks

The search for sites in sedimentary rock is best exemplified by the work that has been carried out by Nagra and ENRESA. A major evaluation report on the properties of the sedimentary rocks in northern Switzerland was presented to the authorities in 1988 (Nagra, 1988). Two sedimentary formations were selected, using the process described below and outlined in Figure 7.1, as having potential for the deep disposal of HLW: the Opalinus Clay and the Lower Freshwater Molasse. This report was reviewed by the government advisory groups during 1990, and in 1991 an initial phase of field work was started with a regional seismic survey to investigate the Opalinus Clay. Information on this formation was already available from some of the deep boreholes into the basement that had intersected it during the investigations of the basement rocks.

Nagra started out with seven different potential host formations, and reduced the number of formations to two and their areal extent to one fifteenth of the original area. The attributes they applied were directly related to those that they felt were most significant in terms of long-term safety, and were determined for each formation as follows:

- lithostratigraphy,
- thickness,
- depth and extent at potential disposal depths,
- *special aspects* (neotectonics, ease of performing a site investigation, presence of mineable resources *etc.*),
- rock mechanics/constructability,
- geological and hydrogeological complexity,
- hydrogeological environment and hydraulic properties of the formations,
- hydrochemical, geochemical and mineralogical conditions and sorption properties of the formations.

Using these attributes, the most favourable Swiss sedimentary formations, the Opalinus Clay and the Lower Fresh Water Molasse (USM), were chosen. Having chosen these two formations, areas of search were progressively reduced by analysing the following factors:

- tectonic conditions and geological predictability,
- depth ranges of the host formations,
- hydraulic and other properties of the host formations,
- hydrogeological environments and predictability of potential future changes in these conditions,
- seismicity and neotectonic conditions and predictability of long-term changes.

Two areas of Opalinus Clay and one area of the USM were eventually chosen for further examination. There were quite extensive data for some of the

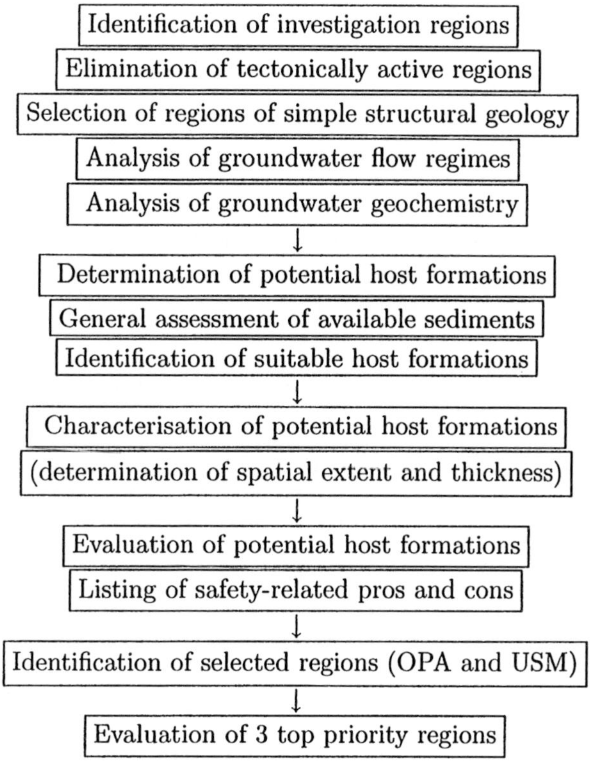

Figure 7.1 The procedure used by Nagra (Switzerland) for the selection of areas for investigation within the Opalinus Clay (OPA) and the Lower Freshwater Molasse (USM).

formations over certain parts of the original areas of search, as significant amounts of hydrocarbon exploration had taken place in parts of the northern Swiss plain, and there were extensive, but relatively simple, hydrochemical data for parts of the USM. An overview of the procedures applied during Nagra's sediment study are shown in Figure 7.2.

Hydrogeological aspects of argillaceous rocks important for waste disposal. The hydrogeological environment of clays is of paramount importance when considering disposal in this type of rock. The following hydrogeological factors need to be taken into account when disposal in argillaceous formations is considered:

Steps in Procedure	Activity	Results	
1 Identification of investigation regions	Selection of tectonically inactive and simply structured investigation areas	Plateau, Tabular Jura in Cantons Aargau and Schaffhausen and Ajoie as sediment investigation area	
	Analysis of flow regime and deep groundwater chemistry (hydrodynamics/hydrochemistry). Attempt to localise important groundwater infiltration areas	Further narrowing-down impossible (insufficient data-base for localising infiltration areas)	
2 Determining potential host rock formations	General assessment of sediment sequences in N Switzerland and identification of suitable host rocks with low permeability and sufficient thickness	Seven potential host rocks	Upper Freshwater Molasse (OSM) Lower Freshwater Molasse (USM) Effingen beds Opalinus clay (OPA) Gipskeuper Middle Muschelkalk (anhydrite group) Permian
3 Characterisation of potential host rock formations, determination of spatial extent	Listing advantages and disadvantages of different host rocks. Defining regions where potential host rocks are 300 to 1200 m thick	Brief description of most important properties. Mapping of spatial extent	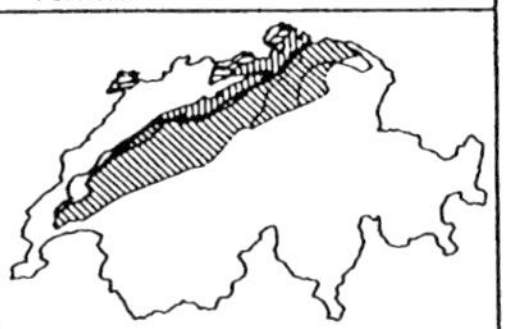
4 Evluation of potential host formations. Listing of safety-relevant pros and cons	First evaluation of significant characteristics. Assessment of selected options from engineering, hydrodynamic and safety viewpoint	Two top priority host rock options: Lower Freshwater Molasse (USM) Opalinus clay (OPA)	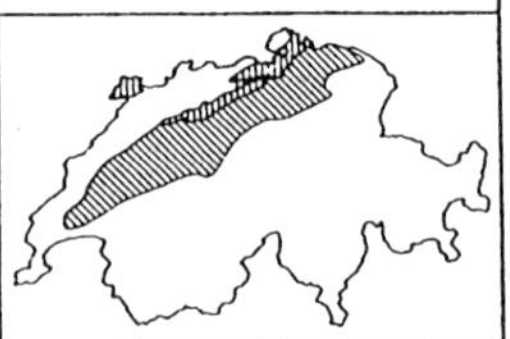
5 Identification of potential siting regions with OPA and USM	Narrowing-down to areas with tectonically inactive conditions and sufficient depth of host rock	Six siting regions with OPA and two with USM	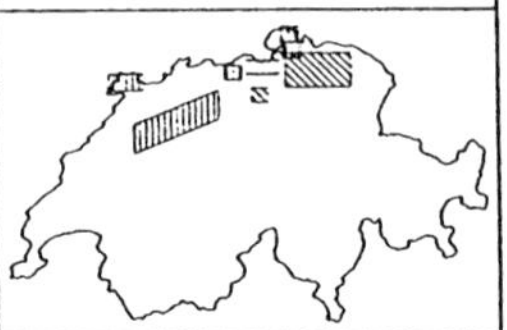
6 Evaluation of three potential top priority siting regions	Description and comparative assessment of the potential siting regions from point of view of geology, tectonics, hydrodynamics and hydrochemistry	Narrowing-down to: - two sites with OPA (Zurich Weinland and area north of the Lägern) - one region with USM (area east of the Limmat)	

Figure 7.2 An overview of the procedures applied during Nagra's sediment study (from Nagra, 1988). Reproduced by permission of Nagra.

1. Groundwater flow in aquifer-aquitard systems with dips that are not too extreme tends to be parallel to the formation boundaries within the aquifer units and normal to the boundaries in low permeability units.
2. Sedimentary rocks, because of their planar sedimentary bedding structures, tend to possess large hydraulic anisotropies. This can result in hydraulic heads being transmitted over long distances, in distinct contrast to crystalline rocks.
3. In a formation with a hydraulic conductivity less than approximately 10^{-9} to 10^{-10} m s^{-1}, (*i.e.* all clays), it is probable that most of the formation will always be in a transient state as regards its hydraulic head distribution. Steady state groundwater flow modelling may well not be appropriate in such a formation.
4. Fracture and fault zones within argillaceous rocks may tend to dominate the regional groundwater movement across the rocks at near steady-state conditions, whilst the response of the rock between the faults will be much slower, and may never be in steady-state and never in equilibrium.
5. The rate of tectonic activity can have a very significant effect on the hydraulic response of argillaceous rocks and can be expected to affect fluid flow in argillaceous rocks with the production of anomalous hydraulic heads.
6. Geological processes can alter the hydrogeological boundary conditions, and this can be important because the hydraulic response time of thick argillaceous rock formations is very slow.
7. Anomalous overpressures may be produced in formations with high clay contents, and osmosis has been invoked (see Chapter 5) as the most probable cause of these pressures.
8. Transport processes in clays are invariably coupled, and this makes the modelling of radionuclide transport through clays considerably more difficult than is the case for non-coupled systems.
9. The flow of groundwater through clays is very poorly understood, particularly at the phenomenological level, and the measurement of hydraulic parameters in clays is fraught with difficulties.

Evaporites

Evaporite formations can be present in a variety of forms, from diapirs, pillows and walls to simpler bedded formations. In many parts of the world bedded evaporites have been deformed, have acted as décollement surfaces or have suffered from a variety of halo- and tectonokinetic effects, so that they are present as pillow structures, in anticlinal cores and as partial diapirs, usually associated with thrusts. Only halite is generally of interest to waste disposal organisations as a host medium for a repository, but halite can often be accompanied by anhydrite, carnallite, sylvite, other minor evaporitic

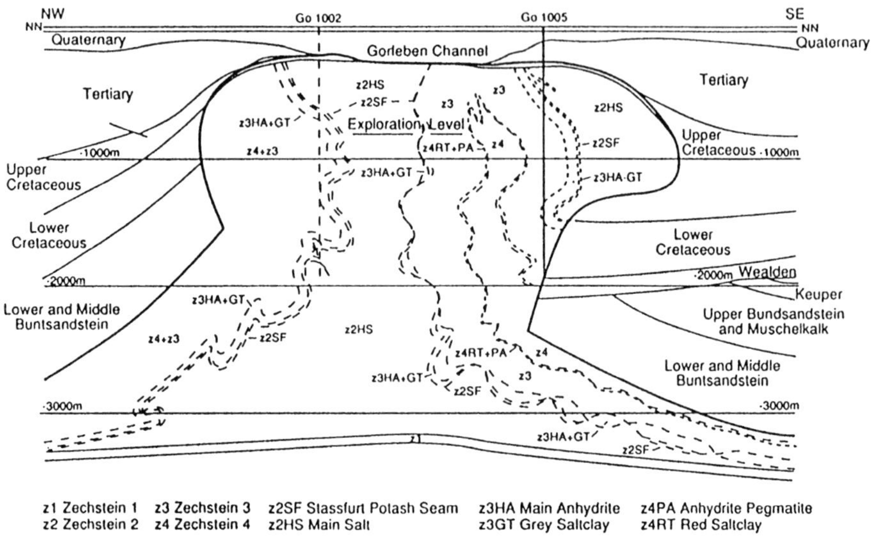

Figure 7.3 Cross-section of the Gorleben salt dome in Germany, which is currently being investigated as a site for a repository for HLW, showing the contorted structure of the original horizontal evaporite and argillaceous beds within the dome. The dome is made up of several different phases of diapiric uplift, with material being derived from several halite horizons, each of which is made up of a variety of evaporitic minerals in addition to halite (such as anhydrite and carnallite). (From Langer *et al.*, 1991). Reproduced by permission of Lawrence Berkeley Laboratory, U.S.A.

minerals and argillaceous horizons. A cross-section of the Gorleben salt dome in Germany, currently being investigated as a site for a repository for HLW, is shown in Figure 7.3.

Very thick units of almost pure halite, in some cases in excess of two thousands metres thick, and with purities in excess of 95 %, are present within Europe and the U.S.A. In areas of low tectonic activity, halite is potentially very stable in the geological environment, contains few if any fractures, is generally self-healing, due to its plastic, visco-elastic and long-term creep properties, and can be mined with ease. The possible problems associated with halite are as follows:

- The sometimes frequent, and often unpredictable, occurrences of highly unstable horizons of minerals such as carnallite, whose locations and geometries can be difficult to predict in diapirs which possess complex internal structures, and which can decompose at low temperatures to produce large quantities of water.

- The presence of anhydrite and argillaceous horizons, which are normally fractured and provide potentially rapid transport pathways.
- The presence of fluid inclusions which can migrate towards a heat source introducing corrosive brines into the repository near-field.
- The risk of flooding an operating repository, a repository in halite being essentially a dry mine operating far below a water table which is present in overlying sedimentary formations, some of which may be significant aquifers.

Many salt domes show no evidence of any contemporary movement, whilst others, in locations such the U.S.A. Gulf States, may be rising at a rate of a few millimetres a year. Salt domes appear to be thermally more stable than bedded salt, but halite is generally an excellent conductor of heat, and temperatures in a repository for heat-emitting waste in halite are predicted to be markedly below those in argillaceous formations.

A suitable, very pure halite formation containing only thermally stable minerals with few hydrous phases would contain only very small volumes of free water, which would be present in the form of intergranular films and as intracrystalline/intergranular and unconnected brine inclusions. Whilst these inclusions tend to migrate up thermal gradients, they would not in the majority of circumstances constitute a radionuclide transport mechanism. In some situations the accumulation of brine around the waste canisters may contribute to corrosion, but there is unlikely to be any mechanism to remove the released radionuclides, except exceptionally slow diffusional transfer. Intergranular diffusion through the salt will occur, but diffusive transfer rates are likely to be many orders of magnitude less than those in free water, and only very local transport through the geological barrier is anticipated.

When a repository in a stable, thick and pure halite has been sealed there is no sensible mechanism that will allow any transfer of radionuclides over more than a few tens of metres from the repository for exceptionally long periods of time. The situation of an *altered evolution* scenario in which radionuclide transport could occur more rapidly obviously must also be considered.

Within Europe, Germany, Spain, Holland and Denmark are considering the deep disposal of long-lived waste within halite. Germany (the former FGR) chose the Gorleben salt dome in 1977, with the then DDR choosing a similar salt structure at Morsleben. ENRESA is currently examining halite as a disposal medium, Denmark has carried out some work on the Mørs salt dome.

The following geological and hydrogeological factors need to be considered in selecting suitable areas of halite for disposal purposes:

1. Lateral *safety* or *exclusion* zones of 200 m need to be applied between the

edge of a repository and the margin of a salt dome. A minimum vertical distance of 300 m between the upper surface of a salt dome and a repository also needs to be applied[1].

2. Within bedded halite the minimum distance from a repository to the top of the deposit should again be 200 m, if the overlying formations have a relatively high hydraulic conductivity, but could be substantially less than this if formations with lower conductivities are present. The distance to the base of the formation could be as small as 50 m. Horizontal exclusion zones should be applied in a similar manner to salt domes.
3. The minimum area of high quality halite required within a salt dome at the disposal depth is of the order of 1 to 3 km^2, with a vertical extent of at least 600 m, if German/Dutch repository designs are assumed.
4. Substantially smaller thicknesses of suitable halite are likely to be required for disposal within bedded salt.
5. Halite deposits exhibiting complex faulted margins and/or cross-cutting faults should be avoided where possible. Salt domes with complex tectonised zones around their margins should also be avoided.
6. Salt domes that contain significant anhydrite, carnallite (and other potassic salts) and associated argillaceous horizons should be avoided where possible. Within salt domes there need to be large exclusion zones around such formations, of the order of 50 m minimum.
7. In bedded halites it is suggested that vertical exclusion zones of up to 200 m should be applied to thick anhydrite formations (>20 m), and zones of 50 m to thinner formations.
8. Salt domes and bedded halites with a purity as high as possible should be sought.
9. Bedded halite formations should not be exposed, and need to be overlain by at least 200 m of sediments. This constraint will not normally apply to salt domes and pillows.
10. The geological environment of the halite formation needs to be as simple as possible on all scales. Salt domes that exhibit current or recent rates of uplift in excess of approximately 0.03 mm a^{-1} should be avoided.

7.2.6 Disposal offshore

In addition to the search on land for suitable environments, a search can also be made offshore, particularly in countries which have large areas of seabed under their control. Selecting areas offshore depends on the information

[1] All these exclusion distances recommended were developed in Germany and are based on German mining regulations, and not on the results of scoping calculations related to radioactive waste disposal. They do, however, appear to have been applied in a similar manner within the various site selection programmes, *e.g.* Germany, Spain, Holland, Denmark, France.

available, and will inevitably be easier in sedimentary environments, as the distribution of fracture zones within hard rocks is not known offshore. In this part of the search process, account can also be taken of:

- areas licensed for oil and gas exploration;
- the extent and thickness of potentially suitable formations, and
- the depth of water, considering the various engineering options available.

Within Europe it would appear that only U.K. Nirex Ltd. have rigorously considered all the options of offshore disposal, and their programme considered the various possibilities of disposal within the U.K. continental shelf (Nirex, 1989).

7.2.7 The disposal of short-lived or shorter-lived waste

Short-lived, or shorter-lived, wastes can be disposed of in a variety of ways. The disposal options depend to some extent on the precise types of waste that are being considered in this category, and in turn depend to a large degree on whether reprocessing is being carried out and whether military-derived wastes are also of concern:

- In above-ground engineered structures (tumuli), with under and side drains, with engineering to limit ingress of water to the waste and probably with anti-intrusion shields (*e.g.* Centre de l'Aube (France), El Cabril (Spain)).
- In engineered structures which are partly, or perhaps mainly above ground, with similar engineered systems (*e.g.* La Manche (France)).
- In trenches, or vaults, which can have a variety of engineered systems (*e.g.* Drigg (U.K.)).
- In underground caverns, tunnels or silos, but at relatively shallow depth.
- Under the sea (*e.g.* SFR (Sweden)), in the side of hills or mountains (*e.g.* proposed Swiss repository of low and short-lived ILW at Wellenberg) or inland in areas with little relief (*e.g.* previously considered Nirex LLW sites).
- At great depth, in conjunction with longer-lived waste (*e.g.* current Sellafield (U.K.) plan).
- At depth in an existing mine (*e.g.* Konrad (Germany)).

All of the above options are either in operation, under construction or under consideration in various countries. There are two different approaches to surface disposal; whether disposal should take place in saturated or unsaturated conditions. In addition, the length of time over which a safety case has to be made for the disposal system has a significant affect on what is considered possible. U.S.A. regulations on surface disposal require that all parts of the repository must lie at all times, except in the case of extreme scenarios, above the water table. In the U.K. the only formations with sufficient unsaturated thicknesses are those associated with major water supply aquifers, and all the sites previously selected by U.K. Nirex Ltd. for the disposal of LLW were in low permeability formations, *i.e.* relatively plastic clays. The geological and hydrogeological attributes of any site for the disposal of short-lived wastes are obviously less important than those for long-lived waste. The types of geological environment considered suitable for disposal also tend to be more dependent on the philosophy of waste disposal, the climate, whether reprocessing is taking place and the type of safety case considered necessary than is the case for deep disposal. A summary of the factors, both geological and non-geological, used by U.K. Nirex Ltd. to select potential LLW disposal sites in the U.K. is presented in McEwen and Balch (1987).

7.3 Non-geological factors

Having defined areas of the country which are believed to have potential for the disposal of radioactive waste, other, non-geological, factors can then be superimposed on the resulting area of search maps, and these factors are normally, at least initially, mapped on the regional scale. There are four issues with respect to the factors that might affect the selection of sites for radioactive waste disposal:

- the influence of the 'Nimby'[2] effect;
- the type of decision process used and its method of employment;
- the overall approach taken to the selection of sites;
- socio-economic considerations.

7.3.1 The Nimby effect or syndrome

One of the most significant factors in the selection of sites for waste disposal is the influence of the Nimby effect. This effect is commonly apparent in the richer countries where the population can afford to concentrate their attention on the *quality of life* (Zetterberg, 1980). The problem with radioactive waste disposal is that, even in countries such as France, where the population apparently readily accepts nuclear reactors in very environmentally sensitive areas, the disposal of HLW is, by contrast, often considered unacceptable. The public's perceptions of radioactive waste disposal can develop slowly or suddenly as a result of a dramatic accident. In addition, the attitude, or perceived attitude, of the government to matters such as the environment can play an important part in the formulation of these perceptions. This subject is discused further in sections 7.3.4 and 7.4.

[2] not in my back yard.

Table 7.1 Summary of factors which are thought to influence the extent of the Nimby effect or syndrome.

The Nimby effect in waste disposal	
Significant factors	**Comments**
1. Population density	• Larger 'backyards' in rural areas. • Low population, scenic areas have larger backyards than those less scenic and/or economically depressed areas.
2. Extent of travel	• Relationship between distance travelled and size of backyard (the less travelled the smaller the backyard).
3. Distance from centres of political power	• Centralised government: – seen as dumping waste on populations with less political influence – local community prevents even initial attempts at investigation • Devolved power structure: – local communities have strong powers of veto – may allow initial investigations as have power to prevent final selection of site
4. Attitude towards administrative regions	• Collective action if site/area chosen anywhere within monocultural area, *e.g.* Scotland. • Perceived 'threat' due to site possibly extends over large area.
5. Topographic effects	• Extensive topography tends to limit effect. • Watersheds perceived as forming potential boundaries to backyard.
6. Environmental /economic matters	• Level of objection to site not necessarily related (proportional) to level of environmental awareness. • Potential economic benefits able to outweigh 'stigma' of accepting site, *e.g.* Switzerland, Spain.

There are certain attributes associated with the Nimby effect which need to be considered during the selection of waste disposal sites, as they are associated with the perceived threat from the site. A person's perception of the size and shape of their 'backyard' and the reasons why they have developed certain attitudes towards radioactive waste disposal are thought to be dependent on the factors listed in Table 7.1. Many of these factors are self-explanatory.

7.3.2 Decision analysis

In order to provide a rational and defensible site selection methodology, especially in the latter stages of any site selection process when the Nimby effect is most pronounced, the methodology itself could be as important as being confident that the answer provided is acceptable. The three main requirements of any decision process are shown in Table 7.2.

The evaluation of areas or sites involves comparing their relative impacts, but this comparison is complicated by several different factors, which are shown in Table 7.3. In order to identify potential sites, screening procedures need to be applied. The formal approaches which have been used for screening purposes can be categorised into three basic procedures: exclusion screening, inclusion screening and comparison screening. Each of these procedures can be used sequentially with different criteria, or concurrently with more than one criterion.

7.3.3 Evaluating procedures for site selection

In most site selection studies, for structures such as power plants for example, there is normally a small number of sites at the end of the screening process. The same is often not the case when selecting radioactive waste disposal sites, and several separate phases of screening may be required, using different factors. At any stage in this programme a more structured and detailed selection process can be applied, in which the details of specific sites are compared.

There are several different approaches which have been used for the latter phase of site selection processes for a variety of structures and they differ greatly in the degree to which the process itself is formalised. Most of them can be categorised into seven basic procedures, in increasing order of formality (Table 7.4), where the important weaknesses in using the first six approaches are also presented. The use of any of the first six processes is unlikely, according to Keeney (1980), to result in the desired requirements of the site selection process.

The one process that does appear to satisfy the desired requirements is known as decision analysis, or MUA. This is a systematic and logical

Table 7.2 The requirements for a high quality decision analysis.

Decision analysis for selecting waste disposal sites	
Points	**Comments**
Requirements of any decision process	• *To promote a high quality analysis:* – require clearly stated guidelines – clear presentation of how information is integrated into overall asessment – should provide useful procedures for conducting sensitivity analyses – should clarify the professional judgements, value judgements and data used – should explicitly indicate factors not considered
	• *To be practical:* – pool of expertise required to implement the analysis – methodology should demonstrably represent concerns of the various parties
	• *To improve perception of the analysis:* – process must be understandable – concerns and impacts must be considered adequately – be able to demonstrate that process was fair, legal and rational • Methodology needs to make value judgements clear and evaluation process transparent.

procedure, based on a set of axioms for rationally analysing complex problems. What makes decision analysis unique is the way in which factors are quantified and incorporated formally into the analysis of the problem. The crucial difference between decision analysis and the other approaches that claim to help the decision maker is that decision analysis provides theoretically sound procedures for formalising and integrating the judgements and preferences of experts and the client in the evaluation of the various possible courses of action.

The Co-operative Siting Policy (CSP) used in Canada is notably different from the approaches listed above, and it may be felt that a less formalised and less structured approach could actually be advantageous (SPTF, 1987; STF, 1990). An extensive search for disposal sites within Ontario for what is termed *historic wastes* (*i.e.* wastes that had already been buried within temporary

Table 7.3 The factors which complicate the comparison of disposal sites.

Potential problems in the comparison of sites	
Important factors	**Comments**
1. Number of sites	• Many potential sites can make comparison difficult and time-consuming.
2. Multiple objectives	• Multiple objectives (long-term safety, waste transport, *etc.*) can make comparison difficult. • Require 'value trade-offs' to optimise necessary compromises.
3. Interest groups	• Several different interest groups (waste producers, local protest groups, *etc.*). • Potentially complex value judgements in determining what is fair.
4. Time scale of impacts	• Some impacts greatly in excess of those associated with other constructions. • Extent of impact depends on when it occurs: – discounted costs – population influx into small community – accidental discharge (more significant early, than later in operation)
5. Uncertainties of impacts	• Uncertainties make dealing with impacts more difficult. • Reasons for such uncertainties: – possibly difficult to collect sufficient data – some data costly and time-consuming to obtain – population around site may change, thus affecting future impacts – priorities of communities change with time – large uncertainties associated with forecasting any future event
6. Attitudes towards risk	• Different types of risk, *e.g.* radiological, economic • Reduce all forms of risk by only selecting sites with simple geologies
7. Government actions	• Uncertainties of future government actions and attitudes. • Danger of pre-empting government decisions. • Attitude and recommendations of regulatory authorities possible to change significantly.

Table 7.4 The seven different procedures that have been used in site selection programmes, not necessarily for radioactive waste disposal sites, but also for nuclear and non-nuclear power plants, chemical works, *etc.* (based on the work of Keeney (1980)).

Approaches to decision analysis and site selection	
Procedure	**Comments**
1. Favourability selection	• Site considered good enough on its merits alone, rather than in comparison with other sites.
2. Qualitative comparison	• All comparative factors used, but are described in terms of a qualitative description of possible impacts.
3. Cost-effectiveness	• Economic concerns carefully analysed, other concerns qualitative or quantitative. • 'Softer' aspects often omitted.
4. Site rating	• A rating is assigned via an informal method. • Qualitative discussion to support ratings given, but necessary value judgements not made explicit.
5. Dominance	• A site is ranked above another site if it exhibits superior behaviour in each of the general areas of concern.
6. Cost benefit analysis	• Weights are assigned to areas of general concern and the several factors within each concern. Overall site rating assigned by multiplying and adding over all concerns/factors.
7. Multi-attribute decision analysis	• Very structured, relatively complex, but defensible, approach.
Potential problems with analytical methods 1 to 6.	• Lack of sufficient data. • Use of unstated and implicit assumptions and value judgements. • Oversimplification of the value judgements

structures that did not meet the requirements for long-term safety) took place in southern Ontario up until 1986. There was extensive public protest, which reached such levels of disorder that the standard Environmental Assessment and Review Process had to be abandoned and it was decided to establish an independent Task Force. The report of this Task Force, entitled 'Opting for Co-operation' (SPTF, 1987) recommended an innovative programme which should include five phases, based on the voluntary participation of communities in collaborative, joint decision making. The first three phases of this process have been successfully completed and the final two are in the course of being applied.

A similar approach is currently being followed in France for the selection of two sites to house Underground Research Laboratories (URLs) for long-term research into the disposal of long-lived radioactive waste. The approach, which is now enshrined in French law (Bataille, 1991, 1993), relies on the co-operation of local communities on a purely voluntary basis.

7.3.4 Socio-economic considerations

Relatively little has been written about the way in which socio-economic factors have been taken into account in determining the location of radioactive waste disposal sites. In general, these indicators operate at two levels, at a strategic level, in which broad areas are defined where further search is likely to prove most fruitful, and secondly at a more local level, where factors such as land ownership and the presence of heritage sites and mineral reserves provide more precise attributes to steer site selection activities.

Two basic approaches have been adopted to manipulate the results of socio-economic analysis. The MUA technique generates an equation from the sum of individual impacts and it is the overall score that is used to evaluate short-listed sites. On the other hand, the CSP technique identifies a series of push and pull factors which tend to be exclusionary or essential to the location of a site. In the latter case, the ranking of different sites takes place in a less formal and systematic manner.

A striking feature of all methodologies which incorporate an analysis of socio-economic factors is the *ad hoc* manner in which certain indicators seem to be identified. Examples of the non-geological factors which have been relied on by various agencies are shown in Table 7.5. The weight given to various considerations tends to vary with the approach taken to site selection. Analyses based on advocacy and those that are prescriptive and technically led generally scan the whole spectrum of different factors and make explicit the weight that is placed on each consideration. In contrast, more pragmatic and centralised methodologies tend to conceal the identity and weighting given to different criteria. In reality, because of widespread public resistance to nuclear facilities, a great deal of emphasis is often placed on land ownership.

Table 7.5 Non-geological factors governing the selection of sites for the disposal of radioactive waste.

Non-geological factors governing the selection of sites	
Main factors	**Issues**
1. Environmental	• Impact on amenity sites: – ecological – historical/cultural • Other impacts: – noise – aesthetic impact/rural intrusion
2. Social	• Proximity to dwellings • Population distribution • Impact on operatives'/public health • Public acceptability/opposition
3. Economic/Land use	• Agricultural land quality • Development costs: – repository, pre- and post-closure – transportation • Impact on mineral reserves • Planning constraints • Availability of local workforce • Access to transportation routes • Land ownership: – availability of government/public utility owned sites – land fragmentation

The approaches can be envisaged as different stages along a continuum of increasing sophistication and openness. This reflects the fact that, generally speaking, the development of site selection methodologies is tending towards more participatory procedures, and more sophisticated, advocacy-led procedures are tending to replace pragmatic, closed site selection exercises.

7.4 The overall approach taken

A key objective in the selection of waste disposal sites has been to devise a methodology for securing public support for locational decisions despite the high profile and perceived risks attached to these facilities. Therefore, a basic problem has been to determine how and at what stage to involve the public in site selection decisions. A further, related difficulty is the need to reconcile the industry's desire for an optimum site, at which point economic and safety considerations come into play, with the public's desire for the best possible location from a safety point of view, even though the public are unaware of the impossibility of finding the *best* site[3].

Against this background, differences in style of government, level of popular support for nuclear power and the consequent legislative framework within which site selection activities are carried out have given rise to a range of procedures characterised by some or all of the following elements:

- the identification of suitable geological and hydrogeological environments;
- sieving of sites against a range of socio-economic criteria;
- ranking of sites and the selection of preferred option(s);
- securing public acceptance of preferred options.

The weight accorded to each of the four elements, the order in which they are carried out and the extent to which they are independent of each other varies considerably. It is possible to construct a typology of different site selection methodologies which may be represented as *model approaches*, which have evolved in response to particular combinations of political and social circumstances. As such, they may act as pointers for the development of best practice techniques, having regard to the circumstances which prevail in different locations. It should be stressed, however, that the examples chosen to illustrate this typology are simply *snap shots* and do not necessarily represent the various authorities' current approaches to this task. The nature of the site selection process has, in most countries, tended to change over time in response to factors such as drawbacks attached to earlier methods, shifts in the public's acceptance of nuclear power, especially in the light of events such as Chernoby, and increasing public interest in environmental issues generally.

7.5 Development of a typology

Generally, categories of site search process can be defined in terms of two key factors: firstly, the extent to which an overtly rational rather than a *pragmatic* approach is adopted and, secondly, the extent to which *satisfactory* technical solutions are imposed rather than advocated and accepted by recipient communities.

Another broad distinction needs to be made between geologically-led and advocacy-led approaches. In this context, geologically-led means that geological and hydrogeological determinants of long-term repository integrity

[3] Parts of this section are based on work by Pieda Ltd. and the arguments are presented in more detail in McEwen and Balch (in press)

are the primary determinants in the first intance of defining areas of search. In contrast, an advocacy led approach would seek to identify volunteer communities and, at this point, geological considerations will be taken into account to exclude those incapable of satisfying long-term safety criteria.

Every country's repository programme has been conducted in a rational manner to the extent that safety factors have been systematically specified and pursued. However, considerable variation has been shown in terms of the degree of sophistication, content and consistency of different authorities' efforts so that, ultimately, in some cases, frequent shifts in the ends and means adopted defy description as *rational* in the sense of a clear, step-by-step, goal-orientated procedure. Generally, four categories of site search process may be distinguished (Table 7.6).

The following sections describe in more detail the characteristics of the four basic approaches to site selection.

7.5.1 Prescriptive/geologically led

In contrast to the situation in U.K., the U.S.A. possesses a well developed legislative framework for the assessment of potential repository sites. The open, decentralised style of government is also coupled with easy access to judicial review of the decision-making process. This environment has encouraged the development of sophisticated, if rather mechanistic, search methodologies, the object of which is to meet objections head on in a rational and demonstrably equitable manner. An example of this approach is the Multi-attribute Utility Analysis (MUA) described by Merkhoffer and Keeney (1987), Keeney (1980) and Keeney and von Winterfeld (1988) that led to the identification of three possible sites for HLW disposal.

This approach is, in fact, geologically-led in that assumptions are made initially as to the most appropriate geological formations. The next stage is to define rigorously repository performance measures in terms of a range of health and safety, environmental, socio-economic and economic impacts and then to compare relative aggregate performance. In particular, it is designed to make explicit assumptions, judgements, preferences and trade-offs in a systematic manner.

7.5.2 Sophisticated/advocacy led

The distinguishing feature of this approach is that it assumes implementation will be most easily achieved by encouraging public understanding of and participation in the search process from the outset so that, ultimately, sites are self-selected by *volunteer* communities. This methodology recognises that, in countries whose populations are well aware of the risks and costs associated with repository development, joint decision-making by waste

Table 7.6 A Typology of methodologies used to select sites for disposal of radioactive waste.

Category	Characteristic environment	Methodological characteristics	Examples
Prescriptive/Geological	• open government–easy access to judicial review of site search process • centralised decision making • detailed technocratic legislative framework. • sceptical public	• sophisticated, mechanistic site selection process, with an emphasis on procedural objectivity and considerations of wider socio–economic and environmental factors. • initial site selection process governed by technical/geological/hydrogeological criteria. • public acceptance exercises engaged in late in the day and focussed at the local level. • siting imposed.	MUA methodology adopted by US US NAS (1986)
Sophisticated/Advocacy	• open government • decentralised decision making • sceptical public	• early consultation over the whole of the potential receiving area • sophisticated, objective site search process and consideration of wider socio–economic and environmental factors. • initial site selection process governed by degree of public support. • siting voluntary/responsive	Canada—-Cooperative Siting Process (SPTF, 1987;) STF, 1990)
Pragmatic/Technical	• centrallised government; little access to judicial review of search process • sceptical public • weak legislative control of search process	• initial site selection process geologically and hydrogeologicaly led (geologically inclusive) • public acceptance exercises engaged late in the day and focussed at a local level • little emphasis on socio–economic considerations	Nirex ILW site search at Billingham/Elstow
Centrallised/Geological	• centrallised government • high level of public acceptance of nuclear power	• siting imposed • site selection geologically and hydrogeologically led • intensive local advocacy following the selection	French LLW repository (Centre da la Manche) (situation pre–Bataille)

disposal authorities and the public may be just as effective if not more so than centralised, imposed decisions.

Two examples of this approach are the Co-operative Siting Process (CSP) developed in Canada to select a site for the disposal of low-level waste in Ontario and the slightly different situation in France where only underground research laboratories (*i.e.* not disposal sites) sites are being selected. The current site selection programmes in Sweden and Finland for potential deep disposal sites may also be following a similar path. A key characteristic is an emphasis on providing mechanisms for community involvement and an exchange of information on risks and benefits at the earliest stage of the site selection process. A special feature is that the costs to the local community are explicitly recognised by making provision for the payment of compensation. A detailed technical analysis is not embarked upon prior to a community's provisional support for a scheme being firmly established. Even then, a wide range of environmental, social and economic studies are used to justify decision-making. But, being aware of a risk (in this case the perceived risk of having a radioactive waste disposal site in the neighbourhood) does not mean that the public will necessarily treat this risk in a rational manner. This irrationality can have quite significant consequences in selecting a disposal site, even when using this very open process.

The approach is relatively new and, at the end of 1993, CSP in Ontario had still to run its full course. The current situation in France and Sweden is rather fluid, and it is not certain what the final outcome might be in each country. It is, therefore, too early to judge the extent to which it may represent a more practical model than the others that have been described.

7.5.3 Pragmatic/technically led

The pragmatic/technically led approach is exemplified by Nirex's search for intermediate-level waste and low-level waste sites during the early 1980's. The key characteristics of this process were an unsophisticated survey, analysis, plan process which focused on geological factors to determine site location, a lack of public consultation prior to site identification and frequent shifts in policy which reflected the difficulty of securing the implementation of development at preferred sites in the light of political events in particular elections.

Underpinning this approach was a general lack of trust and support for the nuclear industry by the public, coupled with well organised and articulate opposition to what were viewed as poorly justified, imposed locational decisions. In the early 1980's, the government's and the nuclear industry's perception of this problem caused them to adopt a secretive and apparently rather arbitrary approach, typified by the sequence of events which led to the selection and sudden rejection of a site at a former anhydrite mine at Billingham in N.E. England – a site selected for the deep disposal of L/ILW. Another feature of the pragmatic approach is the absence of a detailed statutory or non-statutory framework in which to make and justify decisions. Thus, means and ends shifted as the government and Nirex reacted in an *ad hoc* manner to side step opposition rather than enter into debate with a hostile public. The introduction of mandatory environmental assessment procedures and a growing awareness on the part of Nirex of the need to win public support marked the end of such an approach, at least as far as the U.K. is concerned. The relatively simplified multi-attribute approach described by McEwen and Balch (1987) is an example of the way Nirex originally sought to demonstrate that sites are being selected in an objective manner. In addition, the selection of Sellafield and Dounreay as preferred locations for the deep disposal of L/ILW and growing interest in a possible package of community benefits indicate a greater awareness of the need to win over public opinion. The process used in the selection of these two sites is summarised in Nirex (1989) using a far more sophisticated approach than that described in McEwen and Balch (1987). Even this site selection programme was, however, carried out largely in a secretive manner.

7.5.4 Centralised/geologically led

Centralised, geologically led site selection processes are typified by the approach which used to be adopted by the French. The search for both high-level waste and low-level waste repositories was characterised, until very recently, by expert analysis of the geology and hydrogeology of candidate sites to the apparent exclusion of non-geological factors (Numark *et al.*, 1989; ERL 1987). Public relations programmes were generally well developed and, in the past, have focused on securing public support after the identification of a favoured site. Whilst there was provision for the review of siting decisions by means of independent public inquiries, this has not previously posed particular difficulties to the implementation of repository development. A range of factors explains this situation.

Firstly, there is in France a very centralised national decision-making process which concentrates effective power in the hands of national government, so that it is possible to override local and regional objections in the national interest. However, commentators have noted that a general acceptance of technical institutions, a trust and, indeed, deference to government agencies and strong support for the nuclear programme at all levels of society has ensured that projects such as Centre de la Manche and Centre de l'Aube have proceeded relatively smoothly.

In the past, the French have also been able to rely on the availability of land at existing nuclear facilities and it seems likely that the public's consequent familiarity with the industry diminished the level of opposition. More recently, the need to seek sites from further afield has generated a strong public response and the Bataille Report (Bataille, 1991) marked a shift towards a more open, participatory approach. The resulting moratorium on the selection of sites for deep disposal of long-lived waste, the decision to carry out an extensive URL programme lasting 15 years and the participation of local communities in selecting sites for these URLs marks an end to the centralised/geologically-led approach, at least in France (Bataille, 1993).

7.5.5 Conclusions

The approaches which have been described can be envisaged as different stages along a continuum of increasing sophistication and openness. This reflects the fact that, generally speaking, the development of site selection methodologies is tending towards more participatory procedures which specify impacts, acknowledge value judgements and risks and encourage self-selection rather than imposed decisions. Thus, more sophisticated advocacy-led procedures are tending to replace pragmatic, closed site selection exercises with the public's views being canvassed at an earlier stage, as authorities recognise the sensitivity of equity issue (*i.e.* the effect on property values and the level of possible compensation) associated with the siting of nuclear facilities.

However, the extent to which advocacy-based approaches are any more successful in terms of securing the implementation of repository development hinges on the ability to convince recipient communities that they will be adequately compensated. At the present time, this issue is giving rise to considerable controversy: Inhaber's (1991) suggestion, that the search process could be conducted as a reverse 'Dutch auction', where communities bid for facilities in terms of the level of compensation they are prepared to accept, has given rise to charges of bribery. Nevertheless, the approach has been mooted as a way of encouraging the development of repositories in Illinois, New York and California to assist the regeneration of declining rural communities (American News: 19 August, 1991). In other circumstances, advocacy-led approaches appear useful as a means of overcoming initial opposition to schemes, but experience in Canada suggests that, once the selection process focuses on a relatively few short-listed sites, public opinion becomes galvanised and potential opportunities rapidly disappear. A similar effect was seen during the site selection in the U.K. (Nirex, 1989; Chapman and McEwen, 1991) and in Switzerland for the sites for L/ILW disposal (Nagra, 1981, 1983; Kowalski *et al.*, 1986).

Overall, it would appear that, in the absence of widespread public support for nuclear power and governments' ability to impose decisions in the national interest, an early effort must be made to publicise and justify site selection procedures (this is precisely the process now being implemented in France and Sweden *e.g.* Bataille, 1993)). Implementation will depend upon full justification of preferred options and, therefore, careful consideration must be given to the specification of socio-economic as well as technical criteria and the way in which these are weighted when determining priorities.

As a starting point, it would seem that the range of factors identified in Table 7.6 could form a framework for strategic scanning of large areas and more detailed analysis of those which exhibit greatest promise. Subject to the availability of the digital data required for input to geographical information systems (GIS) (now commonly available in many European counties, and in particular Sweden), strategic scanning ought to be capable of being implemented in an economical and effective manner. Once preferred areas have been identified, further consideration will need to be given to the weighting attached to key factors such as land ownership in the light of feedback from public consultation exercises.

7.6 Lessons to be learned

7.6.1 General points, and socio-economic and political matters

1. It is important not to have any rigid criteria in the selection of sites for waste disposal, except where this is absolutely necessary. It is important to develop a site selection process that remains flexible and is based on *guidelines.* These guidelines need to encompass all the important factors, from geological through to socio-economic, that need to be taken into account, but they must not be prescriptive. This is where the U.S. site selection programme has its most serious shortcomings, although it would appear that European cultures are less likely to develop such a prescriptive, legalistic and adversarial system.
2. It is extremely easy to generate intense local public opposition to any proposal to dispose of radioactive waste or to the selection of an area for investigation. Openness by the waste disposal authority is one of the most important issues in this regard.
3. The extent to which a local population is likely to accept the possibility of their area being used for radioactive waste disposal seems to be very dependent on their previous experience of the nuclear industry. This axiom appears to be applicable to many countries, including the U.K., Canada, France, Sweden, Finland, and possibly also to South Korea and Japan. In order to suceed with a waste disposal programme, which will inevitably extend over several terms of government, it is probably important to accept the constraints imposed by political pragmatism and to choose sites for investigation which have some form of public, and therefore political, support. Unfortunately, any waste disposal organisation will be criticised, as Nirex have been, if they select sites which are in the ownership of

the government, the nuclear power company, *etc.* This is in part because the public believe that the *best* site ought to be selected, rather than a radiologically *adequate* site that is actually being sought, and they are likely to give little credence to any arguments that suggest that a site close to an existing nuclear plant is the optimal choice.

4. It is important for the politicians, the public and the licensing authorities to be convinced that the solutions for radioactive waste disposal in their country are not inferior and not substantially different from the solutions being proposed in other countries. Waste disposal organisations also seem to believe that this factor is important, as several of them have arranged to have external and international peer reviews of their programmes carried out. Many such organisations are also often keen to become involved in international research, modelling and validation exercises, as this is seen as a way of providing more credence to their waste disposal programmes, even though such collaboration may not directly influence the choice of a waste disposal site. Such research has been described as producing a 'warm tummy' feeling (Neuman, 1991), and this feeling of confidence in being able to dispose of radioactive waste safely needs somehow to be transmitted to the public.
5. It is tempting for a waste disposal organisation to try and pre-empt the decisions of politicians, and this is particularly relevant for a programme of deep disposal that needs political support over long periods of time if it is to succeed. It is, however, quite dangerous to place too much faith in one's ability to predict the fickle nature of the political process, and it is suggested that whilst a pragmatic attitude should be taken towards possible political influence on a site selection programme, that not too much credence should be given to any predictions.
6. The Co-operative Siting Process (CSP) currently being applied in Ontario, and being copied in some states of the U.S.A., is an interesting way of selecting communities which are supposedly willing to accept radioactive waste. In Ontario, the CSP has reached its more critical phases and it is not at all clear whether any one of the volunteer communities, if it were eventually chosen to house the proposed disposal facility, would still be willing to accept the waste. Similarly, it is not clear what is going to happen in France at the end of the 15 years of the two URL programmes when an actual disposal site needs to be selected.
7. The transport of radioactive waste to a repository can also generate adverse reactions amongst the public and there is an interesting divergence of views on the safest methods for transporting the waste. In the U.K. there is apparently particular public concern regarding the transport of waste by sea, whereas the fact that many sources of the waste in the U.K. are on or close to the coast is seen as an advantage in terms of waste transport by the nuclear industry and by their environmental and transport consultants.

It seems that this is an example of the analysis of the problem by *experts* perhaps not being in tune with the public, who, in certain areas of risk and environmental impact assessment, consider themselves the equal of any experts.

8. The significance of the Nimby effect seems to vary considerably from country to country, and even within different parts of the same country, and it does not seem possible to estimate its significance in advance, except in the most obvious cases. The point in the site selection process at which this effect becomes apparent is also sensitive to the political structure of the country, and particularly to the extent to which the public believe that they can have some influence over the decision making process.

7.6.2 Geological factors

This review has not concerned itself in any detailed manner with specific geological attributes which are relevant to the deep disposal of radioactive waste, for two main reasons:

1. There is a general consensus between the various countries carrying out searches for suitable deep disposal sites on the geological factors which are most likely to influence the suitability of a site, at least during the earlier stage of the site selection process, when limited geological and hydrogeological information is often available. These factors are intimately related to the required performance of the geological barrier in any performance assessment, and are extensively discussed for crystalline rocks in publications such as the Project-90 Final Report (SKI, 1991), SKB-91 (SKB, 1992a, b), Project Gewähr (Nagra, 1985), Kristallin-1 (Nagra, 1994), *etc.*
2. In many countries there may be little to choose between the different types of geological environments available, perhaps up to the stage of defining a short list of potential sites for investigation, as there is often a lack of relevant geological and hydrogeological data. This seems to be a particular problem in crystalline rocks.

It is apparent, however, that there is one factor of the geological, and hydrogeological, environment that is of overriding importance when considering different environments for disposal, and that is the level (or at least the *perceived* level) of geological complexity. This factor is known to have had a major influence in site selection programmes in Finland, Spain and Switzerland, and has been extensively discussed in the U.K., France and Canada. It has been considered under the subject of reducing the uncertainty in the far field (*e.g.* Zuidema, in NEA (1991)) and as enabling a robust approach to be taken to the development of a safety assessment (*e.g.* McCombie *et al.*, 1991). A relatively simple and predictable geological, and by inference a hydrogeological, environment is a desirable prerequisite for the development of a convincing and robust safety case. Similarly, any factor of the geological environment that could be considered to increase the geological complexity of an area should generally be considered as detrimental when scoring sites, even if it is also associated with advantageous properties. A site with a relatively simple geological environment is also more likely to be easier to characterise. The ease of site characterisation, which is directly related to the spatial variability of the site, was considered in the scoring of sites during the final stages of site selection by Nirex (Nirex, 1989; Chapman and McEwen 1991; Billington *et al.*, 1989), is discussed by Nagra in their selection of areas for further study in northern Switzerland (Nagra, 1994) and is also discussed in this context and in the more general perspective of repository performance by Hodgkinson and McEwen (1990). It has been termed the site's *explorability* by Nagra (Zuidema, in NEA (1991)). This aspect of the site selection process is very dependent on the particular country's geological history.

In contrast, an advocacy-led approach would seek to identify volunteer communities and, at this point, geological considerations will be taken into account to exclude those incapable of satisfying long-term safety criteria.

A summary of useful lessons to be learned is presented in Table 7.7.

Table 7.7 Summary of useful lessons to be learned.

Main factors	Subject area	Comments
1. General points	• Methodology	• Important (wherever possible) not to have rigid criteria, *guidelines* are a better option. • Methodology must not be prescriptive, but flexible and pragmatic.
	• Social factors	• Openness by the waste disposal authority is one of the most important issues. • Public acceptance appears to be related to previous experience of the nuclear industry. • Difficult concept to put over that *acceptable/adequate* site, rather than *best* site is being sought. • Solutions for waste disposal need to be similar to those proposed by other countries. • The approach suggested by the Co–operative Siting Policy is interesting and may prove successful. • Waste transport can generate very adverse public reactions. • The Nimby effect varies very greatly within and between countries in its significance and effect.
2. Geological factors	• Applicable to all geological environments	• There is a general consensus on the requirements for suitable geological environments for disposal. • There is often little to choose between many sites, particularly those in hard rock, up to the of short–listing of them for investigation. • The most important geological factor in site selection is the geological predictability/complexity.

7.7 References

Äikäs, T. (1985) The environmental factors to be considered in the site selection studies of the disposal of spent nuclear fuel in Finland. *TVO Report* YJT-85-28, Saanio and Laine Ltd., Helsinki (in Finnish).

Aller, L.T., Bennet, T., Lehr, J.H., Petty, R.J. and Hackett, G. (1987) DRASTIC – A standardized system for evaluating groundwater pollution using hydrogeologic settings. USEPA/600/2-87/035, Washington DC.

Almén, K-E. and Zellman, O. (1991) Äspö Hard Rock Laboratory. Field investigations methodology and instruments used in the pre-investigation phase, 1986-1990. *SKB Technical Report* TR 91-21. Swedish Nuclear Fuel and Waste Management Company, Stockholm, Sweden.

Barton, N., Lien, R. and Lund, J. (1974) Engineering classification of rock masses for the design of tunnel supports. *Rock Mechanics* **6:** 189-236.

Bataille, C. (1991) Rapport sur la gestion des déchets nucléaires à haute activité. *Rapport de l'Office Parlementaire d'Evaluation des Choix Scientifiques et Technologiques.*

Bataille, C. (1993) Mission Report. Mediation mission on siting uderground research laboratories. Ministries of Industry, Postal and Telecommunications and Trade and Ministry of the Environment, France.

Billington, D.E., Lever, D.A. and Wisbey, S.J. (1989) Comparative asessment of concepts and areas for deep disposal (Phase 2): Summary. *UK Nirex Report* No. NSS/A301, UK Nirex Ltd, Harwell, U.K.

Castaing (1982) (1983) (1984) Rapport du groupe de travail sur les recherches et développement en matière de gestion des déchets radioactifs propose par le CEA.

Chapman, N.A., McEwen, T.J. and Beale, H. (1986) Geological environments for deep disposal of intermediate level wastes in the UK. *Proceedings of IAEA Symposium on Siting, Design and Construction of Underground Repositories for Radioactive Wastes*, Hannover, March 1986, pp 311-328. International Atomic Energy Agency, Vienna, Austria.

Chapman, N.A. and McEwen, T.J. (1991) Geological aspects of the British programme for deep disposal of nuclear wastes. In *Applied Groundwater Hydrology – A British Perspective*, (eds. R.A. Downing and W.B. Wilkinson), Clarendon Press, Oxford, p 143-164.

ENRESA (1990) Factores de descalificación y atributos de selección: rangos establecidos. (Disqualification and qualifying selection factors: established ranges.) Proyecto ERA. *ENRESA Report* 46-92-ST-IE-001. ENRESA, Madrid, Spain.

ERL (1987) The disposal of radioactive waste in Sweden, West Germany and France. Report for the County Council's Coalition, UK.

Goguel, J. (1987) Stockage des déchets radioactifs en formations géologiques, critères techniques de choix de site. Rapport du Groupe de Travail présidé par le Professeur Goguel. Ministère de l'Industrie des P et T et du Tourisme, France.

Gustafson, G., Liedholm, M., Rhén, I., Stanfors, R. and Wikberg, P. (1991) Äspö Hard Rock Laboratory. Predictions prior to excavation and the process of their validation. *SKB Technical Report* TR 91-23. Swedish Nuclear Fuel and Waste Management Company, Stockholm, Sweden.

Hodgkinson, D.P. and McEwen, T.J. (1990) Spatial variability: A repository performance perspective. NEA/CEC Workshop on *Heterogeneities in Hydrogeological Systems*, Paris, October, 1990. Nuclear Energy Agency, Paris, France.

IAEA (1977) Site selection factors for repositories of solid high-level and alpha-bearing wastes in geological fromations. *IAEA Technical Report Series* No. 177. International Atomic Energy Agency, Vienna, Austria.

IAEA (1981) Underground disposal of radioactive wastes-Basic Guidance. *IAEA Safety Series* No. 54. International Atomic Energy Agency, Vienna, Austria.

IAEA (1983) Criteria for underground disposal of solid radioactive wastes. *IAEA Safety Series* No. 60. International Atomic Energy Agency, Vienna, Austria.

IAEA (1985) Techniques for site investigation for underground disposal of radioactive wastes. *Technical Report Series* Report No. 256. International Atomic Energy Agency, Vienna, Austria.

IAEA (1989) Guidance for regulation of underground repositories for disposal of radioactive wastes. *Safety Series Report* No. 96. International Atomic Energy Agency, Vienna, Austria.

IAEA (1994) Siting of geologic disposal facilities. *Safety Series Report* (RADWASS) III-G-4.1. International Atomic Energy Agency, Vienna, Austria.

IAEA (1995) Process for selection and characterization of sites for deep geologic disposal of high level radioactive waste. *Technical Report Series.* International Atomic Energy Agency, Vienna, Austria.

Inhaber, H. (1991) Yes, it is possible to site radioactive waste management facilities. Proceedings of 1991 Joint International Waste Management Conference, Seoul, p 201-207.

Keeney, R.L. (1980) *Siting Energy Facilities.* Academic Press, New York.

Keeney, R.L. and von Winterfeld, D. (1988) The MUA analysis and its role for selecting nuclear waste repository sites. *Operational Research '87*, (ed. G.K. Rand), Elsevier Science Publications.

Kowalski, E., Gassner, R. and Kappeler, S. (1986) Constructional concepts and site selection procedures for a low- and intermediate-level waste repository in Switzerland. Proceedings of IAEA Symposium on *Siting, Design and Construction of Underground Repositories for Radioactive Wastes*, Hannover, March 1986, pp 241-166.

Langer, M., Schneider, H. and Kühn, K. (1991) The salt dome of Gorleben – Target site for the German radioactive waste repository. In *Geological Problems in Radioactive Waste Isolation – A World Wide Review*, (ed. P.A. Witherspoon). *Lawrence Berkeley Laboratory Report* LBL-29703. Lawrence Berkeley Laboratory, Berkeley, California, U.S.A.

McCombie, C., McKinley, I.G. and Zuidema, P. (1991) Sufficient validation: The value of robustness in performance assessment and system design. *GEOVAL-90 Symposium*, Stockholm, May 1990. Nuclear Energy Agency, Paris / Swedish Nuclear Power Inspectorate, Stockholm.

McEwen, T.J. and Balch, C. (1987) Geological and environmental constraints on the disposal of low-level radioactive waste in the UK. In *Planning and Engineering Geology*, (eds. M.G. Culshaw, F.G. Bell, J.C. Cripps and M. O'Hara). *Geological Society of London Special Publication* No. 4, pp 507-515.

McEwen, T.J. and Balch, C.J. (in press) Review of the status of siting radioactive waste repositories. *Intera Report* IG3242-1, Intera Information Technologies Ltd, Melton Mowbray, U.K.; *SKB Technical Report (in press)*, Swedish Nuclear Fuel and Waste Management Company, Stockholm, Sweden..

Merkhofer, M.L. and Keeney, R.L. (1987) A multi-attribute utility analysis of alternative sites for the nuclear waste repositories. *Risk Analysis* **7**: 173-194.

Nagra (1981) Die Endlagurung schwach- und mittelaktiver Abfälle in der Schweiz – Potentielle Standortgebiete für ein Endlager Typ B. *Nagra Technical Report* NTB 81-04. Nagra, Baden, Switzerland.

Nagra (1983) Die Endlagerung schwach- und mittelaktiver Abfälle in der Schweiz: Evaluation der potentiellen Standortgebiete. *Nagra Technical Report* NTB 83-15. Nagra, Baden, Switzerland.

Nagra (1985) Project Gewähr 1985. Nuclear waste management in Switzerland: Feasibility studies and safety analysis. *Nagra Gewähr Report Series* NGB 85-09. Nagra, Baden, Switzerland.

Nagra (1988) Sediment Study – Intermediate Report. Disposal options for long-lived radioactive waste in Swiss sedimentary formations (Executive Summary). *Nagra Technical Report* 88-25E. Nagra, Baden, Switzerland.

Nagra (1994) Kristallin-1: Conclusions from the regional investigation programme for siting a HLW repository in the crystalline basement of northern Switzerland. *Nagra Technical Report* NTB 93-09E. Nagra, Wettingen, Switzerland.

NEA (1991) Disposal of high level radioactive wastes: Radiation protection and safety criteria. Proceedings of NEA Workshop, Paris, Nov. 1990. Nuclear Energy Agency, Paris, France.

Neuman, S. (1991) Technical discussion. In *Proceedings of GEOVAL-90,* NEA/SKI, Stockholm, May 1990. Nuclear Energy Agency, Paris / Swedish Nuclear Power Inspectorate, Stockholm.

Nirex (1989) Deep Repository Project – Preliminary Environmental and Radiological Assessment and Preliminary Safety Report (PREA/PSR). *Nirex Report* No. 71. UK Nirex Limited, Harwell, U.K.

Numark, N.J., Paige, H.W. and Wonder, E.F. (1989) Public perception and acceptance of the siting of nuclear waste facilities in seven countries. *Report for ERC Environmental and Energy Services Company.*

Peyrus, J.Ch. (1987) Evaluation de la sureté d'un site de stockage de déchets radioactifs en milieu géologique. Rapport SERE/87, DERS, CEA, France.

Rósen, L. and Gustafson, G. (1993) Possible strategies for geoscientific classification for high-level waste repository site selection. *SKB Technical Report* 93-12. Swedish Nuclear Fuel and Waste Management Company, Stockholm, Sweden.

Salmi, M., Vuorela, P. and Kuivamäki, A. (1985) Geological site selection studies for the final disposal of spent nuclear fuel in Finland. Report YJT-85-27, Geological Survey of Finland (in Finnish).

SKB (1983) Final storage of spent nuclear fuel – KBS-3, Vols. 1-5. Swedish Nuclear Fuel and Waste Management Company, Stockholm, Sweden.

SKB (1989) Handling and final disposal of nuclear waste: Hard Rock Laboratory. Background Report to the R&D Programme 89. Swedish Nuclear Fuel and Waste Management Company, Stockholm, Sweden.

SKB (1992a) Treatment and final disposal of nuclear waste – Siting of a deep repository. SKB Background Report to the R&D Programme. Swedish Nuclear Fuel

and Waste Management Company, Stockholm, Sweden.
SKB (1992b) SKB-91: Final disposal of spent nuclear fuel. Importance of the bedrock for safety. *SKB Technical Report* 92-20. Swedish Nuclear Fuel and Waste Management Company, Stockholm, Sweden.
SKI (1991) SKI Project-90 Summary. *SKI Technical Report* 91:23. Swedish Nuclear Power Inspectorate, Stockholm.
SPTF (1987) Opting for co-operation. Report of the Siting Process Task Force (SPTF) on low-level radioactive waste disposal. Energy Mines and Resources, Ottawa, Canada.
Stanfors, R., Erlström, M. and Markström, I. (1991) Äspö Hard Rock Laboratory. Overview of the investigations 1986-1990. *SKB Technical Report* 91-20. Swedish Nuclear Fuel and Waste Management Company, Stockholm, Sweden.
STF (Siting Task Force) (1990) Opting for co-operation. A Process in Action, the First Phases. Synopsis Report of the Siting Task Force. Energy, Mines and Resources Canada, August 1990. (Available as a report and a synopsis).
TVO (1985) The repository programme for spent nuclear fuel disposal in Finland. *TVO Report* YJT-85-30. The Nuclear Waste Commission of Finnish Power Companies, Helsinki, Finland.
TVO (1992) Final disposal of spent nuclear fuel in the Finnish bedrock. Preliminary site investigations. *TVO Report* YJT-92-32E. The Nuclear Waste Commission of Finnish Power Companies, Helsinki, Finland.
US DOE (1984) General guidelines for the recommendations of sites for the nuclear waste repositories, 10 CFR Report 960, Federal Register 49 (No. 286), 47714-47770. U.S. Department of Energy, Washington DC, U.S.A.
US DOE (1985) Mission Plan for the Civilian Radioactive Waste Management Program, U.S. Department of Energy, Washington DC, U.S.A.
US DOE (1986) A multi-attribute utility analysis of sites nominated for characterisation for the first radioactive waste repository – a decision-aiding methodology. DOE Report DOE/RW-0074. U.S. Department of Energy, Washington DC, U.S.A.
US DOE (1987) Region-to-area screening methodology for the crystalline repository project. Report DOE/CH-1. U.S. Department of Energy, Washington D C, U.S.A.
Vuorela, P. and Hakkarainen, V. (1982) The suitability of Finnish bedrock for the final disposal of high-level radioactive waste. *TVO Report* YJT-82-58, Geological Survey of Finland (in Finnish).
Wikberg, P., Gustafson, G., Rhén, I. and Stanfors, R. (1991) Äspö Hard Rock Laboratory. Evaluation and conceptual modelling based on the pre-investigations 1986-1990. *SKB Technical Report* TR 91-22. Swedish Nuclear Fuel and Waste Management Company, Stockholm, Sweden.
Zetterberg, H.L. (1980) Perceptions of the Nimby effect. In: Proceedings of the 5th International Symposium of the Uranium Institute, Westbury House, UK, ISBN 0-86103-041-9.

8

Site characterisation and URLs

8.1 Introduction

A detailed characterisation of a potential waste disposal site needs to be carried out in order to provide the data necessary for the models used in the performance assessment and so that the repository can be designed and located precisely within the rock mass. Any site characterisation is likely to be performed in phases, which with time will tend to concentrate on diminishing volumes of the rock mass and supply increasingly detailed levels of data. These phases can be described in the following terms:

- Pre-site characterisation phase: Site/area chosen by site selection programme (one or more sites may be selected by this process and be investigated consecutively or sequentially)
- Phase 1: Regional-scale characterisation
- Phase 2: Local-scale characterisation
- Phase 3: Repository-scale characterisation

They can also be conveniently divided into the various stages that are related to the operation of the URL-phase (Underground Research Laboratory) of the site characterisation, and these are discussed in more detail in Section 8.14:

- Pre-shaft construction
- During shaft construction
- Post-shaft excavation and construction of the URL

It would be possible to discard one or more of the sites at any phase during the site characterisation programme. The discarding of a site early in such a programme could only take place based on relatively crude and obvious failings, such as young (perhaps post-glacial) groundwater at the proposed disposal depth. The discarding of a site later in the characterisation programme is more likely to take place due to its comparison with other sites that are deemed to be superior, as a result of early assessment modelling that suggests that some safety target is unlikely to be met, or due to the realisation that the geology, and perhaps the hydrogeology, of the site are too complex and could never be resolved to a sufficient level of understanding.

8.2 Pre-site characterisation

The level of geological knowledge which is available in advance of any site characterisation can be very variable. Site selection programmes will tend to select sites which are associated with simple geological structures, with the hope that these will tend to be associated with relatively simple hydrogeological environments. In sedimentary environments this assumption is more likely to be realised and, in addition, there are likely to be more data on the structure of the rocks than in areas of hard rock. In hard rock areas, it is often difficult to distinguish on geological grounds between many of the areas that are under consideration as, apart from the surface manifestations of fracture zones and their geophysical signatures, little more can be discovered about the sites without deep drilling. Large assumptions have to be made at this stage regarding the probable structures at depth. Some information may be available, sometimes of a proxy nature, on the hydrogeological environment at depth, from the distribution of springs. Figure 8.1 shows the typical block structure of an area of ancient crystalline basement in Finland, with the investigation area selected for future investigation by TVO (TVO, 1992).

The various phases of the site characterisation programme can be examined by reference to the programme of work for the HRL ('Hard Rock Laboratory') at Äspö, Sweden (Table 8.1). The programme for this site is driven, in part, by the desire to validate geological and hydrogeological conceptual models as the programme proceeds. This is done in order to clarify the precision of the early stages of the characterisation programme (*i.e.* to test the assumptions implicit in the development of the conceptual models) and to allow the best decisions to be made on the evolution of the programme and on the design of its later stages.

It is most helpful in this regard if a strategy for examination of the predictions from the various stages of the investigations is set out well in advance of the start of any work in the field, in a similar manner to the programme presented by SKB (SKB, 1989). The process of prediction and validation is illustrated in Figure 8.2, where continuous comparison of predictions with what is subsequently found during the site investigation programme allows for a refinement of the predictive models. The final part of this process of validation is a judgement of whether the models have sufficient predictive capacity within a previously agreed error range.

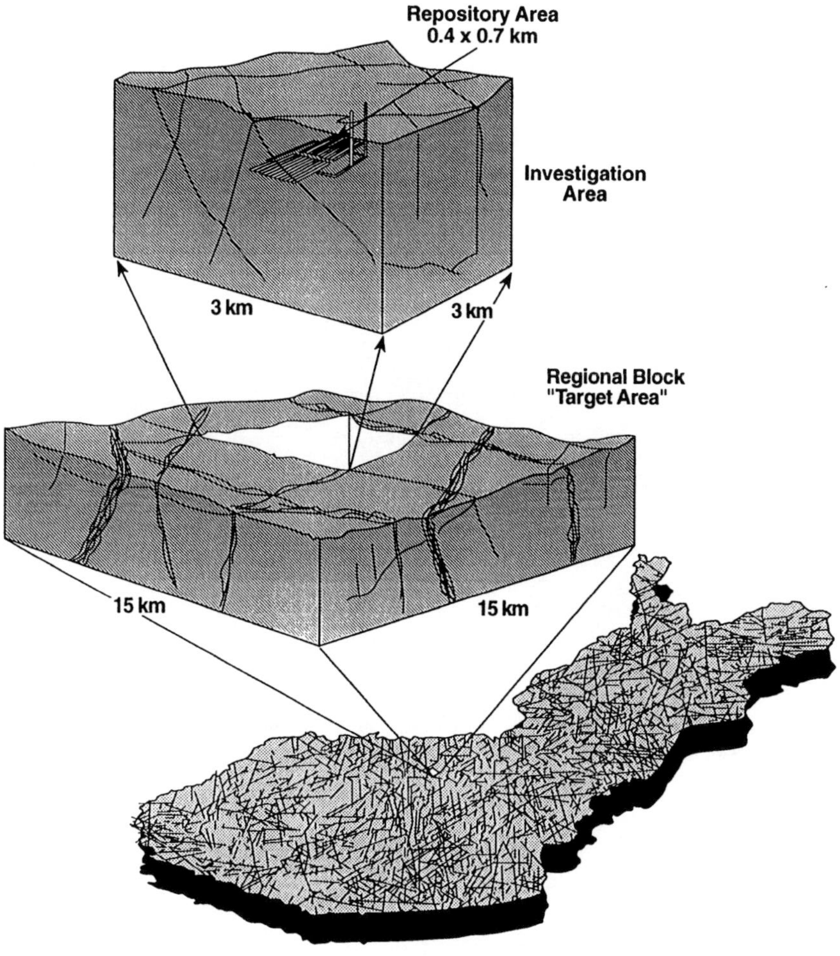

Figure 8.1 Structures within the Precambrian basement of Finland defined mainly from aerial surveys, together with a limited amount of field work, showing the classification of lineaments and the selected area for future investigation (from TVO, 1992). Reproduced by permission from TVO.

Table 8.1 A summary of the four planned phases of the first stage goal of the HRL programme at Äspö, illustrating the staged approach taken to characterising the area and the site. From Wikberg *et al.* (1991). Reproduced by permission of SKB.

First geological and hydrogeological evaluation (1986-1987)
(a) Regional investigations
(b) Evaluation of several target areas
(c) Development of regional conceptual groundwater model
(d) Geological predictions for Äspö area before drilling
The second evaluation (1987-1988)
(a) Development of site specific groundwater model
(b) Results of initial drilling *etc.*
(c) Development of conceptual geological model using new data
The third evaluation (1988-1990)
(a) First comprehensive summary of pre-construction predictions
(b) Basis for all pre-construction predictions
The fourth evaluation (expected in 1994/95)
Compare pre-construction predictions with actual conditions

8.3 Phase 1: regional-scale characterisation

Analysis at this scale is required for several reasons:

1. The site needs to be placed into its regional geological structural setting. The area of interest can vary considerably, but is likely to be larger in sedimentary environments.
2. A regional hydrogeological assessment of existing data will be required, and the size and shape of the area in which this is performed is very dependent on the local catchment area(s) and the hydrogeological regime, *i.e.* the proximity to the coast, to large rivers and to groundwater divides.
3. A regional-scale hydrogeological model is likely to be required, if only to define boundary conditions, so that a smaller local-scale model can be constructed.
4. Data will be required over the region for any eventual analysis of the sesimic risk to the potential site.
5. Any palæohydrogeological studies that are performed may well require data over this area.

It is unlikely to be necessary to drill boreholes over more than a small part of this area, except in the situation where seismic surveys are to be used to define the structures at depth and there is little or no existing borehole

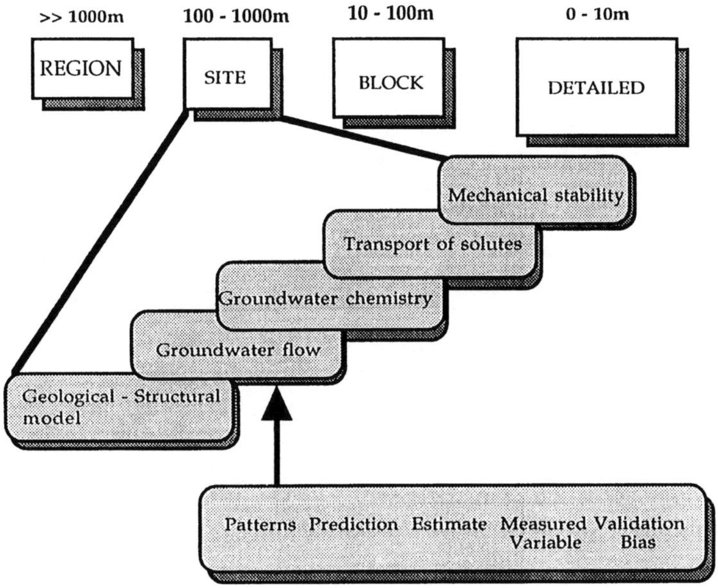

Figure 8.2 The geometric scales considered in the design of the experimental programme at the HRL at Äspö , together with key questions related to the the eventual perfomance assessment of a repository in granite. (From Gustafson *et al.*, 1991). Reproduced by permission of SKB.

control. An example of this might be an area of sediments with gently dipping strata, where the structure is expected to be relatively simple. Undue reliance must not be placed on geophysical interpretations of geological structures, as, without borehole data, they do not provide unequivocal evidence for their existence. (Witness the discovery, during drilling by Nagra, of the Permo-Carboniferous trough in northern Switzerland, that had been interpreted from seismic reflection surveys to be part of the granitic basement.)

A typical regional-scale investigation might include the following types of investigation:

1. Aerial geophysical surveys. The types of survey will vary depending on the geological environment under investigation, though, in general, they are more likely to be of use in hard rock areas. There may also be limitations on the use of some of the surveys, notably magnetic and EM, due to the presence of existing man-made structures. Gamma spectrometric methods are, again, of most use in hard rock areas, but they could also provide background data for anthropogenic radioactive materials.

2. Ground geophysical surveys. Seismic (mainly reflection) is likely to provide the most useful data, though its use in hard rocks can be limited, as they may possess few internal reflectors. The use of this technique was, however, very useful at Äspö (Gustafson *et al.*, 1991) and at Dounreay (Chapman and McEwen, 1989) in delineating low angle fracture zones within basement rocks. Coastal sites can cause problems, as undershooting and/or transition zone seismic surveys, which are expensive and can be difficult to operate, will be required. Magnetic and VLF profiles can often be of use in hard rock areas, as long as they are sparsely populated. Gravity surveys are commonly carried out in conjunction with seismic. EM methods (TEM and magnetotellurics) can be of use for coastal sites as they may be able to delineate the shape of the saline/freshwater interface.
3. A range of geophysical techniques is likely to be employed, and integrated data analysis (*i.e.* the parallel interpretation of different geophysical techniques) is commonly performed. Existing software allows for the development of 2.5-D models with relatively complex structures. Some models allow for the joint 3-D interpretation of gravity and magnetic data. Image processing techniques can be very helpful for multi-parameter data analysis and, in particular, for the delineation of structural trends from gravity and magnetic data.
4. Geophysical interpretations can be linked with surface mapping and air photo and satellite analysis techniques to provide the first indications of the geological structures. Their use is severely limited in the absence of borehole control data, particularly in the case of seismic.
5. Depending on the exposure available, the extent of surface mapping and tectonic analysis will vary greatly. Sedimentary areas, especially those in northern Europe, are frequently covered by glacial and/or recent deposits, which severely limit the opportunities for obtaining useful data on subsurface structures and lithologies. Surface mapping has proved to be of far more use in areas of hard rock in Scandinavia where, away from the cover of sediments deposited by the post-glacial Baltic, good exposure can be obtained and fracture zones analysed.
6. Regional hydrological and hydrogeological studies will allow baseline data on atmospheric precipitation, evaporation, run off, groundwater recharge, groundwater chemical compositions, specific capacities of boreholes, *etc.* to be determined and databased. The extent of pertinent data on the groundwater in areas where there is little groundwater extraction may be very limited.

The regional analysis should allow the following to be determined:

- It should allow the first tentative decisions to be made on the most suitable location for the repository, or perhaps confirm that the pre-investigation interpretations of the geological structures were correct.

- It should allow the next stage of the investigation to be planned, and this is almost certain to include the drilling of the first deep, and perhaps shallow, boreholes. Further, more detailed and more focussed geophysical surveys may be carried out in the next stage.
- The regional hydrogeological analysis, combined with the initial regional structural synthesis, should allow decisions to be made on the boundaries for the regional-scale hydrogeological model and, as a result, the boundaries for any smaller, local-scale models.

The design of the site characterisation programme needs to take into account the important aspect of how many data, and of what quality, it is possible to collect within the time available. The cost of collecting high quality data can be very significant and there needs to be a trade-off between the anticipated level of prediction aimed at, and hopefully achieved, against the cost.

This trade-off is illustrated in Figure 8.3, where it can be seen that the level of prediction, however good, will always be inferior to the actual observations during construction and that there are diminishing returns in attempting to achieve extremely high levels of prediction.

8.3.1 More local-scale characterisation

At the end of the first phase it should become possible to develop a geological and perhaps an initial conceptual hydrogeological model of the region. Although both these models will, of necessity, be simplistic they will allow their predictions to be tested against further geological and hydrogeological investigations.

It is believed that continuous attempts at prediction, *i.e.* simple validation exercises, are a valuable method of steering the investigations and of making most use of the results. The investigations for the Äspö HRL use this type of approach, and these are described in Section 8.14 on the use of URLs in the assessment of sites.

The development of these conceptual models will be linked to the changing scale of the investigation, from regional down to the scale of blocks of rock or rock on the scale of the individual waste canisters or waste disposal vaults, *i.e.* between 10 to 100 m. The scales of the investigation are shown in Figures 8.2 and 8.4, where they are related to the gradual change of the investigation before and subsequent to construction activities, *i.e.* the start of the URL phase of the investigations.

The surveys already performed should allow the optimum locations for boreholes to be determined. The design of the drilling programme will be very dependent on the following factors, outlined in Sections 8.4 and 8.5.

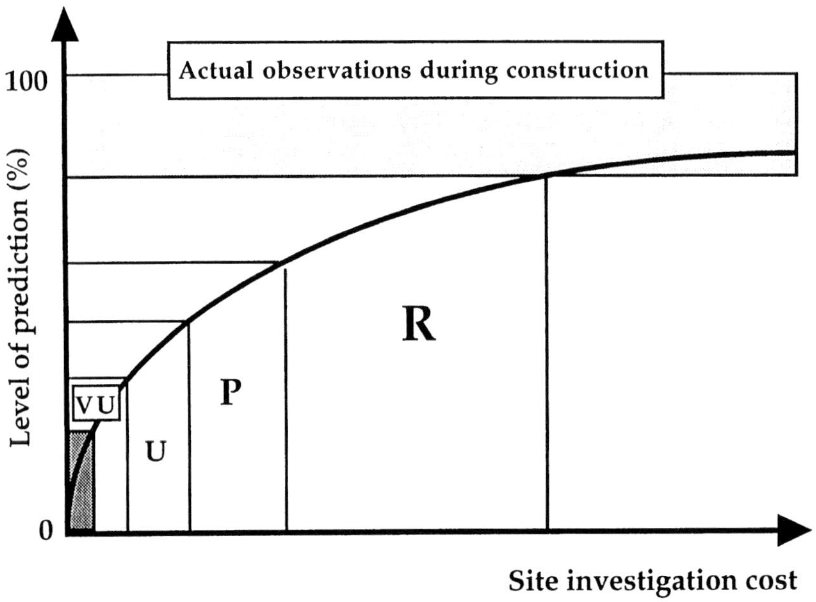

Figure 8.3 The level (accuracy) of prediction versus the site investigation cost. VU = Very Unreliable, U = Unreliable, P = Possible/Probable and R = Reliable. However much investigation work is carried out a 100 % predictive capacity will never be achieved and it is only when underground construction takes place that uncertainties inherent in predictions can be resolved.

8.4 The geological environment

In a relatively simple sedimentary environment, with gently dipping strata and few structural discontinuities, there is likely to be very little need at this stage of the investigations to drill more than perhaps one or two deep vertical, fully-cored boreholes to somewhere beyond the expected repository depth. These boreholes are most likely to be located on existing seismic lines.

In contrast, in low relief granitic terrain, a much larger number of both relatively shallow and deeper boreholes is likely to be required. These are likely to be designed to intersect fracture zones and geophysical anomalies and several of these boreholes may well be inclined and open hole (see Figure 8.9).

Developments, mainly from the oil, and to a lesser extent the mineral, industries, in seismic interpretation techniques and other software products, such as 3-D visualisation and fault analysis systems, now allow far better integrated geological interpretations to be carried out. Developments have been most significant in the following areas:

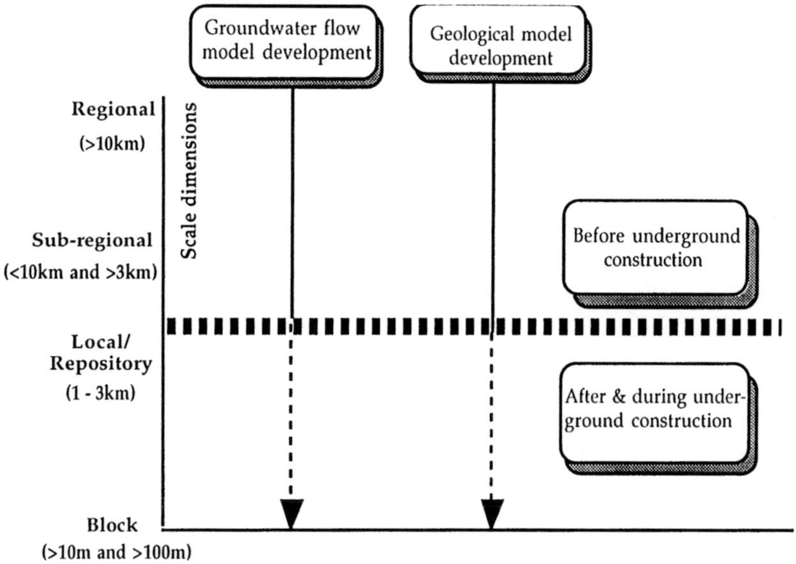

Figure 8.4 The scales of investigation, from regional scale to block scale, and their relationships with the characterisation work in advance of, during, and after underground construction.

- Computer Aided Exploration (CAEX) systems, for use in the interpretation of seismic reflection profiles and the integration of the geological interpretation with all other types of geological and geophysical data (*i.e.* from boreholes and other, non-seismic geophysical techniques, in particular gravity and aeromagnetics). Two major systems are available, GEOQUEST and LANDMARK, both of which run on workstations.
- 3-D visualisation software has been developed that allows the geological data to be presented in 3-D. Three systems are being and/or may be used in this field in radioactive waste disposal site investigations, Stratamodel, Dynamic Graphics Earthvision and Lynx. Earthvision is being used by Nirex at Sellafield, by Nagra at Wellenberg, by SKB at the Äspö HRL and by the USGS and the USDoE at Yucca Mountain. Lynx is also being used at Yucca Mountain.
- Other software products include those designed to examine the validity of fault interpretations from seismic data (*e.g.* FAPS2 (*Badley Earth Sciences Ltd.*)) and section balancing programmes (such as GEOSEC20 (*Cogniseis Ltd.*)).

There are a number of restrictions on any geological investigation which relies on data interpretations made on paper (hard copy). Geological investigations commonly rely upon the integration of large amounts of data, which must be compared and combined continually. It is impossible to achieve this using hard copy of data since all parts of an investigation cannot be compared simultaneously. This is emphasised by considering the offshore seismic database of the 'Sellafield Region' considered by UK Nirex Ltd., which consists of 770 kms of seismic on 58 lines (Figure 8.8 (Nirex, 1993)). With the recent advances in computing technology, the volume of data that geologists can interpret has greatly increased, allowing more rigorous and geologically acceptable interpretations.

The use of such software is important in a number of areas:

- As an aid to understanding the site – the ability to display data in 3-D is very likely to increase the understanding of the geology of the area.
- In the presentation of geological data – a visualisation system (such as one of those listed above) allows increased quality and clarity of presentation of the geological interpretations.
- In the preparation of information to be supplied for modelling of the site – digitised data derived from 3-D software systems can be used directly as input to models of the site.
- In testing geological interpretations – several computer systems are available for interrogation of geological interpretations. These can be divided into two categories. Firstly, visualisation of interpretations can be used to assess if interpretations 'make sense' in three dimensions, (examples include ensuring that fault data coincide with offsets in stratigraphic horizons), these are visualisation software packages (see above). Secondly, software programs are available for assessing interpretations. These are used for ensuring that interpretations are geologically feasible (by balancing the section) and analyzing the quantitative data in interpretations (*e.g.* measurement of fault displacement profiles) (see above).
- Hypothesis Testing – computer software will enable the development of an hypothesis testing programme. This is commonly termed the what-if approach. This will allow testing of the impact of a number of interpretations on the predictions of the sub-surface geology.
- Extraction of additional data – 3-D modelling systems will allow the generation of additional information relevant to the performance assessment of the site, for example the generation of 2-D sections of any orientation, which will provide data related to radionuclide pathways in the geological far-field. In addition, computer software packages can be used to interrogate data and provide quantitative estimates of features such as the fractal dimension of fault patterns and the nature of the sub-seismic fault population.
- Data Management – the incorporation of interpretations into a digital database included within the 3-D visualisation software will enable rapid

access to these data.

8.5 Factors influencing borehole locations

The number and location of deep boreholes, which are drilled as part of a site investigation for the purposes of a repository site are dependent on several different and sometimes competing factors, all of which will need to be taken into account in the design of the investigation. On the one hand there is a need to be able to supply the data for the flow and transport models to a sufficient degree of accuracy and so that the uncertainties inherent in the system will be reduced to below some acceptable level. On the other hand, if too many boreholes are drilled this might compromise the integrity of the host rock and the far-field for any possible use for disposal purposes. A geological environment could be removed from consideration for the purposes of disposal by this very argument, if it could be demonstrated that an unacceptable number of boreholes would be necessary to characterise adequately its properties and to validate the necessary models. The number of boreholes, their locations and the data requirements from these boreholes depend on several different factors:

- the geology of the site and the region,
- the hydrogeological regime on the regional and local scales,
- the acceptable levels of uncertainty associated with the investigation,
- the availability of inferential measurements,
- cost constraints,
- the allowable damage to the potential integrity of the geological barriers,
- the practicality of some of the proposed investigations,
- the iterative nature of the investigation.

A detailed description of the types of boreholes, their depth ranges and the types of testing that could be performed within them, is presented in McEwen *et al.* (1991) for the case of supplying the required data for a performance assessment at a potential repository site. The discussion here is concerned with the general requirements for the number of boreholes, where they might be located and what type of work is likely to be carried out in them with respect to the research programme. A great deal of reliance is placed on the data derived from boreholes, even though such data sample only an extremely small fraction of the rock. This problem of small sampling sizes, and its influence on the type of statistical approach that should be followed, has been discussed by Freeze *et al.* (1988) with reference to safety assessments for repositories.

The extent of the site investigation, and in particular the areal extent of drilling sites, depends to a very great extent on the geological and hydrogeological environment of the site. The area over which drilling will be required can vary from as much as 2500 km^2 in a sedimentary environment, in which groundwater flow cells can have very large lateral extent but where the geological structure could be simple, to a significantly smaller area in crystalline rock, where the flow cells can have less than one tenth the lateral extent but where the inhomogeneities can be of far more significance. Investigation areas of 10 km^2 for research or potential repository sites are common in Sweden and Finland. Areas for regional hydrogeological and hydrochemical investigations in these types of rock can be more of the order of 50-100 km^2 (Stanfors *et al.*, 1991), and a good example of the areal extent of data thought necessary by SKB for their HRL is presented in this report, with accompanying maps that illustrate how the area of interest was reduced in size as the investigations homed in on the Äspö area (Figures 8.5 and 8.6). Recent work by Nirex on the regional hydrogeology and hydrochemistry of Sellafield covered a similar area (Figure 8.7). The locations of the regional-scale boreholes can be related to the more local-scale boreholes of the Potential Repository Zone (PRZ) (Figure 8.7) and can be compared with the areal extent of the much larger investigation areas considered necessary for the geophysical surveys (Figure 8.8).

8.5.1 First phase of the investigation

The first, or reconnaissance, phase of the site investigation needs to examine the regional geological structure and, as a guide, a similar area around the site that is examined in seismic risk assessments for nuclear installations in the U.K., *i.e.* a radius of 60 km around the site, is probably necessary. It is almost certainly unnecessary at this stage to perform any form of detailed analysis at several tens of kilometres from the site, and the examination of the geology at these distances is likely to be confined to a regional structural synthesis. The aim at this stage of the investigation is to place the geology of the site, to depths of perhaps 2-3 km, in its regional context, and to enable initial thoughts to be developed on the structures at depth, especially if little direct geological information is available.

The first boreholes on the site may be positioned by the use of a seismic reflection survey, so that optimal drilling locations can be chosen. Although the regional geophysical surveys are likely to cover areas up to several kilometres from the site, and in the case of aeromagnetic surveys the coverage needs to be at least twice the required depth penetration of the survey for the purposes of interpretation, it is highly unlikely that any boreholes would be drilled at such distances. Any initial groundwater flow modelling that might take place in advance of the availability of any site investigation data could use expert opinion to provide parameter values, and it is only when real data have been incorporated into the models that any decisions can be made logically on the possible locations for further boreholes. The whole

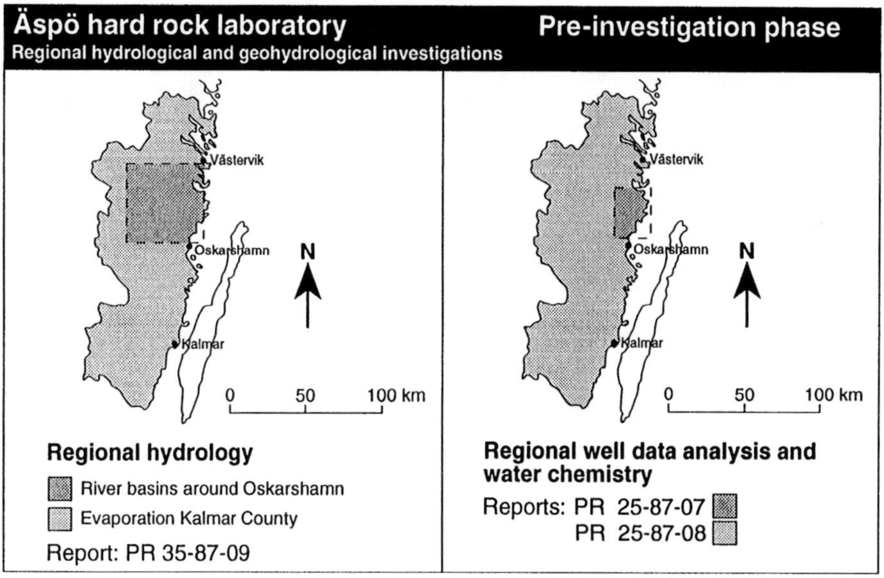

Figure 8.5 Regional-scale hydrogeological and hydrological measurements associated with the development of the HRL (from Stanfors *et al.*, 1991). Reproduced by permission of SKB.

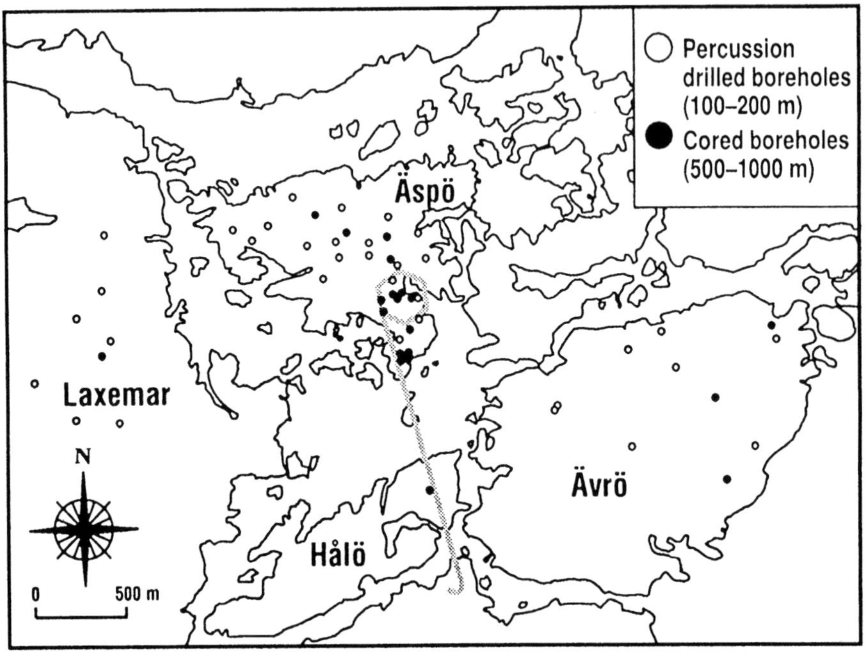

Figure 8.6 Drilling programme around the island of Äspö associated with the site investigation for the HRL (from SKB, 1992b). Reproduced by permission of SKB.

investigation programme cannot be defined at the outset, and must be flexible and able to accept numerous iterations.

This first part of the investigation needs to proceed for at least one year, and probably considerably longer, before any data become available that could be of any use in modelling the groundwater flow regime. Using the U.K. Sellafield and Swedish HRL site investigations as examples, it would appear to take approximately 2 years before a reasonable idea of the more significant aspects of the hydrogeological system can be obtained, though whether this information is really sufficient to select an optimal tunnel or shaft access to the proposed repository is not certain. (It took Nirex approximately 3 years to be able to define the extent of the PRZ – see Figure 8.7).

In this respect, it is important to decide very early in the the site characterisation programme the precise purpose of the URL-phase of the investigation and the approach that will be taken to collecting data. This has particular significance for the following reasons:

1. The types of boreholes that are drilled and the design of the testing

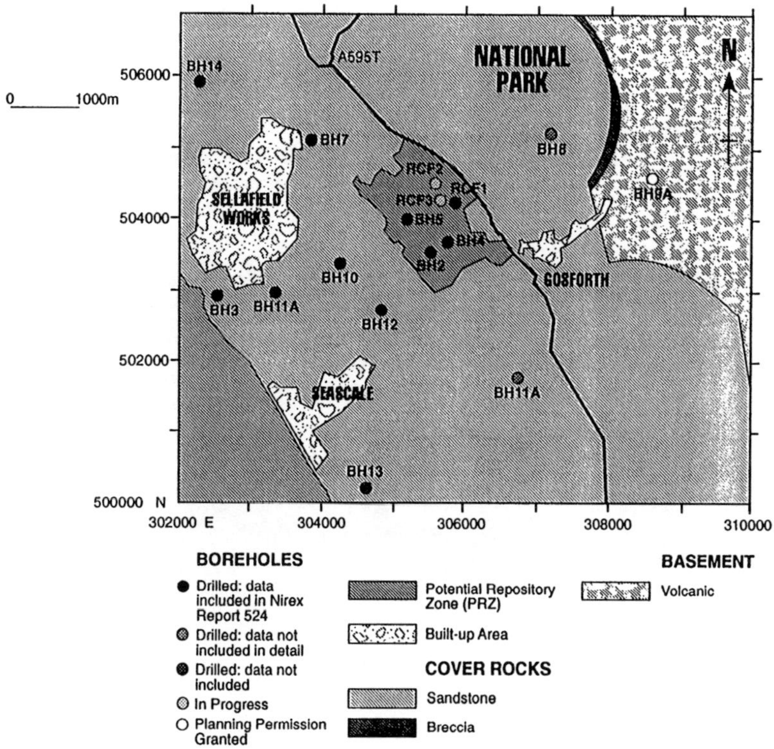

Figure 8.7 Location of boreholes at Sellafield (December 1993; from Nirex, 1993). Reproduced by permission of U.K. Nirex Ltd.

programme within the boreholes are very dependent on the philosophy of the characterisation programme. A decision has to be made on whether to follow the route taken by SKB, which is based on a relatively large number of small diameter (56 mm minimum), fully- or partially-cored, often inclined boreholes (Figure 8.9), and is similar in concept to investigation programmes in the mineral assessment industry, or the route followed by Nagra and Nirex. These two organisations, and in particular Nirex, tend towards fully-cored, larger diameter, vertical boreholes (Figure 8.7), in which standard oilfield geophysical sondes can be run to the base of the borehole (160 mm minimum). In the particular case of Sellafield, where thick sedimentary sequences and the potential problems of gas, made this type of borehole an imperative. Nirex also employed this drilling technique at Dounreay, and have continued to drill large diameter boreholes at Sellafield. The difference in approach has important consequences in terms

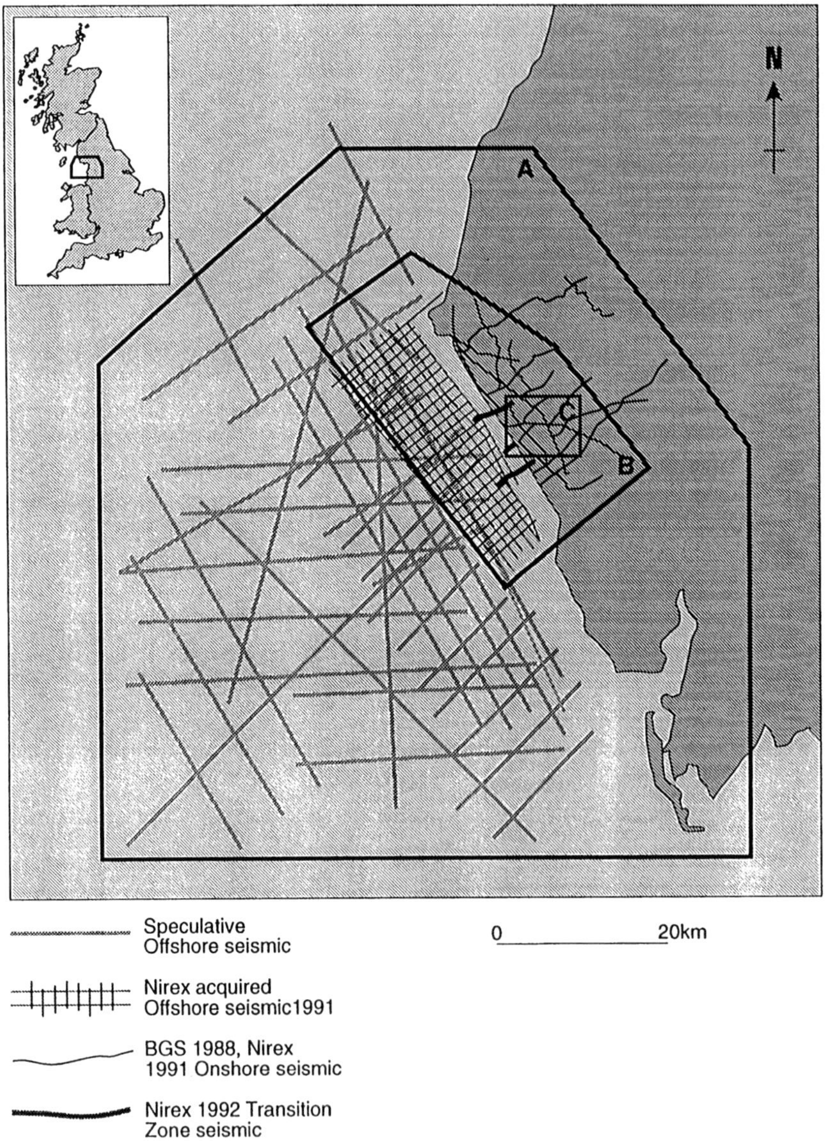

Figure 8.8 Investigation areas around Sellafield: Area A defines the Region-, Area B the District- and Area C the Site-scales (from Nirex, 1993). Reproduced by permission of U.K. Nirex Ltd.

of cost and the time taken to obtain the data:

(a) Fully cored, larger diameter boreholes can provide extremely high quality core to the base of the borehole. Their large size causes problems for their subsequent hydraulic testing, since the volumes of fluid involved are many times greater than those in small diameter boreholes. Larger boreholes will take longer to recover from the drilling activities, and recovery times at depth in low permeability rocks can be very protracted. Standard hydraulic test equipment, of the type routinely employed in the oil industry, is readily available at these diameters, but this equipment is only suitable for initial, somewhat crude hydraulic testing and groundwater sampling. More sophisticated test equipment is normally specially constructed for the purpose, and there is considerable experience of constructing such equipment at small diameters. The cost of a borehole completion also increases rapidly with the borehole diameter.

(b) Such large diameter boreholes are expensive and relatively slow to drill, especially in hard rock at depth. The requirements for full, high quality core and the necessity for running sophisticated borehole geophysical sondes, such as the FMI (Formation MicroImager), to the base need, therefore, to be balanced against the detrimental aspects of such boreholes. A borehole provides only one line source of data, and the potential advantage of having more boreholes, with less detailed information in each borehole, must be compared with more detailed, but potentially isolated, data available from fewer boreholes. In the early stages of the programme the data density will, however, be sparse, regardless of the design of the drilling programme.

2. Fracture zones within both hard rocks and sediments often possess relatively high dips and therefore inclined boreholes could be required to intersect them at acceptable angles for hydraulic testing (*e.g.* Figure 8.9 for boreholes on the island of Äspö (from Wikberg *et al.*, 1991). Drilling only vertical boreholes also introduces the problem of bias in the intersection of fractures and the use of fracture statistics for network modelling and underground design. It is much easier to drill inclined boreholes with the relatively small rigs designed for mineral exploration than it is to deviate boreholes in hard rock when using a large rig. It is also not possible to core whilst deviation of the borehole is taking place, whereas smaller rigs can, if required, be inclined from the surface. Small, quiet electrical rigs are also much less intrusive, and this could be significant in areas in which the local population are sensitive to the site investigation activities.

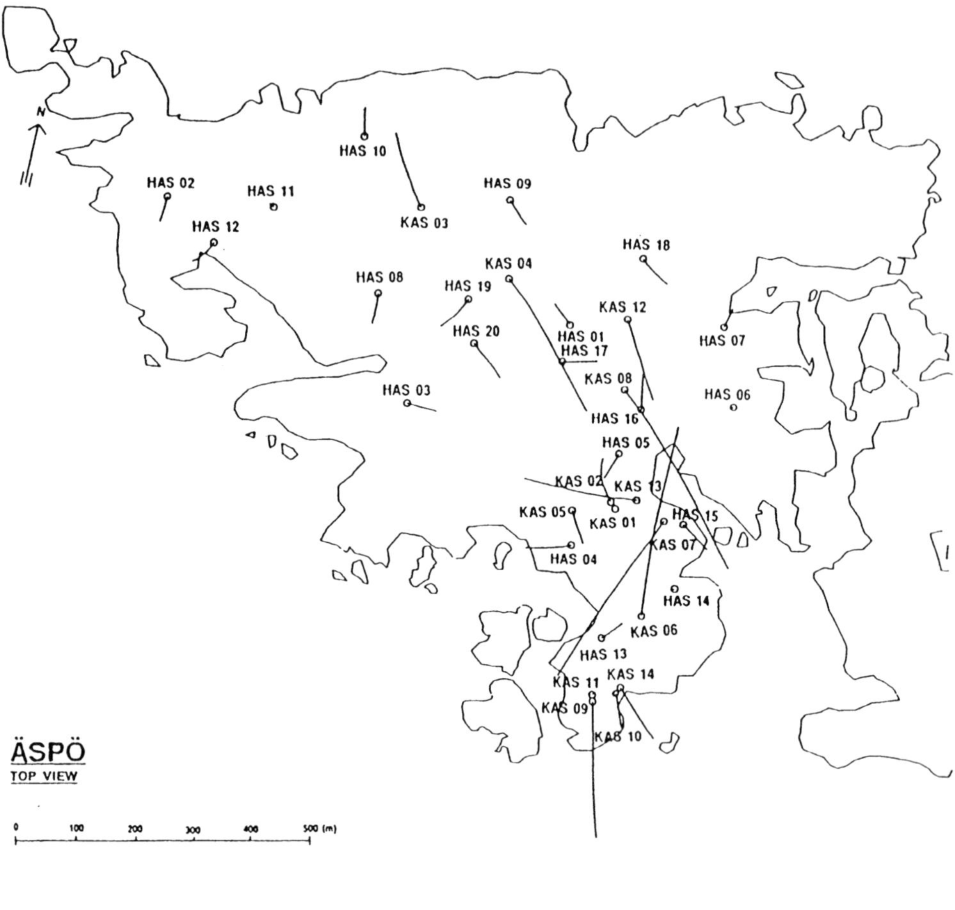

Figure 8.9 Boreholes, mainly inclined, on the island of Äspö. The traces of the boreholes in the horizontal plane are shown (from Wikberg *et al.*, 1991). Reproduced by permission of SKB.

8.5.2 Later stages of the investigation

Having defined the major geological structures of the site and the surrounding area and, having obtained an initial understanding of the hydrogeological environment, some indication of the geological, and hence the hydrogeological, complexity of the site should become apparent. It is this single factor that may be of most significance in the development of the investigation and the future research programme, as it may prove practically impossible to reduce the uncertainties inevitably associated with a complex geological environment below certain levels. Assuming current PRA practice, these uncertainties will be transferred through to the 90 % confidence intervals of the groundwater transit times. This factor is of prime importance within the assessment of a proposed repository site. The best published examples of this approach to estimating the complexity of the geological, hydrogeological and geotechnical system at various scales is probably the predictive work associated with the Swedish HRL, where driving of the inclined access tunnel is complete and the spiral access ramp is under construction (*e.g.* Gustafson *et al.* (1991) for the inclined tunnel and SKB (1993b) for the spiral ramp) and the current work by Nagra in their investigation of the crystalline basement of northern Switzerland (Thury *et al.*, 1994, Mazurek, 1994).

In potentially complex sites it will not be possible, and it may in any case be unnecessary, to measure the hydraulic properties of each of the major fault zones and other significant hydrogeological features. If several of the faults could be shown to be structurally related, in that they formed synchronously in the same stress regime, have similar displacement vectors and displace the same formations, it may be possible to argue, and demonstrate from sensitivity analysis, that only a representative selection of such structures needs examination. If this technique could be used, then the investigation could be made less complicated, and the eventual uncertainties placed within adequate limits. Even so, if there are several different types of hydrogeological discontinuity, each with properties that require assessment, the size of the necessary investigation may still prove to be impracticable. Any site investigation that required the investigation of only a selected, but representative, number of geologically and hydrogeologically significant structures, would itself require some form of validation. Validation of this process of simplification would preferably require the blind prediction, within an agreed error range, of the location, geometry and hydraulic properties of a geological structure that was believed to be hydrogeologically significant. The simplification process could not be justified, even at a research level, until this validation had been achieved.

The complexity of the site may be represented by large spatial variabilities, rather than by large discontinuities. All rocks are heterogeneous and, therefore, one accurate measurement of a given property of a particular rock is insufficient for providing input to a safety case. It is necessary to make several accurate measurements of the same property to quantify the form of the spatial variability and, even having made these measurements, there will still be uncertainty in the effective values that should be used. In this context, it may be possible to construct a superior and more robust model for a site that is highly predictable, and hence spatially uniform, but which is never likely to exhibit very long groundwater transit times than for a site that exhibits spatial variability, but which could perform in an excellent fashion. Any validation that is carried out as part of the site characterisation programme will prove much easier at a predictable site.

8.6 The hydrogeological environment

The hydrogeological investigation at a proposed repository site is likely to comprise three phases, the first phase, one of reconnaissance having been described above. The second phase is likely to involve targeted testing of geological and hydrogeological features that have been detected during the first phase and which appear, from the modelling, to be significant in determining the groundwater flow regime. The third phase will involve testing on the scale of the repository, and is likely to include testing during shaft construction and within a URL. There are likely to be changes in emphasis between hydraulic testing and hydrochemistry as the programme proceeds, and the type of hydraulic testing will also change, from essentially exploratory to more detailed testing required for near-field models.

Hydrogeological investigations need to provide data on the hydraulic properties of the rock mass which can be used in mathematical models of groundwater flow and radionuclide transport. Basic hydraulic parameters include the hydraulic head distribution, hydraulic conductivity, total and effective porosity, storativity and dispersion and other transport characteristics.

8.6.1 First phase of the investigation

In the first phase of the investigations, hydraulic parameters will probably have to be estimated for large parts of the rock mass and only their gross trends can be determined. There needs to be a close relationship between the types of groundwater flow model being used and the data that are required by the models in order to reduce uncertainties below certain critical levels. This level of uncertainty can be translated from the groundwater flow modelling, through transport modelling to probabilistic analysis in order to calculate uncertainties in risk. The first phase of an investigation in an area may involve only a desk study of the geological structure in order to produce a vertical cross-section, together with predictions of the likely hydraulic properties of the rocks involved. This information is then likely to be used to set up a 2-D (and perhaps simple 3-D) porous medium model of groundwater flow. Uncertainties in the input data, which are themselves only educated guesses based on what are believed to be similar hydrogeological environments which have already been examined, will tend to lead to large uncertainties in the model answers.

At the end of the first phase of the investigation, although the geological structure of the site will be known, and not assumed, there will not be very many more data that can be used for modelling and validation purposes than were available for pre-investigation modelling, and the uncertainties in these data may be as large as the uncertainties applied to the assumed parameter values. This has found to be the case after the initial investigations at Sellafield (Nirex, 1992), though the more extensive investigations at Äspö have allowed many estimates to be made of the groundwater flow system prior to the driving of the inclined tunnel (Gustafson *et al.*, 1991; but see the discussion in Olsson *et al.* (1992) on the amount of site investigation carried out for the access tunnel). The way in which the field data at the Äspö HRL have been used in the formulation of models to allow simulations of the fracturing and examine the influence of heterogeneity, anisotropy and uncertainty due to the fractured nature of the rock mass and to study how this influences groundwater flow is illustrated in Figure 8.10 (Axelsson *et al.*, 1990). The use of such modelling approaches could allow predictions of groundwater flow to be made on the scale of several tens of metres, *i.e.* at the scale of structures that could be of interest during shaft sinking and tunnel driving. There will, however, inevitably be a conflict between the requirements of construction and those of scientific research, a problem which is difficult to resolve.

The same situation is likely to exist in the investigation of any geological environment, though the nature in which the uncertainities are resolved and the methods used for the site investigation will vary.

8.6.2 Second and third phases of the investigation

As data begin to be collected from the site, although uncertainties in the geological structure will decrease, there is unlikely to be a parallel decrease in the uncertainty in the model results, at least until a large amount of data has been collected and analysed. With the collection of more data, a 3-D, or perhaps pseudo 3-D (*i.e.* 2.5-D), model of the site may be set up. Although this model is more complex, it is still likely to model any rock mass as an equivalent porous medium, the main change being that the geological structure and its influence on the groundwater flow can be evaluated in a more realistic manner, *i.e.* the hydrogeological discontinuities can be specifically modelled. The hydraulic data needed for this model are likely to be very similar to

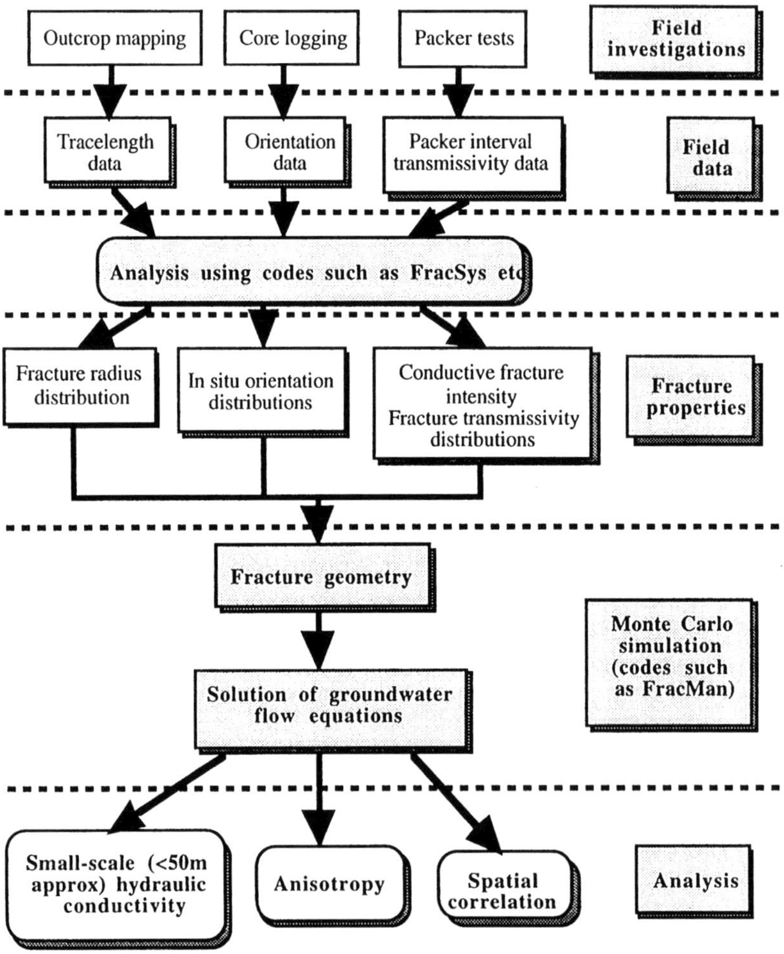

Figure 8.10 Outline of the modelling approach that has been followed at the HRL at Äspö, where 3-D discrete fracture flow modelling has been used at a coarse scale of approximately 50 m to simulate and examine the influence of heterogeneity, anisotropy and uncertainty due to the fractured nature of the rock mass and to study how this influences groundwater flow (From Axelsson *et al.*, 1990). Reproduced by permission of SKB.

those that were suggested for the 2-D model, with the addition of data on the most significant hydrogeological inhomogeneities, which in many geological environments are most likely to be fault and fracture zones.

Figure 8.11 illustrates the types of hydrogeological and geological work that are part of this phase of the site characterisation programme and the way in which they feed into the development of geological and hydrogeological models of the site and the region. Figure 8.12 shows the conceptual model of the groundwater system in the Sellafield area based on hydrogeological data derived from the boreholes shown in Figure 8.7.

A detailed investigation is likely to be required during the shaft sinking operation, and as part of the URL-phase, in order to investigate the effect of shaft sinking on the surrounding rock mass and to examine near-field transport, and this type of more detailed information at the local-scale may be more appropriately used as input to 3-D fracture network models, in the case of crystalline rocks. Each level of the investigation must therefore be carefully planned and co-ordinated to achieve a specific target by the matching of the hydraulic testing to the requirements of the models.

At this stage of the investigation there is a need to define the general properties of the hydrogeological system and determine the hydrogeological factors that could make the site or area unsuitable for long-term disposal. The most obvious factor in this regard would be the presence of young groundwater at depth, *i.e.* within, or close to, the depth range potentially suitable for disposal purposes. Relatively simple groundwater modelling, combined with geochemical studies of the groundwater should allow initial estimates to be made of the likely suitability of the site. It is, therefore, important for this and other reasons that geochemistry is made an important part of the investigations at the earliest possible stage. Useful discussions on the use of geochemistry in radwaste disposal programmes are provided in papers such as those from Andrews *et al.* (1988) and Gautschi and Scholtis (1989), and more recently by Vomvoris *et al.*, (in prep.) and Davis and Nordstrom (1992).

A smaller scale hydrogeological model is likely to be required, at the scale of several km^3 of the rock mass around the most favoured location for the repository.

At the end of the first phase of the investigation, although the geological structure of the site will be known, and not assumed, there will not be very many more data that can be used for modelling and validation purposes than were available for pre-investigation modelling, and the uncertainties in these data may be as large as the uncertainties applied to the assumed parameter values. This was found to be the case after the initial investigations at Sellafield (Nirex, 1992), and the more extensive investigations at Äspö have not allowed many precise estimates to be made of the groundwater flow system prior to the driving of the inclined tunnel (Gustafson *et al.*, 1991). If a geostatistical approach to the problem is to be followed, rather than a

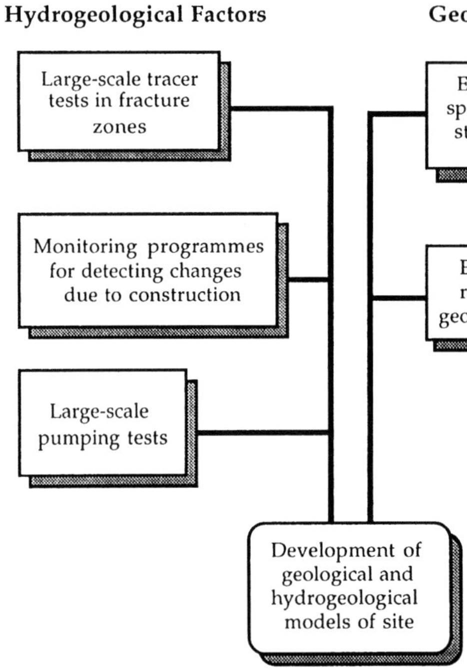

Figure 8.11 The types of hydrogeological and geological work that are part of the later phases of the site characterisation programme and the way in which they feed into the development of geological and hydrogeological models of the site and the region.

more novel technique, such as fuzzy set theory (Shaw and Grindrod, 1988), then a Bayesian statistical method, with an *a priori* assumed distribution of parameter values, based on expert judgement, may be the most useful method of proceeding in this data-sparse environment. The data that are being obtained from the investigation can then be incorporated into the existing distribution. It does not seem sensible to ignore potentially useful expert judgement in place of a few *real* values (Freeze *et al.*, 1988). This approach will also allow validation exercises to take place at an earlier stage that would otherwise have been the case. Nirex have recently followed this approach by the use of data elicitation panels in order to generate a larger volume of hydrogeological data for the initial assessment of Sellafield, and this route was also taken by the U.K. DOE in their Dry Run 3 exercise, in which the Harwell site was used as a model (HMIP, 1992).

It is often assumed that the spatial variations in hydrogeological and geological properties form a correlated stochastic process (de Marsily, 1986),

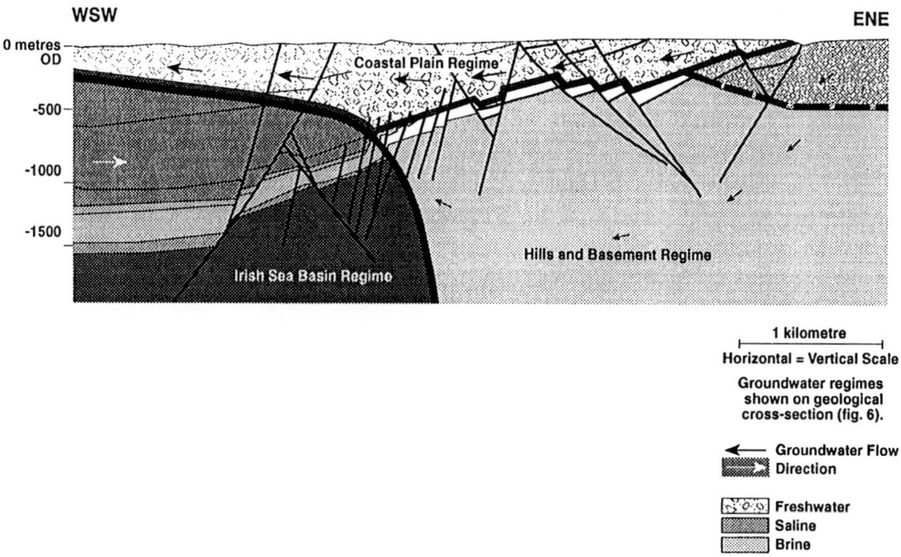

Figure 8.12 Schematic representation of the groundwater system in a 2-D vertical section through the Sellafield area (as of December 1993) (from Nirex, 1993). Reproduced by permission of U.K. Nirex Ltd.

although there is unlikely to be a homogeneous distribution of hydraulic conductivities, with one correlation scale. In order to quantify the spatial variability, the measurement points must cover a region comparable with the length scales of the correlations. If the measurements are too closely spaced, and much closer together than the correlation length scales, a spurious lack of variability will be detected. If the measurement points are too widely spaced, then the variability can be detected but not the correlation structures. The variability of geological and geochemical data is commonly assessed in such areas as mineral exploration and the design of underground openings, and a similar geostatistical approach may be the best method of analysing these types of data. Validation exercises, which measure the predictive capacity of the models derived from these data, will allow a measure of the success of this approach to be assessed.

The previous arguments illustrate that the site investigation has to be approached in an iterative fashion, that the location and number of boreholes have to be considered as a way of decreasing uncertainties and that sufficient data have to be collected to account for the spatial variability and to allow some form of validation, similar to that currently being carried out at the

Äspö HRL (*e.g.* see Gustafsson *et al.*, 1991 for the predictions up to 1990), to proceed as early as possible. The number of deep boreholes is therefore impossible to predict accurately in advance, nor is it possible to calculate the number of hydraulic tests that will be required or the number of groundwater samples that are likely to be analysed.

The iterative approach that is required for the site investigation influences the approach that needs to be followed by any validation exercises that are carried out. In order to demonstrate that the site investigation is in fact reducing model uncertainties, validation exercises will need to be performed to demonstrate that the predictive capacities of the geological and hydrogeological models are improving. The siting of new boreholes, and their testing and monitoring programmes, need to be controlled as much by the need to reduce the parameter uncertainties as by the need to test the current model and provide some measure of validation.

In hard rocks and in also some well indurated argillaceous rocks (*e.g.* the Opalinus Clay of northern Switzerland, the Lower Lias of eastern parts of the U.K.), where groundwater flow is controlled, to a greater of lesser degree, by the transmissive fracture, or fault, network, determining the geometry of the network may be the single most important aim of the site investigation, at least in its initial stages. The detection of fracture zones, which may or may not be significantly more transmissive than the surrounding rock, and the effect that such zones have on the groundwater flow field, may offer the first opportunity for validation, if sufficient is known about the site. In other geological environments heterogeneities, such as pinch-out structures, sandstone lenses and anhydrite beds within halite formations (included in the safety assessment for Gorleben, see also the PAGIS Salt Study (PAGIS, 1984)), can be considered in a similar manner. On the other hand, if little is known about the hydrogeology at depth, several of these zones may have to be investigated before the groundwater flow model can be used in any predictive capacity with regard to, perhaps, an additional fracture zone. By continually attempting relatively simple validation exercises, the confidence in the groundwater flow and geological models, and hence their predictive capacity, can be estimated (this has been attempted by Nagra in their modelling of the crystalline basement of northern Switzerland [Niemeyer *et al.*, 1993]). This can also be attempted by sensitivity analyses in trial performance assessments. This will enable rational decisions to be made regarding the emphasis that should be placed on specific aspects of the investigation and, by implications, the R&D programme. The requirements for validation can therefore help control the direction of the site investigation, and the influence of the requirements for validation on the URL-phase of the programme are discussed in Section 8.14.

8.7 Hydrogeological testing systems

In order to determine the type of testing programme that is necessary, the use to which the data will be put and their accuracy need to be determined. At this stage of the investigations it may be decided that only data from discrete zones are required, with the assumption that the remainder of the rock mass has uniform, or at least predictable, properties, and that any features that are likely to be of hydraulic significance will be detected before testing commences. This strategy is commonly used in the oil industry, but relies on numerous assumptions, which may not be acceptable in terms of the disposal of radioactive waste. In boreholes drilled in unstable sediments, which may require casing immediately after geophysical logging, hydraulic testing will have to be performed through the casing in perforated sections, or in screened sections. Similar problems can arise if the borehole is drilled into rocks where excess gas or oil pressures can be anticipated. In these situations, the testing programme becomes more restricted, because of the overriding need for pressure control measures, and it may be impossible to carry out complex fluid logging or a comprehensive hydraulic testing programme covering a substantial portion of the borehole. The deep boreholes at Sellafield, for example, drilled by Nirex have all had full pressure control systems, so that limitations have been placed on the hydraulic testing and groundwater sampling, at least during the drilling phase itself.

Fixed length testing has been applied for a considerable period in the site investigations for radioactive waste disposal by SKB in Sweden (Carlsson *et al.*, 1988) and has also been used extensively by Nirex at Sellafield in the UK (Nirex 1993). In SKB's reconnaissance surveys the hydraulic conductivity of a borehole was determined using long packer spacings of as much as 50 m (the length also used by Nirex, see Figure 8.13), but more commonly 20 m. This procedure yielded averaged values of hydraulic conductivity. A fixed spacing of as little as 1-2 m was subsequently used by SKB to isolate individual fractures and reveal the heterogeneous nature of the fractured rock mass. This technique was used to determine the statistics of the hydraulic properties of the fracture zones within the rock mass. Initial testing of the complete length of the borehole prevents significant hydraulic features being missed, but the process of using a small packer spacing over the whole borehole length results in high costs and long testing times, as the times required for testing increase more rapidly than the increase in the reciprocal of the packer spacing.

A more cost effective approach to obtaining the hydraulic data needed has been proposed by Black (1988). He describes his approach as focused packer testing, and requires an *active packer string* in which there is the opportunity to change the length of the packer string without removing the whole test string from the borehole. This technique can focus the testing on the more transmissive zones of the rock mass, normally the ones that are of most

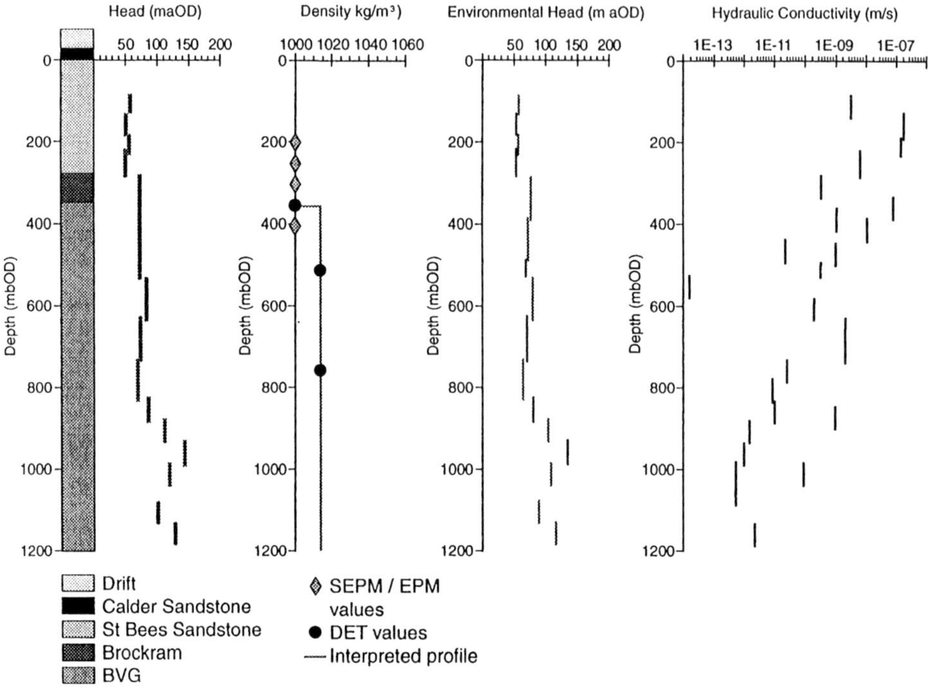

Figure 8.13 Hydrogeological data obtained from one of Nirex's deep boreholes at Sellafield. Hydraulic conductivities and heads have been measured over 50 m lengths of borehole during drilling (from Nirex, 1993). Reproduced by permission of U.K. Nirex Ltd.

interest, and reduce unnecessary testing, without having previously to have made irrevocable decisions on testing strategy. It requires that the tester has previous knowledge of the use to which the data are to be put, and some knowledge of the likely distribution of hydraulic properties of the rock mass under examination. It has been successfully used at Stripa, where time savings of up to 50 % were achieved. A recent review of hydraulic testing methods in fractured crystalline rock (Andersson *et al.*, 1993) critically reviewed the methods available for obtaining the required hydrogeological data. This review discussed the following areas of interest:

- Relevant hydraulic attributes associated with the characterisation of a site and the general principles associated with their measurement.
- The hydraulic and hydrochemical methods used in site characterisation.

- Conceptual problems and experimental errors associated with hydraulic and hydrochemical measurements.
- Suggestions for further improvements in these methods and the development of new measurement strategies.

The improvements suggested for future work within crystalline rocks covered various areas, including borehole logging methods, tracer test methods, groundwater sampling techniques, measurements of hydraulic heads and revisions to measurement strategies.

In this stage of the investigation, testing is likely to be carried out only in single boreholes, with the more sophisticated inter-borehole testing being reserved for the later stages in the programme. The testing equipment can be run on wire line, on an umbilical and on rods, and techniques have been developed over the last few years in both the fields of waste disposal and geothermal energy and in the oil industry. Several of the techniques originally developed for use in the oil industry have been modified for use in crystalline rocks, and a useful overview of such developments and their use in the Swiss waste disposal programme is presented in Grisak *et al.* (1985).

In the oil industry the commonest tool is the Formation Tester. Measurements are made by forcing a cup packer against the borehole wall and monitoring the fluid pressure inside this packer during the test; fluid samples can also be obtained. Newer equipment enables the system to be set several times before it must be removed from the borehole, and this tool is called a Repeat Formation Tester (RFT). This tool is primarily designed for relatively simple measurement of reservoir properties, and in relatively homogeneous sedimentary rock, and is not really suitable in heterogeneous crystalline rocks, nor for sediments with lower permeability, except to provide rather crude, initial hydraulic data.

The majority of testing systems in boreholes use packers, which are now normally inflated hydraulically using gas or fluid pressure, to isolate a test section. Such systems are light enough to be run on wire line and have been designed to be used to depths of 2000 m in boreholes with diameters as small as 56 mm (*e.g.* Almén and Zellman 1991 (Figure 8.14)). Temperatures and pressures, both above and between the packers, are measured downhole and can be read at the surface. Most systems are designed to perform particular types of test, and can be used over a wide range of hydraulic conductivities, from perhaps 10^{-5} to 10^{-14} m s^{-1}.

Wire-line and umbilical testing systems can only be used in vertical or sub-vertical boreholes. In an investigation in a crystalline rock in particular, there is likely to be the need for inclined boreholes, and in these situations the packers must be mounted on drill rods. Almén *et al. (op. cit.)* describe a pipe string designed by the Swedish Geological Company. It has a variable configuration for use in rocks covering a wide range of fracture densities, and

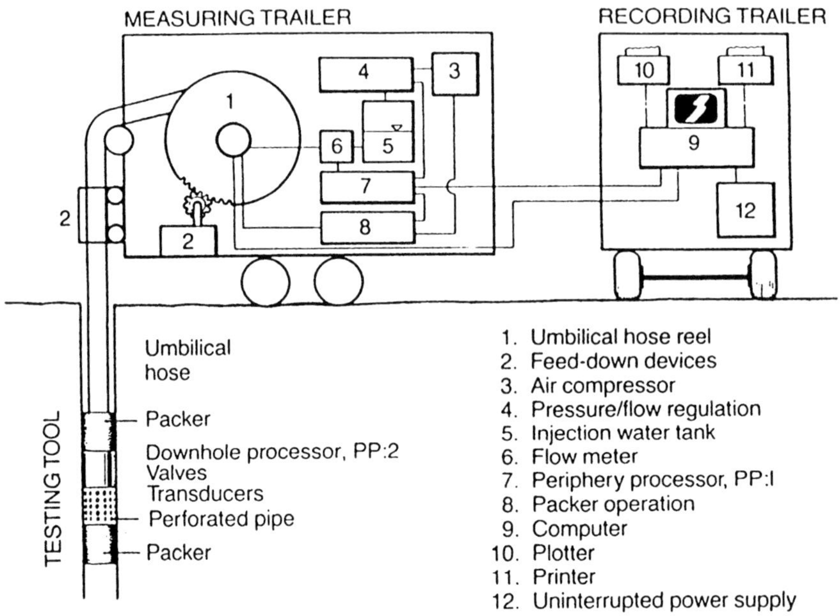

Figure 8.14 Schematic illustration of the umbilical hose system used by SKB to carry out a large number of hydraulic measurements at the HRL (from Almén and Zellman, 1991). Reproduced by permission of SKB.

for tests with variable durations.

The oil industry has used packers on drill string rods for many years, but often at much greater depths than those of interest in waste disposal, and in reservoir rocks with considerably greater hydraulic conductivities than those likely to be of most interest in the site investigation at a potential repository site. More modern systems, such as those described by Grisak *et al.* (1985) and used by Nagra in their deep boreholes in northern Switzerland, incorporate pressure and temperature measurement sensors within the test zone and both above and below the packers (Figure 8.15). They have also been modified to operate in lower permeability rocks and in smaller diameter boreholes, down to 75 mm.

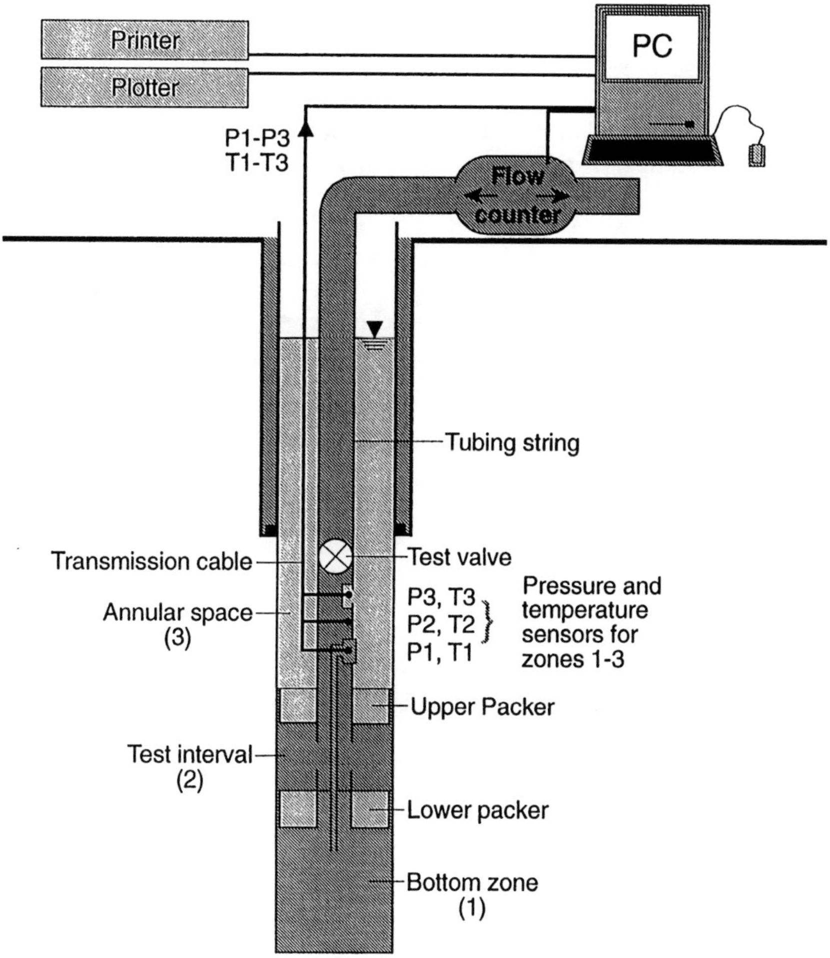

Figure 8.15 Schematic figure of double packer testing system, showing pressure and temperature sensors, test valve and data acquisition system (from Küpfer *et al.*, 1989). Reproduced by permission of Nagra.

8.8 Hydrochemical characterisation

8.8.1 Introduction

An important part of the hydrogeological characterisation of the site will involve the use of hydrochemistry to determine the compositions of the groundwaters, to relate these compositions to the mineralogy of the rocks and to determine the chemical evolution of the groundwater system over long periods of time (see Chapter 5 also). Theoretical arguments are very strong for careful geochemical investigations of sites. The chemistry of the groundwater and the secondary minerals in contact with the water bear an imprint of the present and past hydrogeological system. The imprint may, for example, contain time-related information going back hundreds of thousands of years. This information is contained within the hydrochemical system and does not require the use of the extrapolation of hydrodynamic transport models, which are inevitably dependent on short-term measurements. The limitations of geochemical interpretations are, however, numerous. Firstly much of the information is difficult and expensive to obtain and secondly the results of the analyses are not often definitive and often do not provide absolute numbers required for performance assessments. Errors introduced by poor field and laboratory techniques are probably much more common, though lower in magnitude, than with hydraulic tests and useful geochemical models are not available for the interpretation of some of the isotopic data as well as data on some trace elements and for most trace organics. This is particularly a problem where groundwaters have high salinities (which, for example, is the case at Sellafield (Nirex, 1992) and Gorleben, Germany (Langer *et al.*, 1991)).

Despite these limitations, geochemical work has two very important advantages, one of these is the fact that historical data can be extracted directly from geochemical studies without the uncertainties associated with the long-term extrapolation of hydraulic models, as mentioned above. The second advantage is that a large number of variables can be measured. In normal hydrogeological analysis commonly fewer than six variables are measured, with much of the required information (such as permeabilities) being derived from these variables. Geochemical studies are able to measure at least 100 dissolved constituents in groundwater and at least an equal number of secondary and primary minerals in contact with the water. Most natural systems are likely to provide in excess of 50 constituents which could be measured. Each of these constituents will provide some information on one or more pieces of the overall picture, namely:

- the residence time of the water,
- the thermal history of the water,
- the chemical characteristics of the water at the time of recharge,
- the trajectory of groundwater flow,

- the flow velocity,
- the chemical reactions that have taken place,
- the extent of mixing of different groundwater masses,
- the altitude at the which the groundwater was recharged and,
- the temperature at the time of the recharge.

Given several sampling points within an aquifer, it should be possible to reconstruct the long-term flow system, provided human interference has not been excessive. The use of hydrochemistry within the Stripa programme and the uses that should be made of hydrochemical techniques as part of a site investigation are presented in Davis and Nordstrom (1992).

8.8.2 Groundwater sampling

Introduction

The analysis of groundwaters from boreholes forms an essential element in the site investigation at a potential repository site, and to ensure the collection of representative samples the sampling procedures must be included in the initial design of the borehole programme. It is important that decisions regarding the design of the borehole and the drilling techniques used do not compromise, more than is absolutely necessary, the requirements for taking high quality groundwater samples. Samples can be taken at different stages during the drilling, and each stage has its advantages and disadvantages which must be evaluated in advance.

The field and laboratory tests on the groundwater provide information on the climatic conditions which prevailed during recharge, on the evolution of the groundwater system with time, and allow comparisons to be made between the transit, or residence, times calculated from the measured heads and hydraulic properties of the rocks with those indicated from the geochemistry. The characterisation of the hydrochemical and geochemical conditions is an important component of the safety assessment, as the data are of direct relevance to the behaviour of engineered barriers (Chapter 4) and to the calculations related to the behaviour of radionuclides in the rock mass and their migration rates (Chapter 5).

The groundwater samples providing data for these calculations should be as uncontaminated as possible. Ideally these samples should be at their formation pressure and temperature and be completely unaffected by the drilling fluid. In practice all the samples will be disturbed to a lesser or greater extent, mainly by the following processes:

- contamination with the drilling fluid and the drill cuttings,
- contamination with the lubricating greases (such as pipe dope) and with the borehole cement and its additives, such as accelerators and stabilisers,

- contact with the casing or the tubing and packers of the sampling and hydraulic testing equipment,
- contact with any atmospheric gases,
- pressure release with degassing and potential precipitation of solids,
- cooling of the sample which can cause precipitation and/or dissolution processes.

No single sampling technique is available that prevents all of these disturbances to the groundwater sample. Some effects, such as degassing, can be almost totally eliminated by the use of pressurised sampling systems and by transferring the sample under pressure to the analytical equipment, but such techniques tend to be expensive and time consuming and there are few suitably equipped laboratories capable of analysing what could be a very large number of samples to very high standards. Other disturbances are almost impossible to eliminate, but their influences can be minimised, and a well planned programme will attempt to obtain optimum samples by the use of several different techniques, probably in parallel.

Fluid samples can be sampled during the drilling, either by sampling the drilling returns at the surface or by obtaining samples down the borehole. Both types of sample provide a mixture of the drilling fluid and the groundwater, which will itself be a mixture of all the potentially different groundwaters that are present over the depth range of the open portions of the borehole. If air-flush drilling is being used then the groundwater samples will contain mixed groundwaters and are likely to be considerably oxidised. Regular sampling of the borehole fluid can still provide useful initial data on the chemistries and volumes of groundwater inflows and on the types of rocks which are being penetrated, especially if the groundwater at depth has a high salinity. The collection of such samples, and their rapid analysis on site has minimal cost implications and provides data as soon as possible.

Adding a tracer, or tracers, to the drilling fluid allows the contamination of the groundwater by the drilling fluid to be estimated quantitatively. There is no universally accepted tracer for use in a particular rock type, and a wide variety of different types of tracer has been used in different countries' site investigation programmes. For example, sodium nitrate has been used, as it is argued that nitrate is extremely unlikely to be found naturally at significant depths, and the chemical would not cause any problems when added to drilling mud. However, nitrate may be consumed *in situ* by microbially-catalysed redox reactions. Nagra normally use sodium fluorescein (uranine), for easy analysis, and m-TFMBA (meta-trifluoromethylbenzoic acid), for chemical stability, but the use of a fluorescing tracer in a borehole drilled with full hydrocarbon safety measures is likely to conflict with the use of fluorescing chemicals to determine the presence of hydrocarbons and would therefore not be suitable. Oil field and geothermal drilling in sedimentary basins has often used LiCl and KNO_3. The tracer is added directly to the circulating drilling fluid in the form of dissolved concentrates, and pumping of the drilling fluid tends, and is assumed, to ensure an even dispersal (LiCl is being used by Nirex as a tracer at Sellafield).

After the borehole has been drilled, or in certain circumstances when the first part of the borehole has been drilled, some form of completion will be required, the complexity of which depends on borehole stability considerations and on the requirements for possible future sampling and testing. Sampling can take place before and after this completion, which will vary dramatically depending on the geology intersected by the borehole. In hard crystalline rocks, the borehole is likely to remain open for a considerable period with no support, except for a relatively small section near the surface, and its entire length will remain open for testing and sampling. In many sedimentary rocks the borehole is not likely to remain stable and casing will be required. Access to parts of the rock mass can be made available by the use of screened sections, or by perforations in the casing, though this obviously restricts the possibilities for further sampling and testing.

In sedimentary rocks in particular, where drilling mud is likely to have been used, the borehole should preferably be cleaned before any sampling is attempted. Where safety considerations allow, the borehole will need to be flushed with clean fluid and the mud cake, if any, removed. It is common practice to run geophysical logs before any form of groundwater sampling or hydraulic testing is attempted. Several of these logs can be interpreted to indicate zones of groundwater inflow, and the initial sampling programme is likely to be based on these data. In suitably porous and permeable sedimentary rocks, such as sandstones, a Repeat Formation Tester (RFT) can be run on the logging cable to take fluid samples, though the quality of such samples is not very high because of contamination by the remaining mud cake or by drilling fluid invasion.

More sophisticated sampling devices, which use packers to isolate specific sections of the borehole, require much longer to set up and operate, and large parts of the borehole could remain open for long periods, with the subsequent mixing of the different groundwaters. If this is considered a significant problem, bridging packers can be inserted in order to isolate specific parts of the borehole and prevent mixing. Such packers are commonly used in the oil industry, where several can be installed in the same borehole using a pipe string, and can remain inflated for periods of up to one year. Bridging packers have been used in the Canadian waste disposal programme as part of their site investigation in the East Bull Lake Pluton (Bottomley *et al.*, 1986) and could be considered for the temporary completions of any deep boreholes whilst decisions are made on their future use.

In some of the site investigation programmes for waste disposal, sampling of the groundwaters has taken place a considerable time after the boreholes were completed. Smellie *et al.* (1985, 1987) describe the situation at the several SKB site investigations and also at the Klipperås test site. Bottomley *et al.* (1986) describe the situation in Canada, and Pearson *et al.* (1989, 1990) the situation in Switzerland. The intervening period has normally been taken up with detailed hydraulic testing, so that information is then available on the head distribution and on the positions of significant inflow within the borehole. Sampling of the groundwater can then be carried out in a more controlled manner.

Sampling in these boreholes has either taken place by isolating portions of the borehole with straddle packers or by the use of permanent completions. Groundwater has been sampled mainly in zones of higher hydraulic head, where the contamination by the drilling tends to be minimised, and tracers used in the drilling were used to check for contamination levels. Zones of low hydraulic head at the Canadian URL site were found to remain contaminated with drilling fluid even after extensive periods of pumping, and the measurements of the groundwater chemistry tended to be very variable (Davison, 1985).

There are several different types of sampling equipment for use in a borehole:

- borehole or well bore samplers, which sample the fluid in the borehole,
- formation fluid samplers for use in uncased boreholes,
- formation fluid samplers for use in cased boreholes.

The type of sampler used will depend on the application to which the sample is to be put. In the initial stages of the site investigation and, in particular, during the borehole drilling, it is helpful to be able to obtain samples of groundwater, almost regardless of its quality, so that general ideas regarding the groundwater chemistry can be developed. In the later stages of the investigation, the best possible samples will be required for the analysis of environmental isotopes.

Borehole fluid samplers

These types of samplers are commonly used in the water supply and oil industries, and can vary from the simple grab (Casalla type) sampler, which takes a sample from close to the top of the borehole, to a pressurised fluid sampler which operates on a wire line to depths of several kilometres. Grab samples are also available that allow an 'average' sample to be collected as the sampler is lowered into the borehole, by controlling the rate at which the fluid flows into the container. More sophisticated samplers incorporate valves that are operated from the surface, and these allow samples to be collected at specific depths.

There are several other pressurised sampling systems, with varying degrees of sophistication and reliability, which are normally used on wire-line. They tend to yield small samples (up to a maximum of 1-2 litres) and, if used correctly, can preserve unstable parameters, such as redox, and obtain representative samples of dissolved gases. There is a limited range of commercially available samplers which are probably not sufficiently reliable for use at anything more than shallow depths; and there are purpose built research tools, developed in Sweden and Canada for radioactive waste research (*e.g.* Davis and Nordstrom, 1992), which are notably superior.

Formation fluid samplers

Samplers for use in uncased boreholes have been developed for the oil and water industries, though these tend to be suitable for obtaining only relatively low quality samples, and also for radioactive waste disposal investigation programmes, where sophisticated sampling systems are currently in operation.

The RFT, which is commonly used in the oil industry to sample formation fluids within reservoirs, can only be used successfully in sedimentary rocks, and even in these cases tends to sample a mixture of formation and drilling fluids. Its small packer size precludes its use in fractured hard rocks, as contamination of the resulting sample would be too severe.

Sampling from deep boreholes often requires the use of inflatable packers to isolate a section of the borehole, from which formation fluids can be abstracted using a variety of means. A system, which was designed for use in downhole tracer tests, but which incorporates an electro-submersible pump adjacent to the packers, is described in Noy and Holmes (1986). Other systems, which have been used in the Canadian waste disposal programme, incorporate air lifting or swabbing systems inside the drill pipe (Bottomley *et al.*, 1986). More sophisticated systems are now commonly used, as the importance of obtaining high quality groundwater samples has increasingly been appreciated (*e.g.* Almén and Zellman, 1991). The current groundwater monitoring and sampling network at Äspö is shown in Figure 8.16)

All the systems just described suffer from the same problem, that the fluid collected at the surface is at a much lower pressure than that existing at depth, thus allowing the fluid to degas. Degassing can result in changes to the pH and Eh of the fluid, which in turn can alter some of the chemical measurements. The most significant change, especially in groundwaters from crystalline rocks, where the system is poorly buffered, is likely to be the degassing of CO_2, with a concomitant change in the carbonate content. This will also result in an increase in the pH and a change in the CO_2/HCO_3^- ratio, and the saturation index for calcite will become positive, *i.e.* the samples are apparently oversaturated with respect to calcite, which is assumed to be improbable. However carefully the sampling is carried out, some degassing is likely to occur, and it is normal practice to correct the CO_2 and the

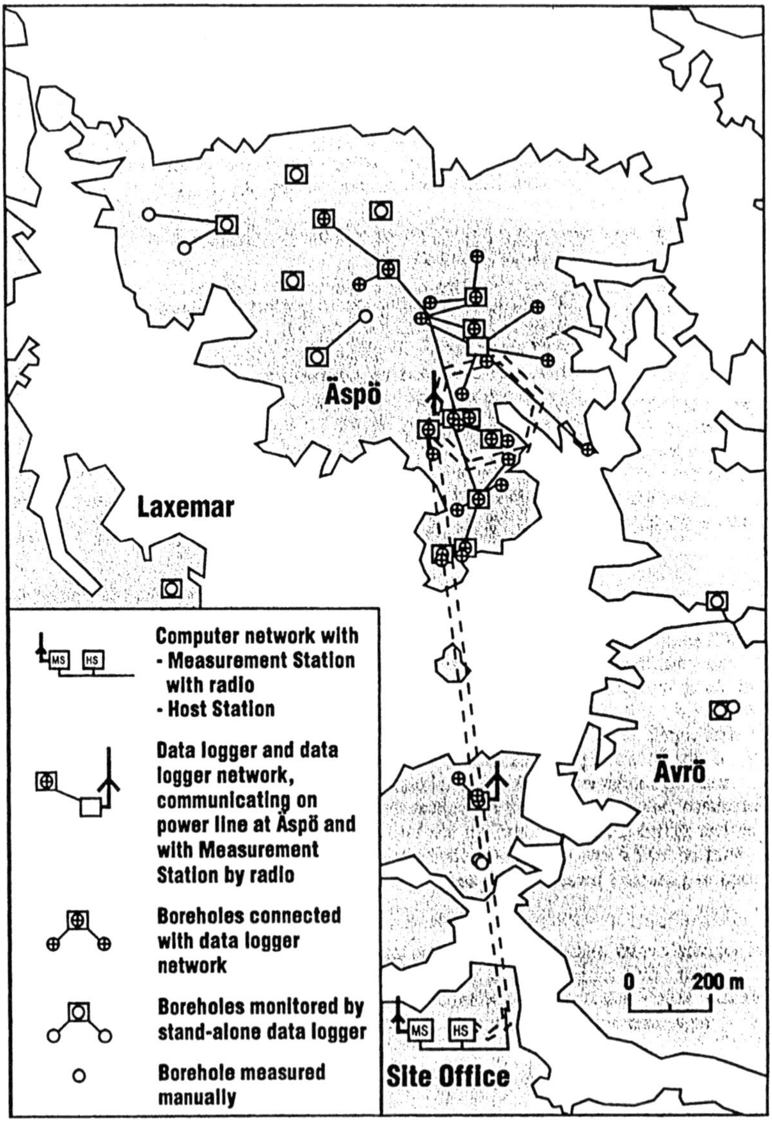

Figure 8.16 Groundwater measurement points and sampling network at the Äspö HRL (From SKB, 1993). Reproduced by permission of SKB.

HCO_3^- contents of the groundwater by using a reaction-path program, such as PHREEQE (Parkhurst *et al.*, 1980). Smellie *et al.* (1985) discuss the problem of degassing in relation to the effect it has on the complexation of actinides, and Wikberg *et al.* (1987) have conducted experiments on the measurement of groundwater at depth and on the surface, and recommend that pH and redox should be determined *in situ* wherever possible.

The most sophisticated sampling and measurement programmes have probably been performed by SKB in Sweden, in the Canadian programme and by Nagra. SKB have developed a complex geochemical testing and sampling system for use in the site investigations in Sweden (Almén *et al.*, 1986a, b). As much as possible is measured and sampled downhole as quickly as possible, in particular the sensitive parameters pH and Eh. Although very sensitive components have been used in the construction of this equipment, there have been various technical problems, in addition to those that one might expect if measurements were being performed at the surface. It appears that it is only in certain circumstances that the effort to measure parameters downhole can be justified. In crystalline rocks, the best example of this being the non-equilibrium Fe^{3+}/Fe^{2+} reactions in the rock affecting the redox potential.

The Swedish sampling programme has highlighted the problems of cleaning out drilling fluid from the formation prior to sampling. In some cases several months of pumping have been required before reliable samples could be obtained, *i.e.* before the drilling fluid tracer concentration dropped below some pre-determined limit, although the decision in the Swedish programme adequately to determine the hydrogeology of the borehole before any sampling took place, can only have exacerbated this problem.

Extraction of fluids from core material

In low permeability rocks, very long periods of sampling may be required before samples of acceptable quality can be obtained. Many of the sampling methods also extract fluids over several metres of the borehole, and therefore tend to smooth out any changes that might be apparent were sampling to take place from a shorter length. In sedimentary rocks, fluid chemistries can change in response to any rapid changes in lithology, and such chemical changes can be significant in developing a model for the evolution of the groundwater system.

Extraction of pore fluids from sedimentary rocks has been attempted on samples with several different lithologies during investigations in the U.K. This technique is only applicable to rocks with relatively large porosities, and therefore cannot be used on crystalline rock samples, and requires that the contamination of the fluid within the core can be determined. Tracers should preferably have been used during the drilling, and the core needs to be well preserved as soon as possible after it has left the core barrel, to prevent any oxidation. The core will require sealing in wax and storing at a temperature of 4 °C to prevent bacteriological action. There are four methods of fluid extraction:

- **Squeezing the core.** Rocks of low permeability and relatively low strength, such as most mudrocks, can have their fluids removed by squeezing the sample in a suitably designed cell. This technique has been successfully utilised by the BGS in the Nirex site investigations at the shallow sites of Bradwell and Fulbeck, on argillaceous core of interest to the Swiss waste disposal programme and on core from the deep boreholes drilled at Harwell (Brightman *et al.*, 1985). BGS have developed temperature controlled cells in which the mudrocks are squeezed at a constant load, and squeezing is also possible in anaerobic conditions. There appears to be a minimum moisture content for a mudrock of about 7 %, below which squeezing at several tens of MPa does not produce sufficient fluid for analysis.
- **Centrifuging.** This method was used extensively throughout the low enthalpy geothermal research programme in the U.K. on several of the potential reservoir rocks, and its application in such rock types is outlined in Edmunds and Bath (1983). A small piece of the core is placed in a centrifuge and is spun at high speed. The fluid removed from the core depends on the spin rate and the pore size distribution of the sample, and the technique is suited to rocks such as the Chalk and to porous permeable sandstones.
- **Immiscible fluid displacement.** A variation of the technique described in (b) is the addition of a chemically inert, heavy and immiscible fluid to the sample in the centrifuge. This fluid displaces the fluid in the core during centrifugation and can increase fluid yields. This technique has been successfully used in the same geothermal programme, and its use is described in Kinniburgh and Miles (1983).
- **Leaching.** Solutes in the pore fluid are removed from the crushed sample by leaching in a known volume of distilled water. There are several different experimental procedures which can be used and this technique is applicable to most sedimentary rocks.

Brightman *et al.* (1985) have reviewed the efficacy of the different techniques, and it appears that in the more permeable rocks the centrifuge technique is likely to be the most appropriate choice, whereas in mudstones, squeezing will yield the best results. Profiles of the pore fluid chemistry in soft plastic and overconsolidated clays have been studied by BGS, and such profiles have been modelled assuming Fickian diffusion (Falck and Bath, 1989). The use of the techniques described above is likely to provide extremely useful data on the evolution of the groundwater system in sedimentary environments, but the interpretation of the data is often not easy.

The use of fluid samples

The requirements for the different uses to which the fluid samples are to be put vary substantially, mainly in terms of the different volumes needed. The requirements are divided into four sections: (a) detailed chemical analysis, (b) natural series U and Th radionuclide analysis, (c) environmental isotope analysis and (d) dissolved gas analysis.

1. **Detailed chemical analysis.** The volumes of groundwater required for the various chemical analyses vary substantially and have important consequences for the design of the sampling programme. A chemical analysis of the major parameters requires a minimum volume of approximately 20 ml of filtered sample, but it is desirable to obtain as full an analysis as possible, including trace metals and REEs, U and Th, and dissolved organic and other colloidal species. The extent of the analytical work increases with the sample volume available and, in particular, the detection limits for certain species may be lowered by having sufficient volume to allow for preconcentration by evaporation or by using an exchange column, *etc.* The minimum desirable volume is approximately 100 ml and this can rise to as high as 1-5 l for a complete analysis. This factor is very important as it is, in many cases, the trace elements, colloids, organics, U and Th and environmental isotopes, which are likely to be of most significance for use in the safety assessment and it is just these parameters that require the largest sample volumes. The problems associated with sampling techniques make it imperative that the raw chemical data are checked before any of the analyses are evaluated in detail. Particular attention should be paid to the extent to which the analytical results are consistent with the chemical and physical conditions at the depths from which the fluids originated.
2. **Natural series nuclide analysis.** U series disequilibrium studies require analyses of the distribution of U, Th and their daughter isotopes in as many solution and mineral phases as possible. Comprehensive analyses of the separate particulate, colloid and solution fractions may require samples of several litres, perhaps as much as 10 l, whereas the minimum sample volume is in the range 100-500 ml. Separation and analysis of some of the Ra daughters and Rn will require special sample handling on site, and special filtration will be required for the colloid fraction.
3. **Environmental isotope analysis.** Measurements of the naturally occurring isotopes in the groundwater provide information in several important areas which cannot be addressed sufficiently by the analysis of the chemistry alone. This applies particularly to the determination of the recharge conditions and the evolution and residence time of the groundwater. The concentration of the stable isotopes deuterium ^{2}H and ^{18}O in the water molecule provide information on the recharge altitude and

the palaeoclimatic conditions. Information on the geochemical evolution of the groundwater can be obtained by determining the stable isotopes ^{34}S and ^{18}O in the sulphate, and ^{13}C in the carbonate. Processes such as precipitation, dissolution, ion exchange and redox reactions can be analysed by studying these isotopes. ^{34}S, ^{18}O, the strontium isotopes, and the ^{4}He concentration, in combination with the ^{3}He/^{4}He ratio, can be used to obtain information on the flow path of the groundwater during its evolution, by providing data on the types of geological formations through which the groundwater passed. Residence times for groundwaters can be provided by measuring the isotopes ^{14}C, ^{39}Ar, ^{3}He, ^{85}Kr and ^{36}Cl and the ratios of ^{18}O/^{16}O and ^{2}H/^{1}H.

Tritium (^{3}H) and ^{14}C provide the most quantitative data on absolute groundwater ages. The former, with a half life of 12.4 years, can be used to detect very recent groundwater ingress and drilling fluid contamination, and ^{14}C, with a half life of 5730 years, can be used for dating groundwater with residence times up to 30 ka. Drilling fluids are likely to contain approximately 90 tritium units (1 tritium unit = one ^{3}H atom to every 10^{18}H atoms), whilst groundwater which infiltrated more than approximately 30 years ago will contain < 1 tritium unit. A detection limit of about 1-2 % allows old groundwaters to be distinguished, but because low levels of tritium close to the detection limit are the norm in present day recharge, it is important that a sufficiently large sample is provided and that contamination is avoided. Conventional ^{14}C analysis requires about 5 g of C, which is equivalent to approximately 50-150 litres of groundwater. Such volumes, of a sufficiently high standard, are only normally available from pumped samples. TAMS (Tandem Accelerator Mass Spectrometry) analyses of ^{14}C are now available in a few laboratories, several of which are in Europe and much smaller sample volumes are required, perhaps a preserved sample of as small as 100 ml. Stable carbon isotopes (^{13}C/^{12}C) are also measured at the same time and require a small sub-sample to be analysed. Stable oxygen and hydrogen isotopes (^{18}O/^{16}O and ^{2}H/^{1}H) are of qualitative value in dating the water, and require only 5-10 ml of untreated water sample. ^{36}Cl, produced mainly by neutron capture within granite, has proved of semi-quantitative value in defining water sources and flow paths in the groundwater flow modelling at Stripa (Davis and Nordstrom 1992). Although sampling for this isotope is easy, the analysis is only possible by TAMS; very few laboratories are capable of the required measurements and several months must be allowed for analyses.

4. **Dissolved gases**. The dissolved gases which need to be analysed fall into two categories, the common gases, N_2, O_2, CO_2, CH_4 *etc.*, which are important for understanding the overall geochemistry of the system, and the inert gases, He, Ne, Ar *etc.*, which are of value in the qualitative dating of the groundwater, and in the case of He, also relate to the distribution of

U and Th. The analysis of ^{222}Rn can be used in fractured crystalline rocks to provide information on fracture widths and to help in understanding the water rock interactions. The gases should preferably be sampled in a pressurised sampler, and whereas for most gases a 100-200 ml sampler is sufficient, good Rn analysis requires as much as 5 l. The inert gases can be analysed using a mass spectrometer, using a sample of only a few ml, sealed in a gas tight tube. The other gases are analysed by gas chromatography, and may require variable volumes depending on their abundance and requirements for separation. Approximate gas concentrations can also be determined from samples pumped to the wellhead and from the drilling returns, although these measurements are only minima.

8.8.3 The correction of raw hydrochemical data

The internal consistency and reliability of the laboratory analyses of the groundwaters and pore fluids can be tested using thermodynamic models. The program PHREEQE, originally developed for the USGS by Parkhurst *et al.* (1980), has been used in several countries' waste disposal programmes in variously modified forms. The program consists of a system of equations which describe the different equilibria and the activities of the components in solution. Saturation indices, which describe the equilibrium state of the water with respect to different minerals, can be calculated from a knowledge of the activities of the individual aqueous species. For minerals such as quartz and calcite, these calculations present no problems; however, difficulties exist with respect to clay minerals, for which there are few reliable thermodynamic data. It is also possible to simulate chemical reactions and changes in concentration which occur when components are added or removed from the solution.

In many hydrogeological environments, it has been found that the concentrations of major ions, even in some crystalline rocks, are controlled to a large extent by the carbonate geochemistry (*i.e.* the dissolution and precipitation of calcite in a closed system). It is difficult to make a correct assessment of the carbonate system in deep boreholes, because the carbonate equilibrium is very sensitive to the pressure and temperature changes associated with the sampling process. PHREEQE has been used to correct the CO_2 contents and the HCO_3^- concentrations in the hydrogeological characterization of groundwaters found in the basement of northern Switzerland, by assuming that all the waters are saturated with respect to calcite (Gautschi and Scholtis, 1989). In weakly mineralised groundwaters, such as those found in some granites, it cannot be assumed that only OH^-, CO_2 and HCO_3 have a significant effect on the alkalinity of the groundwaters, and components such as silicate and borate need to be considered.

The redox potential is an indication of the state of equilibrium between the activities of the oxidising and reducing species in solution. In order to correct the redox potential from that measured, even in situ, it is necessary, in order to obtain the best hydrochemical results, to determine a range of redox potentials for different ion pairs (see comments in Chapter 5 also). There are several different redox couples, and calculations can also be based on the dissolved U concentrations. Gautschi and Scholtis (1989) report for the Nagra sampling programme that the calculations based on U values tend to represent an upper bound to the redox. This is because the redox potential for the solution is calculated assuming that the dissolved U is in equilibrium with uraninite. The values calculated this way, as well as those for As(V)/As(III), are to some extent higher than those for other ion pairs. The H_2O/H^2 pair, with the dissolved hydrogen in the water probably being formed as a by product from the anaerobic corrosion of the steel components in the drill string or the casing, tends to represent a lower bound. These calculated redox potentials can then be compared with those measured using Pt electrodes, which suffer from the problem that, as all groundwater samples are contaminated to some extent with the drilling fluid, the redox potentials are shifted to higher values. Pt electrode measurements should therefore be taken as an upper bound.

Review papers have been published over the last few years, compiling several years work in Switzerland (Pearson *et al.*, 1989 and 1990) and in Sweden (Andrews *et al.*, 1988; Davis and Nordstrom, 1992) on the analysis and interpretation of deep groundwaters sampled as part of these countries' waste disposal investigations. There is now a much better understanding of how groundwaters evolve and of the importance of such an understanding of this evolution to performance assessments for a potential repository.

8.9 Repository location

Another aspect that can begin to be considered at this stage is the location of the repository. In hard rocks it will be the location, orientation and estimated significance of fracture zones that will be the main determinants in selecting the most suitable volume of rock. In salt, in either bedded or domal form, the precise location may well be of less concern, as the volumes of potentially suitable rock are likely to be large. In argillaceous rocks there may also be a significant degree of freedom, except with respect to the maximum depth, which may be severely limited for geotechnical reasons. Depth constraints in the other two geological environments are unlikely to be a problem in high quality rock, at least to depths approaching 1000 m. Unsuitable geotechnical conditions could, however, be a second reason for discarding a site, especially if large volumes of waste are being disposed of, as the requirements will be for large underground openings or greatly increased lengths of tunnel.

In hard rocks, in particular, boreholes may be required specifically to investigate the fracture zones of most significance to this repository location, as it is likely to be these fracture zones that will provide the potential fast pathways to the biosphere.

The general constraints on the location of a repository in hard rock were discussed in Chapter 7. The actual constraints for any particular repository are likely to be site-specific and may comprise a variety of factors in addition to geological ones, such as land ownership, for example. The constraints for a deep repository at Sellafield (see Figure 8.7 for the location of the PRZ), for example, comprise the following:

- Geotechnical constraints on the maximum depth for cavern construction in the basement rocks (caverns are needed in this case for the large volumes of ILW that require disposal).
- Planning constraints imposed by the presence of the Lake District National Park.
- Land ownership constraints.
- Geological constraints imposed by:
  - — the depth to the top of the basement rocks, in which the repository would be located,
  - — the presence of large faults, to the west and to the south, that downstep the basement by substantial amounts,
  - — the presence of a permeable sedimentary unit that is thought to be important in controlling the geometry of the groundwater flow system.

The area believed potentially suitable for repository construction is defined by the PRZ and the location of a repository within this zone is likely to be determined by the location, orientation and hydraulic properties of fracture and fault zones. The volume of rock between these potentially more transmissive zones will determine whether the repository will be constructed as a single layer or as multiple layers. Similar arguments are presented by Nagra (1994) and SKB (1992a).

8.10 Repository-scale characterisation

At the end of the earlier stages of an investigation there will probably be insufficient geological, and particularly hydrogeological, information available to reduce the uncertainties in the groundwater flow and transport models below those which might have been developed, based on expert judgement on the assumed geology and hydrogeology of the sites. The specifications for more detailed phases of site characterisation work will be dictated to a large extent by the results of the earlier phase, but it is possible to identify some likely components. It is possible that more detailed reflection seismics will be carried out to provide an infill to any existing reflection survey(s), and this additional work is most likely in the area which exhibits most promise as a potential area for a repository. The experience gained from the first phase seismics may allow more appropriate acquisition parameters to be used, especially if it is proposed to image steeply dipping structures within the basement. It may be decided to conduct a full 3-D seismic reflection survey, which would, because of its cost and complexity, necessarily only be conducted over relatively small areas, such as the proposed locations for the repository, or over local structures, adjacent to the proposed repository location. A seismic refraction survey may be considered, especially at sites located on the coast, and it may be possible to fire all the shots at sea with receiving stations on shore.

There is likely to be a greater concentration at this stage of the investigation on drilling and testing anomalies identified during the earlier surveys. Again standard geophysical surveys are likely to be used, but with the addition of more sophisticated inter-borehole and borehole to surface techniques, in particular seismic and radar tomography. There is an overlap here with what might be considered as borehole geophysics. This phase of the investigation is likely to lead up to the geophysical work associated with shaft sinking, which is considered in Section 8.11.

The data derived from the geophysics surveys are, in many cases, the only data that are likely to be available on the structure of the rock mass between the boreholes. There are limits to the number of boreholes which could be drilled on any site, for reasons which will be discussed in the next section, and the investigation programme needs to be designed with the limits of the various geophysical techniques borne in mind. For example, there are maximum distances over which seismic and radar tomography can be used, and this depends not only on the requirements for certain levels of uncertainty in the interpretations of the geological structure, but also on the attenuation properties of the rock mass. The inter-borehole techniques may be carried out in parallel with hydraulic testing, and this may pose additional constraints on the separation and depth of the boreholes. No one geophysical technique provides a unique solution to the geological structure, and it is often better to perform complementary geophysical surveys in order to reduce uncertainties in the predictions. For example, if seismic tomography, electrical and electromagnetic surveys are performed between boreholes and are interpreted to give similar structures, there will be more confidence in the final combined interpretation.

Uncertainties associated with seismic interpretations can be investigated in several ways, for example by analyzing the assumptions as the distribution of seismic velocities with depth. An analysis of this type has been carried out by Nirex at Sellafield. TVO, in Finland, are using what are known as *image point techniques* in developing 3-D models of the fracture zones within the basement rocks, based on seismic data.

8.11 Shaft construction investigations

At some stage in the site characterisation programme direct access will be required to the rock at depth and some type of access shaft will need to be constructed. An outline of the type of work that is likely to be required during the pre-shaft phase of the investigations has been described above. Although described as the pre-shaft phase, this work is likely to continue during and long after the shaft(s) have been constructed. In essence, this phase of the characterisation is concerned with two subjects:

1. Demonstrating that one has a good understanding of the groundwater flow regime over the regional scale (with dimensions of several kms to perhaps several tens of kms) and over the scale of the proposed repository (with dimensions of perhaps 1-3 km). In order to demonstrate that a sufficient understanding has been achieved, it will be necessary to demonstrate that groundwater conditions can be predicted with reasonable accuracy with the existing borehole testing programme. If a sufficient understanding has not been achieved, more site investigation work may be required; perhaps more boreholes and/or different and/or additional testing.
2. The drilling, testing and the installation of monitoring equipment in boreholes which need to be drilled relatively close to the shaft(s)/ access tunnel in order to monitor the response of the rock mass to the shaft construction. Monitoring the response of the rock mass, and comparing this response to predictions of the effect of the shaft sinking on the local, and perhaps regional, groundwater flow and on the changes in *in situ* stress and time dependent rock deformation. This should present an excellent opportunity to demonstrate an understanding of the geological and hydrogeological environment and should allow for some testing or validation of:

- The groundwater flow model on the scale of the repository, and possibly at larger scales.
- The groundwater flow model at smaller scales within the rock mass, related to the fracture network, the movement and dilation of fractures, and the effect that these movements have on hydraulic apertures, hydraulic connectivities and porosities.
- The geotechnical model of the rock mass, in particular movements on fracture planes and the time-dependent response of the rock close to the shaft and tunnel walls.

8.12 Geological and groundwater flow models

The largest uncertainties associated with the site characterisation are likely to be with the predictions of the groundwater flow and the resulting radionuclide transport models. This is mainly because geological and hydrogeological environments are inherently uncertain, and defining the geometry of the flow paths and their transport properties can be extremely difficult. It is very important that this fact is not used to distract from the potential benefits of the geological barrier and to concentrate an inappropriate amount of effort on engineering aspects of the near-field. The purpose of the geological barrier is to provide a slowly evolving environment at depth and one which limits the ingress of groundwater into and through any potential repository volume. Engineered barriers in themselves cannot provide the necessary long-term isolation in the absence of the geological barrier.

In order to collect the necessary information for the development of these hydrogeological and transport models boreholes will need to be drilled. The number of these boreholes, their locations and the measurements that are required within them have been discussed in Sections 8.5 and 8.6, and the subject is discussed further below. The development of conceptual geological and hydrogeological models is discussed in Sections 8.4, 8.5 and 8.6.

8.13 Providing data for models

The data requirements that can be met by the site investigation techniques can be divided into those that can be readily satisfied by the use of:

- currently and readily available site investigation techniques;
- more esoteric techniques, which are perhaps still in the development stage or whose interpretation philosophies have not yet been universally agreed;
- those that can only be satisfied by the use of inferential interpretation, or for which it is believed that, based on the current availability of the investigation techniques and interpretation philosophies, expert judgement will have to be used in the majority of cases.

The data requirements can also be described in terms of the accuracy that is necessitated by the different types of model used, and can be placed in the context of the requirements from the models that the uncertainties in the results decrease to acceptable levels as more data are collected.

Although there is general agreement amongst geologists as to the investigation techniques that have to be used to define the geological structure of the site, there is often no universal agreement on the interpretation of data on such matters as the geochemistry of deep groundwaters and on the way in which fracture networks can be modelled. In both these cases there is in any case unlikely to be a 'right answer'. In cases such as these the different interpretation techniques may well influence the types of data that are designed to be collected. This could be particularly significant if data collected for one interpretational or modelling approach proved insufficient or unsuitable for a different approach, especially if further data could not be collected, due perhaps to operational reasons.

Some of the data that are required by the models cannot be obtained satisfactorily from site investigations. Completely new techniques are unlikely to be developed in time for any investigation at a proposed repository site, and the best advances are likely to be made here in the use of natural analogues. Although in many cases the data derived from the natural analogue studies cannot be used directly as input to the transport models, a better understanding of the processes involved in radionuclide transport is likely to allow some form of validation to take place (the use of natural analogues is discussed in Section 8.14.8). It must be appreciated that some of the data used in the safety assessment of a site may not be derived from the site, or region, itself, and it may not be possible to provide some of the data without some recourse to inferential measurements. In this respect a repository cannot be treated like a normal engineered nuclear structure, in particular because some of the uncertainties inherent in geological and hydrogeological interpretations can sometimes never be resolved.

The data will need to be supplied over a range of scales, from the scale of major geological structures, in particular those that are related to hydrogeological discontinuities, down to the scale of single fractures and their associated alteration zones or perhaps fine-scale sedimentary laminations. Although limited data are often acceptable for groundwater flow calculations, transport modelling requires far more detailed information.

In considering the requirements for data from site investigations that arise from the use of the various codes, it is clear that very little data from a site investigation are directly input to the codes. Rather, the site investigation gives a database from which the code data must be derived after an appropriate conceptual model has been formulated.

A groundwater flow and transport code, such as NAMMU, can model at many levels of detail, so the question to be addressed is not simply which items of data are required but at what level of detail they are required. This question can only be answered in the overall context of the full safety assessment, since it is affected by many diverse factors, including the relative importance of the geosphere in the safety case, the degree of uncertainty which can be tolerated and cost of acquiring the data.

Since the data are really required for the conceptual model rather than the code, most of the issues can be discussed in a general way, and are likely to be equally relevant to any code, although there will be some issues peculiar to specific codes. The main data requirements from the site investigation as regards modelling can, therefore, be described under three headings:

- knowledge of significant geological structures,

- average hydrogeological (and transport) properties of the rock types,

- secondary data for validation/calibration, such as geochemical information.

Data are also required for the construction of the repository and the operation of the URL-phase of the site characterisation. In those hard rocks which are considered suitable for waste disposal there are unlikely to be any geotechnical problems, at least for the relatively small underground openings that are necessary for the disposal of HLW. The larger volumes of ILW and LLW, will require considerably larger galleries or vaults. Much more consideration will have to placed on their stability, particularly if long stand-up times are envisaged and if complete backfilling is not planned. Considerably more problems can be present if disposal is planned to take place at depth in many types of low permeability sediments.

Significant depth constraints are apparent for many clays, in some cases easy construction may not be possible at depths much in excess of 300 m. In this type of environment support costs are likely to be high and waste packing densities low. More indurated clays and marls can be stable up to much greater depths (*e.g.* the Opalinus Clay in Switzerland), and the ironstone mine at Konrad appears to have been stable for several decades, even at depths up to 1300 m. Depth constraints can be less of a problem in rock salt as, except in some more restricted bedded salt formations, there is often a large depth range of possible depths for disposal, and support problems are not as significant as in clays. Disposal of the larger volumes of ILW and LLW may be impractical in many of the weaker formations, including salt, for geotechnical reasons, even though these sediments may otherwise be suitable. Geotechnical aspects of repository design and their relationships with backfilling are also discussed in Section 8.14.6.

8.14 A URL in research and characterisation

8.14.1 Introduction

Underground Research Laboratories (URLs) associated with research into the disposal of radioactive waste were first operated as part of Project Salt Vault in the U.S.A. in the early 1960's. Large-scale *in situ* experiments took place in bedded salt; these were mainly of a geotechnical nature and were concerned with enhancement of room and pillar closure rates under thermal loading. Underground experiments at Asse started in 1965, but URLs in general have, however, only been an important part of the radioactive waste disposal scene since the 1970's, with experimental programmes being started in several countries, such as Sweden (Stripa), the U.S.A. (WIPP, G-tunnel, Climax and Rainier), Belgium (Mol), Canada (URL), *etc.* Some of these URLs were sited at potential, or proposed, repository sites, whereas others, such as the Canadian URL, were specifically sited in areas where no waste disposal would be allowed. In addition to research into the feasiblity of disposal of long-lived radioactive waste, it is now generally appreciated that the proper assessment of a site for the long-term disposal of these wastes needs to include a stage of underground experimentation and validation. Rather than the term URL, which implies *research*, a more appropriate term for this phase of the assessment of a site might be UCVS (Underground Characterisation and Validation Stage), or an acronym with a similar meaning (Nirex, for example, call their URL a RCF (Rock Characterisation Facility)).

A URL can, therefore, be developed at a proposed repository site, or can be considered as existing only for the purpose of research, and thus could be located independently from any site selection process for a deep repository. An example of this second approach is provided by the selection of the site for the HRL at Äspö (Figure 8.17), which took place in 1987/88, compared with the process of site selection in Sweden for a deep disposal site for spent fuel which commenced in 1992. An example of the first is provided by the WIPP. At present, underground research and validation work at proposed, or possible, repository sites is limited to the WIPP, Mol (Figure 8.18) and Konrad (Germany). To this date more work has, in fact, been performed in generic URLs than in these proposed repositories.

A URL which is designed never to be associated with waste disposal can be operated in a more relaxed manner, since the long-term influences of experimental programmes and extensive drilling do not have to be considered. This type of URL can be operated more in terms of a research station, whereas a URL which is developed as an integral part of the assessment of a proposed disposal site needs to consider problems which are directly applicable to the assessment of the site and probably will tend to be more constrained by the requirements of the assessment process.

The early stages of an investigation programme at a potential repository site are designed to increase the confidence in the safety assessment and the geotechnical design work to a level at which a choice can be made on the general suitability of the site. In the situation in which more than one site is being investigated, it should allow comparison to be made between the various sites on offer. The emphasis of the investigation will then change, becoming more detailed in order to fit the repository carefully into the rock mass and hone its design. An important feature of the next stage of the investigation is the demonstration that the models and concepts used in the assessment to that date, and those that are likely to be used later in the assessment,

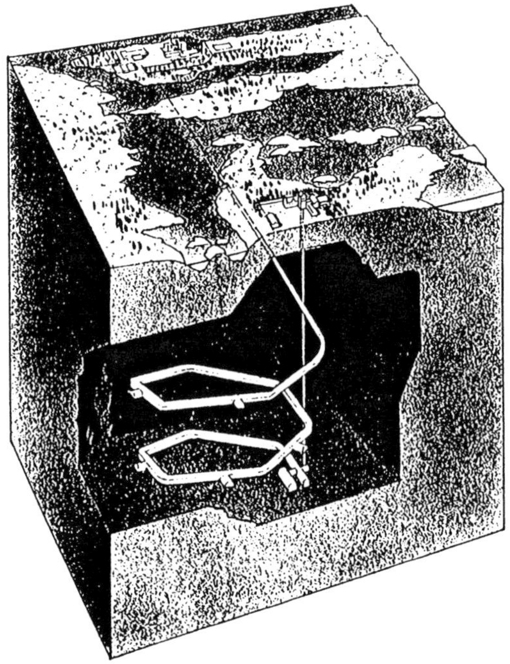

Figure 8.17 The general structure of the HRL at Äspö (From SKB, 1992b). Reproduced by permission of SKB.

are valid for the specific characteristics of the selected site that will emerge from the underground construction. Validation at the site, rather than generic validation of modelling approaches, is likely to be a key component of this stage of the assessment and subject to intense peer review.

This stage of the site investigation is likely to comprise a period of shaft and initial repository tunnel construction, together with underground exploration in and around the actual volume of rock considered most suitable for the repository. This period, which will merge progressively into the main construction phase, will provide an opportunity to use these facilities, together with monitoring boreholes from the surface, to validate crucial aspects of the safety assessment modelling. It also provides an opportunity to make full-scale tests or demonstrations of various engineering aspects of the repository design, such as waste emplacement, backfilling and localised sealing techniques.

Validation is an internationally accepted process which is integral to the development and presentation of a safety case for the deep disposal of long-lived radioactive waste (IAEA, 1981, 1988; NEA/SKI, 1990; Niederer, 1991), and is an important and integral part of national disposal programmes in

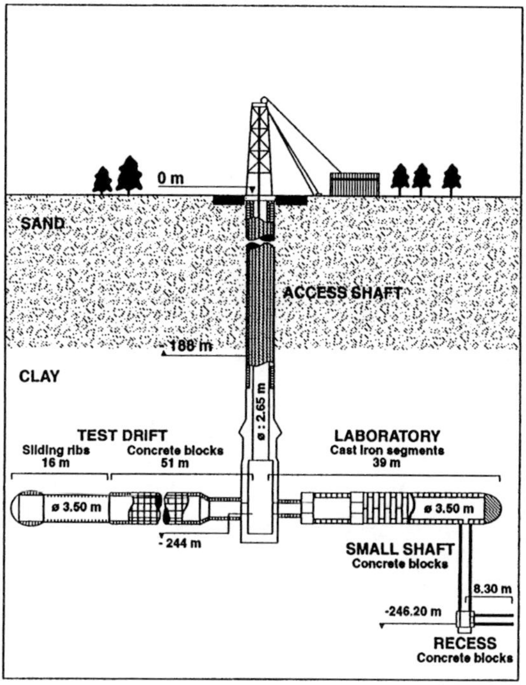

Figure 8.18 Schematic view of the URL at Mol, Belgium (From Bonne, 1991). Reproduced by permission of Lawrence Berkeley Laboratory, U.S.A.

most countries planning deep waste disposal. Models of radionuclide transport and other processes that are important in analysing the performance of a repository, cannot be validated in the generic sense, but only on a site specific basis, and validation is therefore important to the presentation of any safety case, regardless of any previous validation work that might have been carried out in a similar rock type at another site. The validation of models is therefore not an optional extra or a luxury that can be foregone in order to accelerate a repository development programme and is likely to form a major part of any programme of work within a URL.

Before any work can be performed underground at the potential repository level, access has to be provided by a shaft, whether it be inclined or vertical, and the process of shaft construction allows a large amount of data to be collected on the rock mass, which cannot easily be obtained from boreholes (*e.g.* Olsson, 1992). In addition, the construction of the shaft will perturb the groundwater flow regime and will redistribute the stresses close to the shaft walls, and these changes will provide the first opportunity to test the groundwater flow models on a large scale. By drilling, testing and monitoring boreholes in the area of the proposed shaft for a sufficiently extensive period before any construction commences, the changes due to the shaft construction should be adequately quantified (*e.g.* Gustafson *et al.*, 1991). During shaft construction, *in situ* testing of the shaft walls to examine for spalling and groundwater inflows *etc.* will allow data relevant to near-field modelling to be obtained. Depending on the nature of the rock through which the shaft is being sunk, the amount and type of data that can be directly obtained will vary considerably. In competent hard rocks, where the shaft walls are potentially stable, *in situ* monitoring of changes such as wall movement could be performed relatively easily. A model of effects, such as the hydraulic skin and the excavation disturbed zone around the shaft, could then be set up based on the monitored changes and theoretical failure mechanisms. In more transmissive formations there is perhaps less that can be obtained from the shaft itself, as the shaft cannot be constructed without some form of temporary support, such as an ice curtain or chemical and/or cement grouting, which is designed to prevent groundwater ingress and shaft wall failure before permanent linings can be installed.

The URL, or validation, phase of the assessment can, therefore, be considered to commence before any shaft is constructed, as monitoring boreholes are required significantly in advance of excavation. The time necessary for drilling, completion and monitoring of these boreholes is likely to vary considerably depending on the hydrogeological environment at the site and in particular its complexity. In fact, this phase of the assessment can itself be divided into three phases:

- pre-shaft construction,
- during shaft construction,
- post-shaft excavation and initial stages of repository construction.

These phases are not necessarily temporally distinct, since shaft construction is likely to continue during the initial stages of repository construction, and will take place concurrently with other aspects of the site investigation which are not associated with validation work. Therefore, although the site investigation can be nominally divided into three phases, there will in fact be a large overlap between the second and third phases, and this will be mirrored by the spread of validation work across a large part of the investigation programme. The third phase of the investigation, which will be mainly concerned with work underground, and which can be described in terms of detailed site characterisation and validation work in the near-field, cannot be divorced from the work which is planned to take place from the surface. The programme for a URL, therefore, needs to consider not only the underground investigation work, but also that work which will be required considerably in advance of any shaft sinking. Work at this pre-shaft stage of the site investigation is concerned with three main subjects:

1. The drilling, testing and long-term monitoring of boreholes to examine the effects of shaft sinking on the hydrogeology and geotechnics of the rock mass.
2. The continuous validation of groundwater flow models, by a series of relatively simple experiments, and as soon as data become available, in order to obtain a measure of the understanding of the hydrogeological environment and to examine the assumptions inherent in the design of the investigation strategy. Such validation work could include the use of large-scale tracer tests, in fracture zones, for example, if the groundwater flow modelling suggests that such zones are significant and if they are amenable to analysis from the surface.
3. How the hydrogeological evidence available from the early parts of the site investigation can be used to make logical decisions on the locations of the shafts, and the implications, in terms of long-term safety and ease of validation, of specific shaft and repository locations.

8.14.2 Operating and planned URLs

There are several URLs, or related underground experimental operations, that are either operating at present or are planned to be in operation within the next few years. The URLs in this category are:

Those in hard, fractured rocks

- HRL, Äspö, Sweden (currently under construction; no waste disposal permitted within the HRL, but a *demonstration facility* containing radiactive waste could be located in the vicinity).
- Underground work associated with the assessment of one or more sites in Sweden for spent fuel disposal (from approximately 2000?).
- Canadian URL (operating now; waste disposal not allowed).
- Grimsel, Switzerland (operating now, no proposed waste disposal).
- The Sellafield RCF, U.K. (Initial boreholes for shaft design and local detailed geology and hydrogeology of the shaft area now complete. Monitoring boreholes being drilled. Shaft construction possible during 1995, underground gallery construction possible by 1998).
- Underground investigations at one site in Finland, (possibly after 1998?, for spent fuel).
- Yucca Mountain, U.S.A. (Initial horizontal trial gallery construction. Proposed repository for spent fuel and heat-emitting ILW and HLW).
- A URL in a granite in France (programme to commence after site has been selected in 1994; possible eventual use of the same area for HLW and heat-emitting ILW).
- Tono and Kamaishi, Japan (Both experimental sites in granite).

Those in other rock types

- Mol, Belgium (plastic clay; possible repository for HLW).
- WIPP, U.S.A. (bedded salt; probable repository for military-derived TRU wastes).
- Konrad, Germany (disused iron ore mine; planned repository for LLW and non heat-emitting ILW).
- Asse, Germany (disused salt mine; previously used for LLW disposal).
- Gorleben, Germany (salt dome, planned repository for HLW and heat-emitting ILW).
- A second URL in France (possibly in clay; programme to commence after site has been chosen in 1994, similar conditions to granite URL site).

Additional sites

There are, and have been, other small underground experimental sites for radioactive waste research, such as Climax, U.S.A., and Fannay Augères, France in hard rocks and, in sediments, Tournemire (a disused railway tunnel in marl) and Révin (in Cambrian schists within a hydroelectric scheme) (both in France). El Berrocal, Spain (in granite) is currently being used for research, but is not a true URL. Other countries, such as the Czech Republic, are currently considering setting up URL programmes.

URLs or experimental operations in a variety of rock types (*e.g.* WIPP, Canadian URL, Stripa, Mol, Grimsel *etc.*) have been operating for several years and can be of use for illustrating the temporal, spatial and logistical constraints relevant in operating a URL or carrying out experiments *in situ*. In particular, many lessons can be learnt from the work that has been performed at Stripa, Sweden. This research is at a particularly interesting stage, as the final conclusions regarding this long period of work have recently been prepared (The first of these concluding reports to be published was Olsson *et al.* (1992)). Many useful lessons can, therefore, be learnt regarding the practical aspects of performing experiments, particularly those in hard rock and salt, together with the associated modelling.

Information can also be obtained from other operations, which were not URLs, but mainly large-scale field experiments. An example of this type of operation is the Finnsjön fracture test site in Sweden.

8.14.3 Work carried out in URLs

An examination of the work which has been performed or is proposed in these underground laboratories, proposed repositories and test sites listed above illustrates the diversity of experiments that could be envisaged as part of a URL programme. The types of underground experiments that have been carried out can be placed into the following categories (Figure 8.19):

1. Groundwater flow model testing at different scales, including the use of tracer tests with conservative tracers to measure flow.
2. The testing of geochemical transport models using sorbing tracers and the analysis of radionuclide migration within fracture zones combined with an analysis of matrix diffusion.
3. The evaluation of the evolution of redox conditions in situ.
4. Colloid transport (experiments on colloid transport are likely to be included within the testing of geochemical transport models, as part of a suite of tracer tests).
5. Gas generation and migration experiments.
6. Repository construction tests – response of the rock to different construction techniques (excavation response).
7. Sealing tests for the shaft, exploration boreholes, tunnels and, perhaps, galleries.
8. Evaluation of thermo-hydromechanical effects on radionuclide transport.
9. Repository engineering systems, *e.g.* waste emplacement, backfilling and sealing tests *etc.*

The types of information that are potentially available during the URL construction activities are illustrated in Figure 8.20. Complications are introduced into a URL programme in the situation where a URL site could, eventually, be considered for disposal purposes, but where the URL is being constructed and operated before any decision is made on the suitability of the rock mass for waste disposal. The complications introduced are as follows:

- The siting of the URL needs to consider the requirements of the repository, in terms of volume of rock required, separation from fracture zones *etc.*, and the relationship (*i.e.* separation) of the URL with the repository
- The site investigation programme, the number of deep boreholes and their locations, needs to consider at the outset the requirements for eventual borehole sealing.
- The tunnel and/or shaft access to the URL needs to be designed and located in such a way that the integrity of the rock mass for disposal purposes is not compromised.

The majority of these factors would not have to be considered if it were known from the start that the site would be used only for research purposes, or if it could categorically stated that no repository would be positioned where the presence of a URL could affect its long-term safety. Similar complications should also not arise where a URL is constructed and operated as part of a well-planned assessment of a site, but could arise in situations where the investigations are too hurried and, in particular, where the repository is located before sufficient data are available.

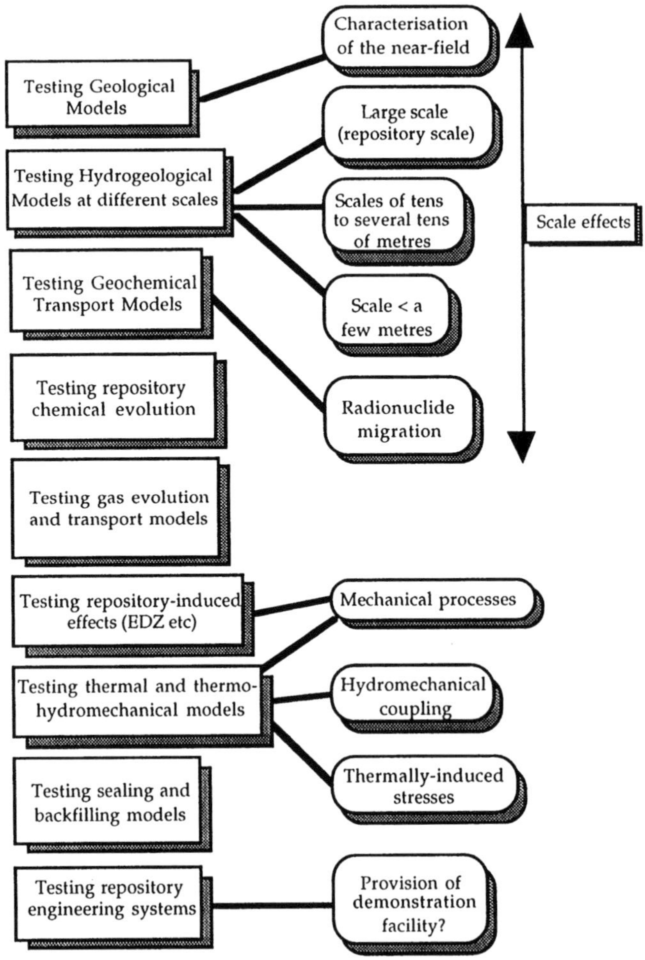

Figure 8.19 The general categories of research and testing that can be carried out after the access shafts to the URL have been constructed.

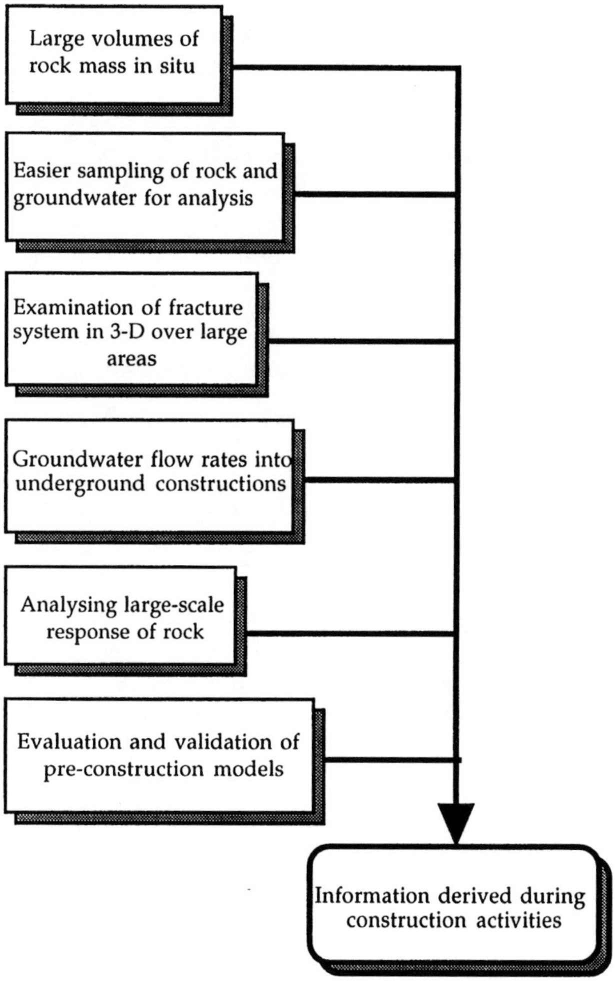

Figure 8.20 The types of information that are potentially available during the the URL construction activities.

8.14.4 Temporal co-ordination of URL work

The temporal co-ordination of work within and associated with a URL is one of the most important factors that needs to be taken into account in the design of a research programme. The best published information on this co-ordination relates to the planned work at the Swedish HRL, and SKB's proposed programme provides a very good introduction to this subject (SKB, 1989; Gustafson *et al.*, 1991).

The original plan for this work is described in several SKB reports, which describe in turn the various aspects of the programme, including the evaluation of the conceptual models developed from the data collected in what is known as the *pre-investigation phase*, *i.e.* in advance of the construction of the HRL (SKB, 1989; Wikberg *et al.*, 1991; Stanfors *et al.*, 1991; Almén and Zellman, 1991; Gustafson *et al.*, 1991).

Phases of work in the Swedish HRL programme

The programme of work associated with the HRL can be related to what are termed *Main Goals* and *Stage Goals*. Whereas the three main goals are general in nature and, in part, are intimately related to other parts of SKB's research programme, the five stage goals are more temporally related and are more specific to the operation of the HRL. These stage goals are as follows:

1. To verify the pre-investigation methods.
2. To finalise a detailed investigation methodology.
3. To test models for groundwater flow and solute transport.
4. To demonstrate construction and handling methods.
5. To test important parts of the repository system.

Prior to the siting of the final repository site for HLW in Sweden the activities at the HRL are designed to demonstrate that investigations at the surface and in boreholes can provide sufficient data relevant to safety-related properties of the rock mass (Stage goal 1) and refine and verify the methods and the technology required for the detailed characterisation of a rock mass at the repository scale (Stage goal 2).

In order to optimize the final repository design and to provide sufficient data for input to a safety assessment it is necessary to test models for groundwater flow and solute transport (Stage goal 3) by refining and testing on a large scale and at expected repository depths methods and models for describing groundwater flow and solute transport in the rock mass.

In preparation for the construction of the final HLW repository it is necessary to demonstrate that construction and waste handling methods are available (Stage goal 4). In order to achieve this, access has to be provided to underground areas where methods and technology can be refined and tested, so that a sufficiently high quality can be guaranteed during the design, construction and operation of the final repository. In addition, it will be necessary to test important parts of the repository system (Stage goal 5) at full scale, particularly those that are of most significance to its long-term safety.

8.14.5 Pre-URL construction

There are important implications related to the requirements for extensive investigation, testing and monitoring before any URL is sited and its construction commences. There are also equally important implications regarding the rate of construction of the URL, particularly as regards the amount and quality of information that is available during this process. These implications are significant as regards the timing of various aspects of the URL programme. The general sequence of events in the early stages of a URL programme in hard rock are outlined in Table 8.2. The general aspects of the programme are likely to be similar in other geological environments, though the relative times taken to carry out the various parts of the site investigations are likely to differ.

Table 8.2 illustrates the extensive times required before any URL should be constructed. If underground construction were to start before the types of work outlined in this table had been carried out and interpreted, many of the potential benefits available from a URL programme would be lost. The temporal implications for the URL programme, as regards the requirements for R&D and validation work during construction, are illustrated in Table 8.3. These two tables illustrate the extensive time from the start of the site investigation programme (7.5 to 9 years), that is likely to be required before any underground experiments can take place. There would seem to be few opportunities for accelerating this process if as much as possible is to be made of the data potentially available from a URL programme. The temporal constraints applicable to other geological environments are likely to be similar.

8.14.6 The purpose of *in situ* work within a URL

Over the last two decades there has been a large amount of site investigation, particularly in hard rock areas, for the deep disposal of long-lived radioactive waste. This work has supplied extremely useful data, which has been of use in the generic assessment of most potential rock types for disposal purposes. Research programmes have also taken place in existing mines, such as Asse, Stripa and Fannay Augères, and for the examination of specific areas of interest, such as the Finnsjön fracture zone. In some areas, an equal amount of work has taken place in purpose-built URLs, most notably at the Canadian URL and Grimsel, and considerable more of this type of work is planned over the next few years in the HRL.

Table 8.2 Outline of possible programme in advance of URL construction to illustrate temporal constraints.

1. Regional analysis (12-18 months)
(a) Specific area within granite selected for analysis (30×30 km approx) using data already available and new desk surveys (b) Region for hydrogeological reconnaissance survey defined (c) Region for hydrochemical reconnaissance survey defined (d) Regional geological mapping (if required) (e) Regional geophysical surveys (f) Simple initial modelling
First report produced, with geological and hydrogeological predictions
2. Siting stage (12-18 months)
(a) Narrowing down of area of interest (b) More detailed geophysics, including localised seismics (c) More detailed geological mapping (c) Initial drilling programme (d) Simple hydraulic testing (e) More detailed modelling (f) Comparison with initial predictions
Second report prepared (recommendations for siting URL)
3. Detailed pre-construction investigations (18-24 months)
(a) Second drilling phase (b) More detailed hydraulic testing and monitoring (c) More detailed hydrochemical analysis (d) Numerical hydrogeological simulations of URL construction *etc.* (e) Monitoring boreholes drilled, tested and completed (f) Detailed plan for URL construction developed (g) Plans for testing and monitoring during construction developed
Total time required pre-construction = 3.5 to 5 years

Table 8.3 A general outline of the construction of the URL in hard rock illustrating the temporal constraints.

1. Location of URL (18-24 months) (Equivalent to end of Siting Stage (Table 8.2))
(a) Decision on location of URL (b) Decision on location and design of access shaft/tunnel (c) Modelling of influence of URL on groundwater system (d) Drilling of monitoring boreholes for URL construction (e) Testing and completion of boreholes (f) Monitoring of boreholes for >12 months before construction (g) Predictions of excavation effects and inflow zones
2. Initial construction phase (24 months)
(a) Construction of access shaft commences (b) Forward drilling and testing, installation of monitoring equipment (c) Validation: findings continuously compared with predictions (d) Updating of geological and hydrogeological models as required (Tunnel drivage rates purposely slow to maximise data input and allow opportunities for validation exercises.)
3. Construction of URL galleries (24 months)
(a) Continued monitoring of effects of URL construction (b) Continued validation exrecises (c) Final, detailed design of experiments (d) Temporal and spatial separation of experiments and/or construction
4. Comparison of pre-investigation predictions with those found

The construction of a URL and the experiments that can subsequently be carried out underground will enable the results obtained from the surface explorations to be verified, and will allow validation of the geological and hydrogeological models to be carried out to greater and greater levels of sophistication. This, in turn, should increase confidence in the ability to model the performance of the potential repository system and to predict the response of the geological and hydrogeological systems. With increasing experience of carrying out such *in situ* work, it should become easier to demonstrate to the regulatory authorities, to the politicians and to the public an increased understanding of the disposal environment. This increased understanding should enable better technical decisions to be made regarding the eventual choice of disposal environments and the engineering aspects of the disposal system.

Another important aspect of a URL programme is the development of experience of *in situ* work, which will be of direct relevance to the work required in an actual repository. It takes many years to develop the necessary expertise to carry out underground experiments, and good examples of the time required are provided by the experience from Stripa and Grimsel. It is important in this regard that there is continuity in the URL programme, so that the experience of the teams can be developed in a rational manner, and that hiatuses in funding *etc.* are avoided. It is also important that contact at the working level is made with other URL programmes and that co-operative research is also carried out with these programmes. Both these factors are likely to impart a greater credibility to a waste disposal organisation's programme.

The main objectives of a URL programme can be divided into the following subject areas:

- The demonstration and development of a site characterisation approach
- The testing of available techniques
- The demonstration that techniques can be applied routinely
- The scaling up from borehole data
- The provision of data for performance assessment
- The validation of models
- The demonstration of waste emplacement and repository operation and sealing

Demonstration and development of a site characterisation approach

Part of the research programme at the URL site will be to demonstrate that an adequate site characterisation approach can be developed, so that a future repository site in the same, or similar, geological environment can be sufficiently well characterised. In order to demonstrate adequacy, validation of the geological and hydrogeological models will be required, and true validation can only be attained by attempting exercises in blind prediction.

The R&D programme for the HRL in Sweden is a good example of this type of approach, which is being assessed during the construction of the laboratory, with the intention of applying similar techniques to an actual deep repository for HLW (SKB, 1989). An extremely well managed database system, together with accompanying QA, is required in order to perform predictive assessments of the type described above. It is believed that such a predictive capacity, together with a demonstration that the hydrogeological evolution of the site is well understood, are likely to provide the most convincing evidence for validation purposes, even if, in the case of the analysis of the hydrogeological evolution, much of this information cannot be used directly in determining the possibilities for the safe disposal of radioactive waste in this type of environment.

As part of this approach it is necessary to develop a procedure for the construction of conceptual models of the site at the various scales necessary. An example of this is shown in Figure 8.21 for the case of the geological structural model within the SCV Project at Stripa, which was concerned with the detailed examination of a small volume of rock at several hundred metres depth. Following on from that is the development of flow and transport models (Figure 8.22), where the structural geology of the system, in this case the fracture system geometry, was input from the geological structural model.

The testing of available techniques

Specialised techniques will be required in order to characterise large volumes of the rock mass at depth. Many of these techniques can only be tested in conditions representative of a real repository, and it is only after demonstrating the efficacy, reliability and accuracy of such techniques that they could be used to supply data acceptable in a performance assessment. Much of the work at Stripa, Grimsel and the Canadian URL has been involved with the development of techniques for supplying the data necessary for modelling purposes and demonstrating that such techniques do, in fact, supply the required data. Many of the techniques are generally applicable to hard rocks, and their use in a specific hard rock does not require any validation. Development work has also taken place on similar techniques for use in argillaceous rocks and evaporites, and it would, therefore, seem sensible for any organisation to co-operate with other URLs in similar geological environments in any programme of technique testing and development that was believed to show promise.

Demonstrating that techniques can be applied routinely

Unlike the fairly relaxed and flexible programme that can be accepted in the operation of a purely research-based URL, the construction of the actual repository is likely to be carried out on a much tighter schedule, and it is important, in advance of this construction, that site investigation techniques can be routinely applied in an efficient manner. In order that this can occur, the reliability of the measuring systems needs to be proved and the efficient operation of the equipment needs to demonstrated. Carrying out testing during construction is expensive, and it is a more sensible use of resources if money is spent on experiments within the associated, or another, URL which could result in the more efficient use of investigation techniques within the repository.

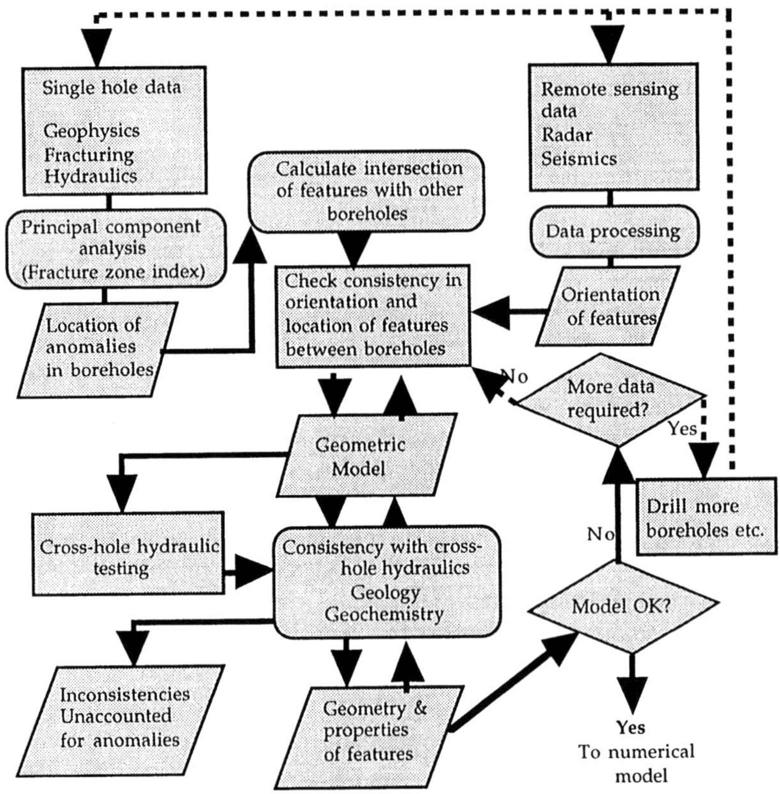

Figure 8.21 Outline of the procedure used for the construction of the geological conceptual model for the SCV (Site Characterisation and Validation) site at Stripa (modified from Olsson *et al.*, 1992). Reproduced by permission of SKB.

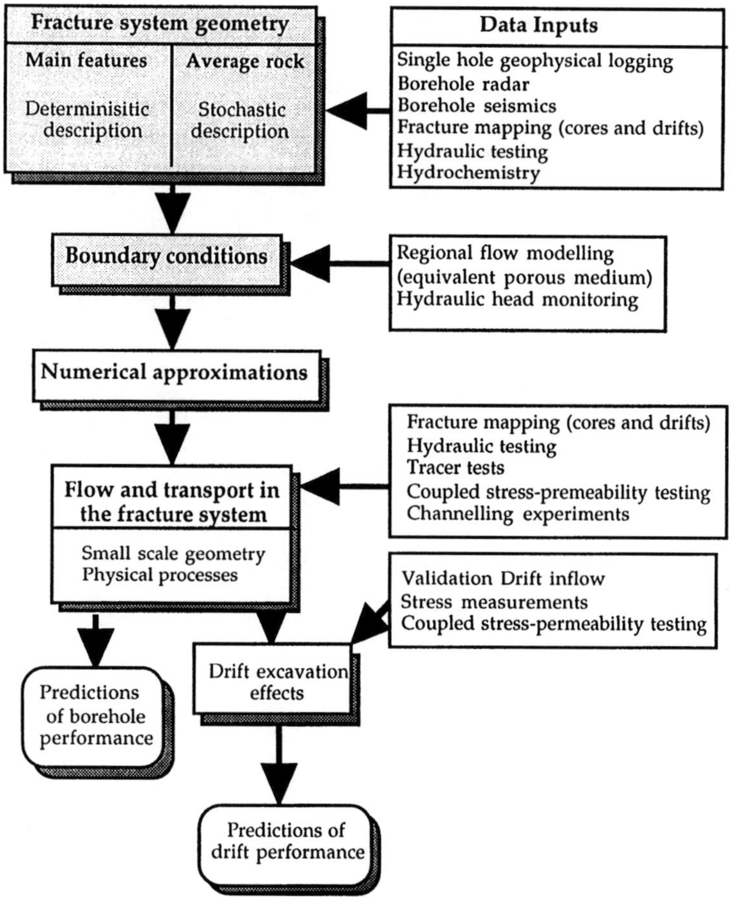

Figure 8.22 The basic components of the flow and transport models that were set up as part of the SCV Project at Stripa and the data required to construct them. The components belonging to the geological structural model (see Figure 8.21) are shown shaded (modified from Olsson *et al.*, 1992). Reproduced by permission of SKB.

Demonstrating that techniques can be routinely applied in an efficient and routine manner will allow an investigation methodology to be developed, in the knowledge that this methodology can be tailored to the requirements of the geological environment of the repository. The development of such a methodology, and the improvement in data collection that should result is likely to increase confidence in the data quality and, thereby confidence in the ability to understand the repository system.

Although it is in the investigation of the near-field that techniques need to be applied repeatedly and routinely during the construction phase, there is a great deal of work that also needs to take place from the surface during the characterisation of a site. It would be sensible also to develop a characterisation methodology at a URL, rather than wait until investigations are taking place at a repository site. This is one of the main aims of the current work at the HRL and requires that the site investigation for the URL and the subsequent work from the surface are carried out with this subject in mind.

Scaling up from borehole data

The early stages of site characterisation programmes in all rocks involve drilling and testing deep boreholes and carrying out surface and aerial geophysical surveys. The geological and hydrogeological models which can be developed from these data are of a preliminary nature, and it is only when underground access has been obtained that sufficiently detailed models can be set up. There is a need to demonstrate the ability to extrapolate from the larger-scale borehole-derived data to the more local geological and hydrogeological structures on the scale of a URL and a repository. The success of this process of extrapolation can be used as a measure of one's understanding of the geological environment, and the process of underground construction allows numerous possibilities for the continuous updating and validation of geological and hydrogeological models.

In order to demonstrate ability to carry out this extrapolation and to demonstrate a predictive capacity, validation exercises are required at several stages during the site characterisation phase. By far the best example of this approach is the work that has and is still being carried out at the HRL. The first few stages of the HRL programme have all been centred around the verification of the geological and hydrogeological predictions derived from previous phases of the investigations (Table 8.2) and the validation of geological and hydrogeological models of the site and the surrounding region.

The provision of data for performance assessment modelling

One of the most important reasons for carrying out in situ work is to provide the data necessary for performance assessment modelling. In a hard rock repository for HLW much of the barrier to radionuclide transport is provided by the ability to isolate the waste in low permeability, sorptive engineered barriers and the time taken to reach any major fracture zone, through what is often termed *averagely fractured rock* (the *respect distance* of SKB (SKB, 1983)). In both evaporites and in low permeability argillaceous rocks similar arguments can apply, but in these cases the more transmissive pathways can be of a different nature, *i.e.* lithological. A large amount of data, of a very detailed nature, is therefore required, particularly in the repository near-field, in order to provide sufficient input to the performance assessment models. An example of this detailed level of work is provided by the migration fracture test at Grimsel, where a fracture zone was intersected by several boreholes drilled from an experimental tunnel (Figure 8.23 (from Frick *et al.*, 1992)). It is proposed to excavate this fracture zone when all the tracer testing has been competed. A URL provides the best opportunity to collect these data before they are required in earnest in a repository environment. The lessons learnt, in particular at Stripa, regarding the best methods of obtaining these data are very helpful in designing the experimental programme for any future URLs. Olsson *et al.* (1992) provide a useful discussion on this subject with respect to the characterisation of the SCV zone at Stripa.

The validation of models

As discussed in previous chapters, a safety assessment for a radioactive waste repository will involve a chain of mathematical models which describe the evolution of the various barriers, both engineered and natural. As part of this process, detailed research models are currently being constructed in various national radioactive waste disposal programmes, based on a theoretical framework which is supported by laboratory and field studies. These research models are likely to be simplified and grouped together to form assessment models which are linked into a model chain in order to describe the overall performance of the repository system. Having verified the research models by testing their numerical accuracy against exact solutions or other independent models, it is necessary to validate these and, in some respects, the assessment models. In order to be able to validate models, there needs to be a good understanding of the physico-chemical processes that occur in rocks, particularly their radionuclide transport properties. A distinction therefore needs to be made between the work that is still necessary in order to develop and refine models and site investigation strategies and validation work that is site specific.

A process can first be identified and described in terms of a conceptual model, which is then translated into a computer code. The model can then be applied to experimental data and its results compared with measurements. If successful, it can be described as being validated with respect to this particular

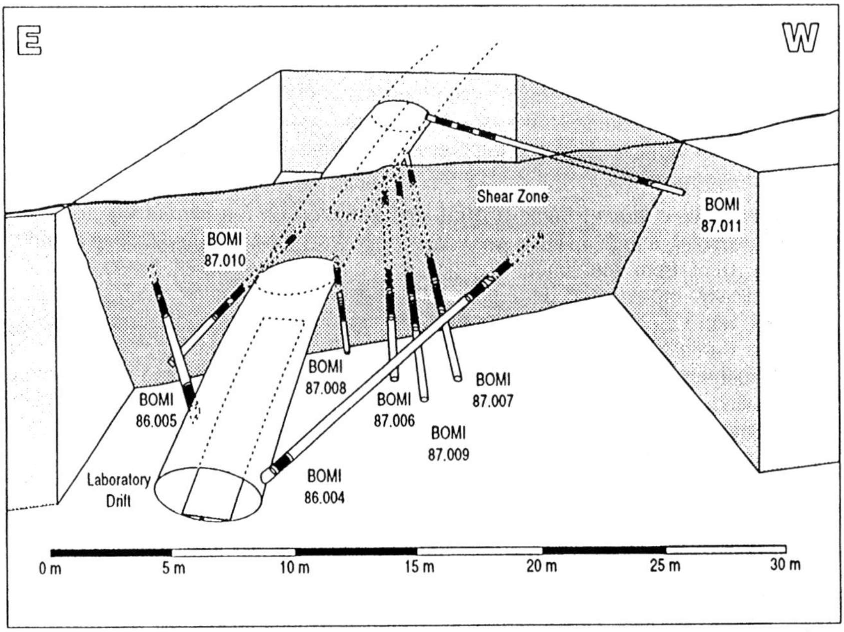

Figure 8.23 3-D perspective of the migration fracture experiment at Grimsel indicating the location of the exploration boreholes (from Frick *et al.*, 1992). Reproduced by permission of Nagra.

process. For a site specific environment, the modelled system is likely to be composed of a number of processes and structures which are linked. In order to perform any site specific validation, validation of the relevant processes must already have taken place and, once the separate parts of the system have been identified, the site specific group of models can be used to simulate and be compared with field observations. The group of models can be said to have been validated with respect to a particular site within the range of the observations made in the field experiments. Validation for conditions outside this range or in another geological environment (*i.e.* another site) will not have been achieved.

Assessment models, together with their associated databases, have varying requirements for validation which depend on the complexity of the system that is to be modelled, the time over which the assessment is required, the spatial extrapolation of the models used and the sensitivity of the overall performance of the system to the processes being modelled. A robust assessment is preferable, and perhaps a necessity, if permission for disposal is to be permitted (McCombie *et al.*, 1991). For a model to be described as being robust, it is necessary to have confidence that the results of the model are either correct, that is they are sufficiently realistic, or that they are conservative.

The process of model validation is to demonstrate that the model can sufficiently describe the processes involved, and an adequately validated model is by definition robust. A model that is consistently conservative, and may, therefore, be consistently incorrect, may still be robust and is likely to be acceptable. An advantage of this robust approach, which is advocated particularly strongly by Nagra (McCombie *et al.*, *op. cit.*) and was used in part by Nirex in their procedure for selecting suitable sites for investigation, is that the requirements for validation are likely to be reduced by this simplification of reality. A relatively simple, predictable system, whether it is in the near- or far-field, together with a conservative approach to the choice of models used, will place fewer demands on the process of validation. In order to take this robust approach, the geological and hydrogeological environments of the site need to be simple.

The problem of validation becomes apparent when the complexity of many natural systems are considered and when the requirements of the model, in terms of its predictive capacity over large temporal and spatial dimensions, are examined. Suitable models need to be able to make predictions of future behaviour, and part of the validation process should be to attempt to check that such predictions are in fact correct. The temporal and spatial scales can be so large that opportunities for testing models are very limited, but in any case models can only be shown to be invalid, rigorous validation being impossible. Validation is therefore concerned with building sufficient confidence in the predictions of a model. Predictions derived from assessment models that concern the arrival times at the surface of specific radionuclides at long times into the future, cannot of course be validated in any definitive sense. It seems that it is only by an analysis of the palæohydrogeology of the site, which could allow some form of back-analysis of the transport of dissolved constituents through the repository volume, that even an approximate validation of such a model could be attempted.

Demonstration of waste emplacement, repository operation and sealing and near-field response

Another major part of any waste disposal organisation's R&D programme for a URL is demonstrating that waste can in fact be emplaced and backfilled *in situ* to the design standards, that a repository could be constructed and operated as planned in this geological environment and that backfilling and sealing of vaults, tunnels and shafts can be shown to meet previously agreed performance standards. This work is equivalent to the proposed fifth stage goal of the proposed programme for the Äspö HRL (SKB, 1989; Gustafson *et al.*, 1991).

The design of a repository will be very dependent on the types of waste that are to be disposed. In the type of high quality hard rock suitable for disposal purposes there are likely to be few geotechnical problems related to the construction of a HLW repository at depths up to 1000 m.

The inclusion of much larger volumes of ILW reprocessing and decommissioning waste may require the construction of large galleries, vaults or silos, and the geotechnical problems with these structures will increase in proportion to their dimensions. In both cases, however, an analysis will be required of the effects of the EDZ (Excavation Damage Zone), and it is also important to consider the long-term stability of the large openings and the influence that their evolution or potential collapse could have on the release of radionuclides (Figure 8.24). The evolution of the ILW waste packages, the type of mass backfill that could be considered in the ILW vaults and the interaction between the evolution of the vault structure with these two factors needs to be analysed conceptually and the concept examined in situ.

Repositories in sediments or evaporites are associated with different geotechnical problems, and the problems in plastic clays could be very significant. Again, the placement of the waste, its interaction with the backfill and any support or linings are important. The effects of elevated temperatures and gas generation within these low permeability rocks are likely to be of considerably greater significance than they are in hard rocks, and experimental programmes to examine these effects are likely to be required.

It is important at the outset to distinguish between the various possible repository options, so that the volume of rock that would be required for each option is known and so that the underground experiments in the URL are tailored to the requirements of testing and validating models relevant to each option. The possible presence of large underground openings for ILW disposal requires very different thinking as to the suitable conditions for the disposal of long-lived waste from the requirements imposed by the disposal of much smaller volumes of HLW. Much of the R&D work to date in URLs has concentrated on the problems associated with the much smaller volumes of HLW or spent fuel and has tended to ignore the potential problems associated with larger volumes of lower level, but still long-lived, wastes.

The backfilling of the repository is also very dependent on the types of waste to be disposed. HLW canisters can be placed in the tunnels themselves or in boreholes drilled from tunnels, and surrounded by backfill, such as compressed bentonite. Each canister, or group of canisters, can be effectively isolated, and any R&D programme in this area is likely to follow a similar path to that of SKB or GFS/BGR (Germany). For this type of disposal, in situ experiments are likely to be the most helpful. Such waste isolation and backfill and sealing opportunities are not possible for the much larger volumes of ILW, and the

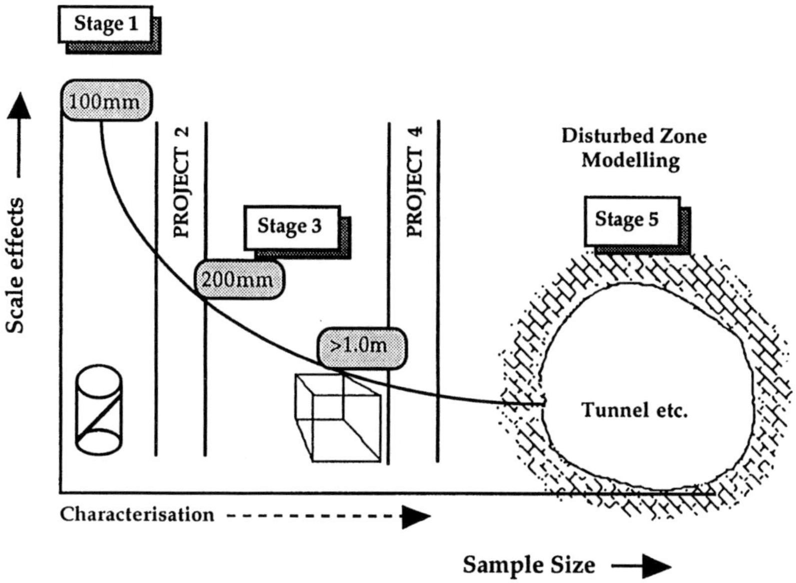

Figure 8.24 The problem of scale effect in relation to the testing of rock samples illustrated with respect to the development of a model of the EDZ (Excavation Damage Zone), with the ultimate fracture size being determined by the natural *block size* in the EDZ zone (modified from Barton *et al.*, 1992). The *Stages* and *Projects* in the figure are related to the staged programme of work within the Stripa Project concerned with the analysis of the geotechnical and hydrogeological effects of the EDZ.

backfilling of large waste vaults will require a very different set of experiments, many of which cannot take place underground. In this area of concern, Nirex's R&D backfill programme provides the most relevant example.

Much of the research within URLs in all rock types has been concerned with the near-field response. Validation of near-field models as part of the R&D programme needs to investigate some or all of the following factors:

- the temperature increase caused by radiogenic heating, and whether the temperature rise is as predicted, and the effect that this temperature rise has on the performance of the rock mass;
- the chemical conditions within, and the physical properties of, the backfill surrounding the waste. Whether the pH and redox conditions, the porosity and permeability (to groundwater and gas) of the backfill, the voidage

within the backfill and its strength and deformation properties are as predicted;
- the interaction, if any, between the backfill and the support requirements for the tunnels and the disposal boreholes. This concerns the potential problems associated with any rock support/backfill interface and potential highly conductive pathways in this area;
- the efficacy of grouting and sealing of fracture zones in tunnels and vaults and perhaps more importantly, in the shafts. Long-term hydraulic testing and subsequent exhumation of such zones will be required, all of which could be carried out in experimental galleries away from the shafts;
- the efficacy of emplacing tunnel and vault backfill and examination of the voidage, the bonding to the walls and the packing density using different techniques.

Some of the work in this area is more related to engineering design and operational requirements.

8.14.7 Work in the URL itself

Most of the work is likely to take place after the shafts have been constructed and, in the case of the URL being part of the assessment of the site, whilst the first part of the repository, perhaps a peripheral or spline tunnel, is being mined out. The construction of such a tunnel provides one of the best opportunities for validation and detailed characterisation of the rock mass within the repository volume. The types of experiments that are likely to be required as part of the validation exercise are listed below:

Work at this stage of the URL programme can be divided into the following general categories of collecting suitable data and testing predictive models:

1. Models of the geology on the scale of the repository down to the scale of single fractures.
2. Models of the hydrogeology at similar scales.
3. Models of geochemical transport models at similar scales.
4. Models of colloid transport.
5. Models of the geochemical evolution of the repository.
6. Models of the production and migration of gas from the waste.
7. Models of sealing and backfilling systems for shafts, boreholes, tunnels and vaults.
8. Models of the EDZ and other repository-induced effects.
9. Models of thermal and (thermo)-hydromechanical effects on radionuclide transport.
10. Engineering systems, such as waste emplacement.

8.14.8 The use of natural analogues in a URL programme

It is important to consider the issue of how evidence from the natural geological and geochemical environments of a URL site could be useful both in helping to test and validate models, particularly geochemical models, and in supporting the findings of the site characterisation. This type of work should be started as soon as possible after the site has been selected, as some analogue studies can take considerable periods of time to evaluate.

Evidence from local man-made sources around the site may also also be available. The use of natural evidence is generally termed the 'natural analogue' approach but this only covers one aspect of its usage. Much natural analogue information has, so far, been applied in a largely generic sense to the performance assessment of disposal concepts (*e.g.* Chapman, *et al.*, 1984; McKinley, 1989), rather than at specific sites. Four types of natural evidence is likely to be useful (Figure 8.25), all of which could be used in various ways to help in the testing and validation of models, in demonstrating that a good understanding of the evolution of the site has been obtained and in illustrating that the natural system at the site evolves only slowly at depth :

1. Generic, concept related natural analogues: these address key processes in the safety assessment. Information can come from very many sources, and from environments quite dissimilar to those of the two sites in question.
2. Site-specific natural analogues: where processes of interest in the safety assessment can also be found to have been active at the site.
3. Palæohydrogeological evidence; information on the rates of past fluid fluxes throughout the site which can be used as an indicator of rates and styles of future movements, for direct comparison with the simple radionuclide transport models used in safety assessments.
4. Geochemical cycling studies; where evidence on the large scale, long-term sources, fluxes and sinks of naturally occurring radionuclides or their chemical analogues can be used as an alternative end-point with which to compare releases from the repository.

Each of these approaches needs to be included in a thorough presentation of a safety case for a repository, regardless of their use in validation exercises, and could form a very constructive part of a URL programme.

Information on large scale geochemical cycling (essentially a description of how much comparably hazardous natural material is moving through the environment over the same timescales for which we calculate radionuclide releases) is most useful as a qualitative comparison when presenting the release/dose results of a performance assessment for a repository. It might be regarded as purely presentational or, if some weight is to be placed in a repository safety assessment on not increasing local background levels of radioactivity, in a more quantitative sense (see Chapman and Côme, 1989).

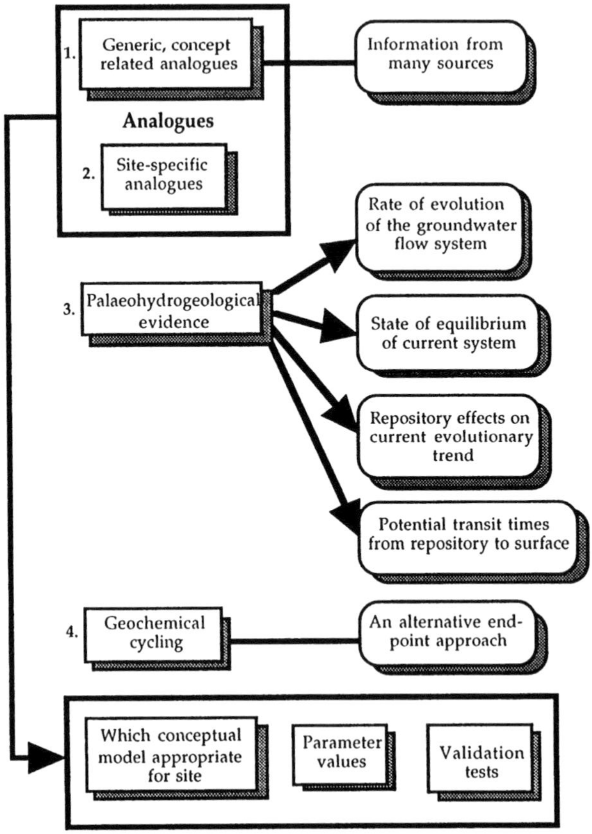

Figure 8.25 The four types of natural evidence that are likely to be useful in various ways to help in the testing and validation of models, in demonstrating that a good understanding of the evolution of the site has been obtained and in illustrating that the natural system at the site evolves only slowly at depth.

Debatably, too much emphasis is currently placed on individual risk targets. If this aspect were well presented, it might prove very useful for demonstrating the safe disposal of radioactive waste, using the URL site as an example of the use of this technique.

8.14.9 URL summary and conclusions

A URL-phase is now generally considered necessary as part of a comprehensive assessment of a potential waste disposal site for long-lived waste. In addition, an independent URL can be of use in the development of site characterisation methodologies, in testing modelling techniques, in demonstrating waste handling and emplacement techniques and in backfilling studies.

A URL programme actually starts well in advance of any underground construction and in situ experimentation, as the construction of the URL, and its influence on the surrounding groundwater regime and in situ stress field, provides some of the best opportunities for attempting validation exercises. In order to make best use of a URL, whether it be an integral part of the assessment of a disposal site or operated independently, sufficient time must be allowed for the following:

- sufficient site characterisation to allow optimal siting of the URL;
- construction of specialised testing and monitoring equipment, much of which may need to be developed and built specially for the URL programme;
- monitoring programmes to be in place before construction;
- pre-construction modelling to allow eventual comparison with results obtained post–construction;
- relatively slow construction rates;
- testing during construction;
- good interaction between modellers and experimenters;
- sufficient volume of available rock at depth to prevent interaction between experiments.

As has been indicated, a URL programme has to be viewed as a long-term venture, with investigations and modelling taking place over a period well in excess of a decade. Any attempt to speed up this process is only likely to result in inconclusive results, and may well prove to be counter-productive.

In addition, collaboration between different URL programmes, to prevent duplication of effort and to allow cross-fertilisation of ideas and techniques, is almost certain to be profitable.

8.15 References

Almén, K-E., Andersson, O., Fridh, B., Johansson, B-E., Sehlstedt, M., Gustafsson, E., Hansson, K., Olsson, O., Nilsson, G., Axelsen, K., and Wikberg, P. (1986a) Site investigation equipment for geological, geophysical, hydrogeological and hydrochemical characterization. *SKB Technical Report* TR 86-16. Swedish Nuclear Fuel and Waste Management Company, Stockholm, Sweden.

Almén, K-E., Andersson, J-E., Carlsson, L., Hansson, K. and Larsson, N-A. (1986b) Hydraulic testing in crystalline rock. A comparative study of single-hole test methods. *SKB Technical Report* TR 86-27. Swedish Nuclear Fuel and Waste Management Company, Stockholm, Sweden.

Almén, K-E. and Zellman, O. (1991) Äspö Hard Rock Laboratory. Field investigations methodology and instruments used in the pre-investigation phase, 1986-1990. *SKB Technical Report* TR 91-21. Swedish Nuclear Fuel and Waste Management Company, Stockholm, Sweden.

Andersson, P., Andersson, J-E., Gustafson, E., Nordqvist, R. and Voss, C. (1993) Site characterisation in fractured crystalline rock – A critical review of geohydraulic measurement methods. *SKI Technical Report* 93:23. Swedish Nuclear Power Inspectorate, Stockholm, Sweden.

Andrews, J., Fontes, J-C., Fritz, P. and Nordstrom, D.K. (1988) Hydrochemical asessment of crystalline rock formations for radioactive waste disposal – The Stripa experience. *Stripa Project Report* TR 88-05. Swedish Nuclear Fuel and Waste Management Company, Stockholm, Sweden.

Axelsson, C., Jonsson, E-K., Geier, J. and Dershowitz, W. (1990) Discrete fracture modelling. *SKB Technical Report* PR 25-89-21. Swedish Nuclear Fuel and Waste Management Company, Stockholm, Sweden.

Barton, N., Makurat, A., Monson, K., Vik, G. and Tunbridge, L. (1992) Rock mechanics characterisation and modelling of the Disturbed Zone Phenomena at Stripa. *Stripa Project Report* 92-12. Swedish Nuclear Fuel and Waste Management Company, Stockholm, Sweden.

Black, J.H. (1988) Focused packer testing: a radical methodology to improve single borehole hydraulic measurements. In *Proceedings of the International Association of Hydrogeology, Symposium on Hydrogeology and Safety of Radioactive and Industrial Hazardous Waste Disposal*, 7-10 June 1988, Orleans, France.

Bonne, A. (1991) Geoscientific investigations in the Belgian R&D programme concerning the disposal of radioactive waste in clay. In *Geological Problems in Radioactive Waste Isolation – A World Wide Review*, (ed. P.A. Witherspoon). *Lawrence Berkeley Laboratory Report* LBL-29703. Lawrence Berkeley Laboratory, Berkeley, California, U.S.A.

Bottomley, D.J., Gascoyne, M., Ross, J.D. and Ruttan, J.T. (1986) Hydrochemistry of the East Bull Lake Pluton, Masset, Ontario. *AECL Technical Report* TR-382. Atomic Energy of Canada Limited, Manitoba, Canada.

Brightman, M.A., Bath, A.H., Cave, M.R. and Darling, W.G. (1985) Pore fluids from the argillaceous rocks of the Harwell region. *British Geological Survey Fluid Processes Group Report* FLPU 85-6. British Geological Survey, Keyworth, U.K.

Carlsson, L., Winberg, A. and Arnefors, J. (1988) The Swedish final repository for reactor waste - Part I - Hydraulic testing and characterisation of the bedrock. In *Proceedings of the International Association of Hydrogeologists, Symposium on Hydrogeology and Safety of Radioactive and Industrial Hazardous Waste Disposal*, 7-10 June 1988, Orleans, France. Vol 1, pp 211- 225.

Chapman, N.A. and Côme, B. (1989) Long term predictions: Making proper use of geological evidence. *International Symposium on the Safety Assessments of Radioactive Waste Repositories*, NEA/OECD, Paris, October 1989. NEA/OECD, Paris, France.

Chapman, N.A. and McEwen, T.J. (1989) Geological aspects of the British programme for deep disposal of nuclear wastes. *International Geological Congress*, Washington DC, July 1989.

Chapman, N.A., McKinley, I.G., Smellie, J.A.T. (1984) The potential of natural analogues in assessing systems for deep disposal of high level radioactive waste. *Nagra Technical Report* NTB 84-41. Nagra, Baden, Switzerland.

Davis, S.N. and Nordstrom, D.K. (eds.) (1992) Hydrogeochemical investigations in boreholes at the Stripa Mine. The Hydrochemical Advisory Group and Their Associates. *Stripa Project Technical Report* 92-19. Swedish Nuclear Fuel and Waste Management Company, Stockholm, Sweden.

Davison, C.C. (1985) Far-Field hydrogeological monitoring at the site of Canada's Underground Research Laboratory. *Proceedings of CEC-NEA workshop on Design and Instrumentation of In Situ Experiments in Underground Laboratories for Radioactive Waste Disposal.* Brussels 15-18 May, 1984. Commission of the European Communities, Brussels, Belgium – Nuclear Energy Agency, Paris, France.

Edmunds, W.M. and Bath, A.H. (1976) Centrifuge extraction and chemical analysis of interstitial waters. *Environmental Science and Research* **10:** 467-479.

Falck, W.E. and Bath, A.H. (1989) Evaluation of vertical changes in groundwater chloride and stable isotope compositions in coastal London Clay. *British Geological Survey Fluid Processes Group Report* WE/89/90. British Geological Survey, Keyworth, U.K.

Freeze, R.A., de Marsily, G., Smith, L. and Massmann, J. (1988) Some comments about uncertainties in nuclear waste disposal. *Proceedings of DOE-AECL Symposium on the use of Geostatistics,* San Francisco, Sep. 1987.

Frick, U., Alexander, W.R., Baeyens, B., Bossart, P., Bradbury, M.H., Bühler, Ch., Eikenberg, J., Fierz, Th., Heer, W., Hoehn, E., McKinley, I.G., Smith, P.A. (1992) The radionuclide migration experiment – Overview of investigations 1985-1990. *Nagra Technical Report* 91-04. Nagra, Wettingen, Switzerland.

Gautschi, A. and Scholtis, A. (1989) Sampling and analysis of groundwaters from deep boreholes. *Nagra Bulletin Special Edition*, December 1989, p 50-66. Nagra, Baden, Switzerland.

Grisak, G.E., Pickens, J.F., Belanger, D.W. and Avis, J.D. (1985) Hydrogeologic testing of crystalline rocks during the Nagra Deep Drilling Program. *Nagra Technical Report* NTB 85-08. Nagra, Baden, Switzerland.

Gustafson, G., Liedholm, M., Rhén, I., Stanfors, R. and Wikberg, P. (1991) Äspö Hard Rock Laboratory. Predictions prior to excavation and the process of their validation. *SKB Technical Report* TR 91-23. Swedish Nuclear Fuel and Waste Management Company, Stockholm, Sweden.

HMIP (1992) Dry Run 3. A trial assessment of underground disposal of radioactive wastes based on probabilistic risk analysis. Volumes 1 to 10. *UKDOE Report* No. DOE/HMIP/RR/92. UK Department of the Environment, London, U.K.

IAEA (1981) Safety assessment for underground disposal of radioactive wastes, procedures and data. *IAEA Safety Series Report* No. 56. International Atomic Energy Agency, Vienna, Austria.

IAEA (1988) Radioactive waste management glossary, 2nd ed. IAEA-TECDOC-447, International Atomic Energy Agency, Vienna, Austria.

Kinniburgh, D.G. and Miles, D.L. (1983) Extraction and chemical analysis of interstitial water from soils and rocks. *Environmental Science and Research* **17:** 362-378.

Küpfer, Th., Hufschmied, P. and Pasquier, F. (1989) Hydraulic testing in the Nagra boreholes. *Nagra Bulletin Special Edition*, December 1989, p7-23. Nagra, Baden, Switzerland.

Langer, M., Schneider, H. and Kühn, K. (1991) The salt dome of Gorleben – Target site for the German radioactive waste repository. In *Geological Problems in Radioactive Waste Isolation – A World Wide Review,* (ed. P.A. Witherspoon). *Lawrence Berkeley Laboratory Report* LBL-29703. Lawrence Berkeley Laboratory, Berkeley, California, U.S.A.

de Marsily, G. (1986) *Quantitative Hydrogeology.* Academic Press, 440pp.

Mazurek, M. (1994) Geology of the crystalline basement of northern Switzerland and derivation of geological input data for hydrodynamic modelling and safety performance analysis. *Nagra Technical Report* NTB 93-12. Nagra, Wettingen, Switzerland.

McCombie, C., McKinley, I.G. and Zuidema, P. (1991) Sufficient validation: The value of robustness in performance assessment and system design. *Proceedings of GEOVAL-90 Symposium,* Stockholm, May 1990. Nuclear Energy Agency, Paris / Swedish Nuclear Power Inspectorate, Stockholm.

McKinley, I.G. (1989) Applying natural analogues in predictive performance assessment. *Nagra Technical Report* 89-02. Nagra, Baden, Switzerland.

McEwen, T.J., Chapman, N.A. and Robinson, P.C. (1991) Review of data requirements for groundwater and solute transport modelling and the ability of site investigation methods to meet these requirements. *UK DOE Report* No. DoE/HMIP/RR/90/095. UK Department of the Environment, London, U.K.

Nagra (1994) Conclusions from the regional investigation programme for siting a HLW repository in the crystalline basement of northern Switzerland. *Nagra Technical Report* NTB 93-09E. Nagra, Wettingen, Switzerland.

NEA/SKI (1990) The International INTRAVAL Project, Background and Results (1990). Nuclear Energy Agency, Paris / Swedish Nuclear Power Inspectorate, Stockholm.

Niederer, U. (1991) In search of truth: The regulatory necessity of validation. In *Proceedings of GEOVAL-90*, NEA/SKI, Stockholm, May 1990. Nuclear Energy Agency, Paris / Swedish Nuclear Power Inspectorate, Stockholm.

Niemeyer, M., Hugi, M., Smith, P.A. and Zuidema, P. (1993) Kristallin-1 performance assessment – First results from sensitivity analysis studies. *International IAEA Symposium on Geological Disposal of spent fuel, high-level and alpha-bearing wastes*, Antwerp, Belgium, October 1992. IAEA, Vienna, Austria.

Nirex (1992) The geology and hydrogeology of Sellafield. *Nirex Report* No. 263, 2

Volumes, (ed. U. McL. Michie). UK Nirex Limited, Harwell, U.K.

Nirex (1993) Nirex deep waste repository project – Scientific Update 1993. *Nirex Report* No. 525. UK Nirex Limited, Harwell, U.K.

Noy, D.J. and Holmes, D.C. (1986) A single hole tracer test to determine longitudinal dispersion. *British Geological Survey Fluid Processes Research Group Report* FLPU 86-5. British Geological Survey, Keyworth, U.K.

Olsson, O. *et al.* (eds.) (1992) The Stripa Project. Site Characterisation and Validation – Final Report. *Stripa Project Report* TR 92-22. Swedish Nuclear Fuel and Waste Management Company, Stockholm, Sweden.

Olsson, O. (1992) Characterisation ahead of the tunnel front by radar and seismic methods – a case history from the Äspö HRL. *SKB Progress Report* PR 25-92-01. Swedish Nuclear Fuel and Waste Management Company, Stockholm, Sweden.

PAGIS (1984) Performance Assessment of Geological Isolation Systems – Phase 1 report for the Salt Option. CEC R&D Programme on Radioactive Waste Management and Storage.

Parkhurst, D.L., Thorstenson, D.C. and Plummer, L.N. (1980) PHREEQE - A Computer Program for Geochemical Calculations. *US Geol Survey Water Resources Investigation* No 80-96, p187; (revised and re-printed 1985). United States Geological Survey, Reston, Virginia, U.S.A.

Pearson, F.J., Lolcama, J.L. and Scholtis, A. (1989) Chemistry of waters in the Böttstein, Kaisten and Leuggern boreholes: A Hydrochemically Consistent Data Set. *Nagra Technical Report* NTB 86-19. Nagra, Baden, Switzerland.

Pearson, F.J., Balderer, W., Loosli, H.H., Lehmann, B.E., Matter, A., Peters, T.J., Schmassmann, H., and Gautschi, A. (1990) Isotope Hydrogeology of Northern Switzerland. *Nagra Technical Report* NTB 88-01. Nagra, Baden, Switzerland.

Shaw, W. and Grindrod, P. (1988) Investigation of the potential of fuzzy sets and related approaches for treating uncertainty in radionuclide transfer predictions. *CEC Report* EUR-12499. Commission of the European Communities, Brussels, Belgium.

SKB (1983) Final storage of spent nuclear fuel – KBS-3, Vols. 1-5. Swedish Nuclear Fuel and Waste Management Company, Stockholm, Sweden.

SKB (1989) Handling and final disposal of nuclear waste: Hard Rock Laboratory. Background Report to the R&D Programme 89. Swedish Nuclear Fuel and Waste Management Company, Stockholm, Sweden.

SKB (1992a) SKB 91 – Final disposal of spent nuclear fuel. Importance of the bedrock for safety. *SKB Technical Report* 92-20. Swedish Nuclear Fuel and Waste Management Company, Stockholm, Sweden.

SKB (1992b) Treatment and final disposal of nuclear waste. The Äspö Hard Rock Laboratory, Background Report to the R&D Programme 92. Swedish Nuclear Fuel and Waste Management Comany, Stockholm, Sweden.

SKB (1993) Äspö Hard Rock Laboratory. Annual Report 1992. *SKB Technical Report* 93-08. Swedish Nuclear Fuel and Waste Management Company, Stockholm, Sweden.

Smellie, J., Larsson, N-A., Wikberg, P. and Carlsson, L. (1985) Hydrochemical investigations in crystalline bedrock in relation to existing hydraulic conditions: experience from the SKB test sites in Sweden. *SKB Technical Report* TR 85-11. Swedish Nuclear Fuel and Waste Management Company, Stockholm, Sweden.

Smellie J., Larsson N-A., Wikberg P., Puigdomenech I. and Tullborg E-L. (1987)

Hydrochemical investigations in crystalline bedrock in relation to existing hydraulic conditions: Klipperås test site, Smaland, Southern, Sweden. *SKB Technical Report* TR 87-21. Swedish Nuclear Fuel and Waste Management Company, Stockholm, Sweden.

Stanfors, R., Erlström, M. and Markström, I. (1991) Äspö Hard Rock Laboratory. Overview of the investigations 1986-1990. *SKB Technical Report* 91-20, Swedish Nuclear Fuel and Waste Management Company, Stockholm, Sweden.

Thury, M., Gautschi, A., Mazurek, M., Müller, W., Naef, H., Pearson, F.J., Voborny, O., Vomvoris, S. and Wilson, W. (1994) Geology and hydrogeology of the crystalline basement of northern Switzerland. *Nagra Technical Report* NTB 93-01E. Nagra, Wettingen, Switzerland.

TVO (1992) Final disposal of spent nuclear fuel in the Finnish bedrock – Preliminary site investigations. *TVO Report* YJT-92-32E. The Nuclear Waste Commission of Finnish Power Companies, Helsinki, Finland.

Vomvoris, S., Voborny, O., Wilson, W., Hürlimann, W. and Andrews, R.W. (in prep.) Hydrodynamic synthesis of crystalline rocks of northern Switzerland: Summary of the results relevant to safety analysis, reference area West. *NagraTechnical Report*. Nagra, Wettingen, Switzerland.

Wikberg, P., Axelsen, K. and Fredlund, F. (1987) Deep groundwater chemistry. *SKB Technical Report* TR 87-07. Swedish Nuclear Fuel and Waste Management Company, Stockholm, Sweden.

Wikberg, P., Gustafson, G., Rhén, I. and Stanfors, R. (1991) Äspö Hard Rock Laboratory. Evaluation and conceptual modelling based on the pre-investigations 1986-1990. *SKB Technical Report* TR 91-22. Swedish Nuclear Fuel and Waste Management Company, Stockholm, Sweden.

9

Performance assessment

9.1 Introduction

In the field of radioactive waste disposal there is a profusion of different types of performance assessment. There are three main questions to ask when considering what type of assessment is required:

- **who** is carrying out the assessment?
- **why** is the assessment being undertaken?
- **how** is the assessment to be carried out?

With regard to **who**, the person or organisation carrying out the assessment can be very important in determining the type of assessment. At the risk of oversimplifying, four categories of interests will be considered:

- politician/strategist;
- developer;
- waste producer;
- regulator.

In general, the **politician/strategist** will be concerned with the impact of the performance assessment on policy matters.

The **developer** needs in the first instance to identify a suitable site. Once a site has been selected he has to convince the regulator that he should be allowed to proceed with repository development. The **regulator** may consider that he needs to undertake an independent performance assessment in order to be able to judge the case made by the developer.

The viewpoint of the **waste producer** may be rather different from that of the developer. He is primarily concerned with being able to dispose of the wastes he produces (partly to avoid expensive, and possibly hazardous, storage on his site), and is less concerned about the wider issues of repository development. He will, of course, want costs to be kept as low as possible, and may want to resist any moves by the regulator or developer to widen the range of wastes which can be accepted at the repository if this results in a cost penalty for his own wastes.

In terms of **why** an assessment is carried out, most will be associated with one of the following categories:

- strategic studies;
- siting criteria and site selection;
- site characterisation and repository design;
- waste acceptance criteria;
- regulatory submission.

There are inevitable overlaps between these categories, and slightly different categorisations are possible (*e.g.* McCombie and Zuidema, 1992).

By its very nature, radioactive waste management covers very long timescales and it is worth considering in a little more detail the various strategic issues which need to be addressed from the production of the waste to its final disposal. Some examples of such issues are:

- identification of feasible disposal options;
- choosing between direct disposal of spent fuel and reprocessing;
- determining the period of interim storage prior to final disposal;
- encapsulating wastes so as not to prejudice final disposal.

Repository performance assessments have a major part to play in the determination of important strategic issues. The objective of the assessments in this context is to provide information to guide strategy development. For these very general studies few data are likely to be available, since only an outline of the disposal system will be employed; the assessments will be 'generic' in nature.

Once the search starts for a repository for a particular class of radioactive wastes, criteria have to be set to limit the geographical areas considered, and these will generally be based on geological considerations. Potential sites have then to be identified and compared. In both these stages performance assessments play an important part. The whole subject of site selection has been dealt with in detail in Chapter 7, but it is important to bear in mind that the developer is not looking for the safest site from a post disposal point of view, but one which is at least adequate. The post disposal safety performance of a repository will have to be weighed against other considerations, such as the ease of repository construction and transport of waste to the repository, and socio-economic factors. At this stage, although the performance assessments will be site specific, detailed geological and hydrogeological data may not yet be available for those sites.

When a particular site has been chosen for further investigation, performance assessments will be an important input to site characterisation and repository design. Results from the assessments will help guide the placement of priorities. Where are better data required from the site characterisation programme in order to be able to make the safety case more robust? What are the most cost effective improvements in repository design? Performance assessments at this stage will help to identify research needs. There is a continuing need for iterations between the site characterisation team, repository designers, research scientists and those carrying out the performance assessments.

Whilst repository development is underway, waste producers have to continue to manage their wastes until a disposal route is available. For many wastes there are various management options, and the waste producer needs to be assured that when choosing between such options he is not prejudicing the suitability of the final waste form for disposal. As the repository design develops it will become easier to identify those waste management options which are compatible with the disposal concept. The determination of waste acceptance criteria is bound to change as more data about the repository become available.

Before repository construction can proceed, and before operations can commence, some sort of regulatory submission will be required of the operator, the details depending on the legislation in the country where the potential repository is sited. The performance assessments carried out at these stages need to have adequate data to be able to make the safety case required by the criteria applicable in that country.

How a performance assessment is undertaken can be affected by the following considerations:

- the regulatory regime;
- end points to be evaluated;
- timescales to be considered;
- the target audience.

The first of these is often the most important. In Chapter 1 regulatory criteria were discussed. These vary greatly from country to country and can directly influence the other considerations listed. The regulations may dictate, at least to some extent, what end points are considered in a performance assessment: individual doses, risks, radionuclide fluxes or something else? They may determine the timescales for which calculations are undertaken; is there a cut off time after which there is no need to evaluate the end points?

The last consideration, the target audience, is very important, and this is often not fully appreciated. The way in which the safety of a repository is to be demonstrated depends very much on who the audience is. A full and technically complex set of calculations may be needed to obtain regulatory approval, but this is unlikely to be understood by decision makers and members of the public. It is vital that the lessons learnt from detailed performance assessments are translated into robust arguments that can be understood by other groups of people.

The remainder of this chapter will consider the 'how' question in more detail. In this regard, there are various approaches and techniques that are available for organising and analysing information for repository performance assessments. A general theme is the need to follow systematic procedures in order to develop a defensible train of argument. The major components of a comprehensive assessment may consist of:

- Scenario Development,
- Conceptual Model Development,
- Mathematical Model Development,
- Consequence Analysis,
- Uncertainty and Sensitivity Analysis,
- Confidence Building.

Depending on the purpose of the assessment, some of these stages may be omitted or severely curtailed. While it is convenient to discuss these topics in a sequential manner, this is not meant to imply that performance assessment is a one-pass process. To the contrary, there is ample evidence to demonstrate that assessments benefit considerably from iteration between the above activities and with other aspects of repository development such as site characterisation.

9.2 Scenario development

A key element of assessments of post-closure safety and performance of underground repositories is the identification of the possible future evolutions of the disposal system – this is the process called scenario development or scenario analysis.

There is a large and unavoidable uncertainty about the long-term future state and/or evolution of radioactive waste disposal systems. This has traditionally been treated by carrying out performance assessment calculations for a small number of stylised descriptions of possible future states or evolutions termed *scenarios.* Structured scenario development procedures are necessary to identify and define the scenarios to be treated. Estimation of scenario probability or *degree-of-belief* causes particular difficulty. *Environmental simulation* techniques can provide assistance to scenario development and may provide a more satisfactory basis for probabilistic safety assessment of radioactive waste disposal systems.

This section presents a summary of the state of the art in scenario development for performance assessment and is based in part on Sumerling and Hodgkinson (1992).

9.2.1 Background

A particular feature of the performance assessment of geological disposal of radioactive waste is the very long timescale over which assurance of safety is sought. For example: the U.S. Environmental Protection Agency have developed standards that set quantitative limits for releases that may occur within *10,000 years after closure*; the U.K. Department of the Environment have adopted a risk target that should not be exceeded *at any time* (in the future).

Over such timescales, both the natural environment and the engineered waste repository will change due to natural processes, human actions (unrelated to the disposal), and the interactions of natural environment and repository. Even given a well characterised disposal site and completely specified waste and repository, there will be large and unavoidable uncertainty about the future state or evolution of the system due to:

- uncertainty about the importance (impact) or rate of various natural processes which will act on the system, *e.g.* degradation, geochemical and biological processes;
- the timing or frequency of certain natural phenomena, *e.g.* seismic events and climatic change;
- essentially unpredictable human activities in the far future, *e.g.* groundwater and mineral resource use.

In the assessment of post-closure impacts of radioactive waste disposal, the uncertainty in future states has traditionally been handled by carrying out performance assessment calculations for a number of simplified or stylised conceptual descriptions of future state or evolution, termed *scenarios.* Scenarios first entered common parlance in the area of predictive modelling of the future via the work of Kahn (1967) in the context of predicting the consequences of thermo-nuclear war. For that purpose, scenarios were defined as:

> '... hypothetical sequences of events constructed for the purpose of focusing attention on causal processes and decision points. They answer two types of questions: (1) Precisely how might some hypothetical situation come about, step by step? and (2) What alternatives exist, for each actor, at each step, for preventing, diverting, or facilitating the process.'

For the purposes of evaluating the future behaviour of a waste repository the same approach might be used, changing the second question to *What is the likelihood of this sequence of events occurring and what are the consequences?* The answers to these questions would then allow judgements to be made about the safety of a repository system, and these might be reflected in the design of the system, the siting of the repository, *etc.* Since the work of Kahn, the idea of using scenarios or possible 'futures' has become widely used in business and industry as a planning and brain-storming tool.

Recent work by the Swedish Nuclear Power Inspectorate (SKI) has proposed a definition of scenarios more directly appropriate to radioactive waste disposal:

> A scenario is a hypothetical sequence of processes and events, and is one of a set devised for the purpose of illustrating the range of future behaviours and states of a repository system, for the purposes of making or evaluating a safety case. It is not necessary, or indeed possible, to describe individually all possible scenarios. However, consideration of the complete set should provide an adequately robust test of repository safety by addressing the most likely futures together with less likely futures which exhibit features of widely acknowledged concern.

The process of eliciting the scenarios most important to the performance assessment, for which detailed calculations of performance should be undertaken, is termed scenario development. Scenario development should also provide a logical framework by which the individual scenarios can be organised or related, enabling a coherent use and presentation of results. More recently, the developing disciplines of climatology and geomorphology have been invoked to provide quantitative *environmental simulation* models which have been used as an aid to scenario development and, also, to generate input to radiological assessment models directly.

9.2.2 Scenario methodologies

The Nuclear Energy Agency (NEA) co-ordinated a Working Group with the objective of considering, at an international level, issues related to the identification and selection of scenarios for performance assessment of radioactive waste disposal. A review of approaches to scenario development examined by the working group was presented at an NEA/IAEA/CEC Symposium in Paris in 1989 (Hodgkinson and Sumerling, 1990). The stages generally included in a scenario development are:

- identification and classification of all phenomena relevant to the performance of the repository and site;
- screening of phenomena according to well-defined criteria;
- identification and grouping of scenarios relevant to the performance of the repository and site;
- specification of scenarios for consequence analysis.

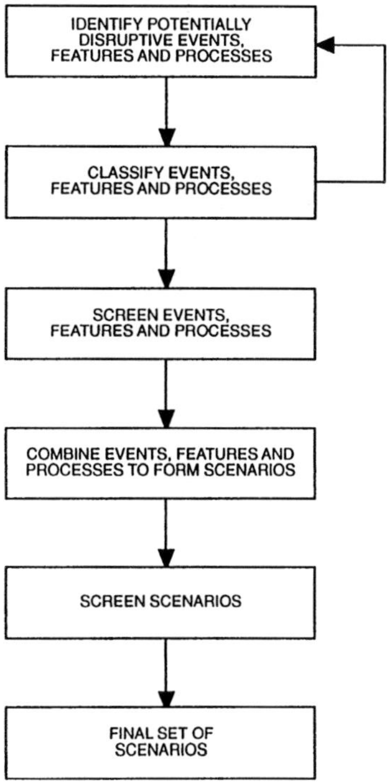

Figure 9.1 The Sandia/NRC scenario selection procedure (from Cranwell *et al.*, 1982). Reproduced by permission of the Sandia Corporation, U.S.A.

The Sandia/NRC procedure

More than a decade ago Sandia National Laboratories developed a structured approach to scenario selection for radioactive waste disposal for the United States Nuclear Regulatory Commission (Cranwell *et al.*, 1982) see Figure 9.1. This has been applied to hypothetical HLW repositories in unsaturated tuff (Guzowski *et al.*, 1986) and in basalt (Hunter, 1983), and to proposed disposals of transuranic wastes at a bedded salt site (Guzowski *et al.*, 1990).

Expert opinion is used to identify the relevant phenomena. The methodology focuses explicitly on disruptive events, features and processes, that is, phenomena that will lead to changes in the repository and/or natural environment such as to initiate or modify radionuclide release and transport.[1]

[1] The terminology of 'features, events and processes' or 'FEPs' is now widely used in scenario development work. Broadly, FEPs can be defined as:

- **Features:** of the disposal system which are not generally accounted for in performance assessment of an unperturbed repository system, such as undetected geological structures (solution cavities, breccia pipes, *etc.*), or unplanned aspects of the repository system (such

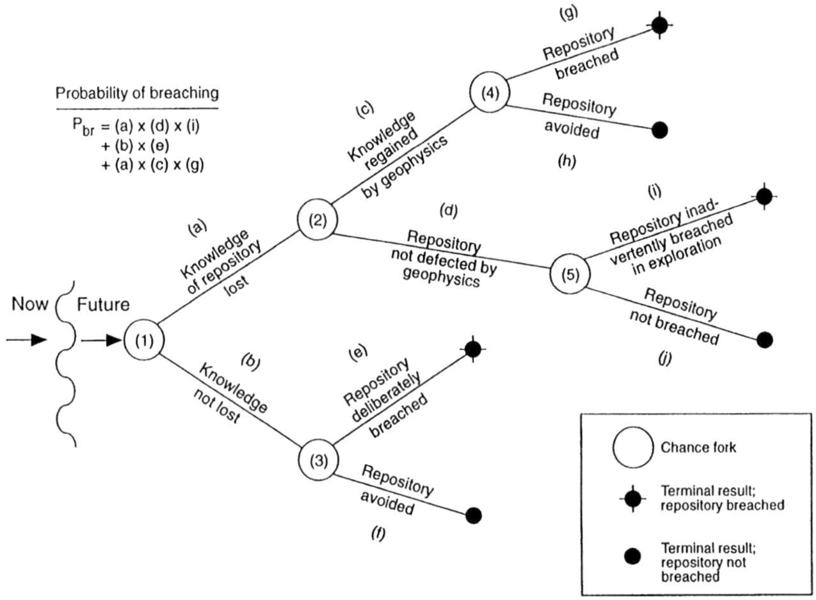

Figure 9.2 Part of an event tree to assist in estimating probability of human intrusion (from Hunter and Mann, 1989). Reproduced by permission of the Sandia Corporation, U.S.A.

Screening criteria can be based on: physical reasonableness of phenomena, low probability of occurrence of an event, and/or low potential consequences of an event or process, for the specific site and disposal system under consideration. Phenomena remaining after screening may be lumped into a single composite phenomenon if they have similar effects. Scenarios can then be constructed by taking all physically meaningful combinations of occurrence or non-occurrence of remaining phenomena to form an *event tree.* Hunter has developed the use of event trees, *e.g.* see Figure 9.2, to analyse quite complex combinations of phenomena (Hunter and Mann, 1989).

The scenario in which no initiating or modifying phenomena occur is

---

as unsealed exploratory boreholes, inadequately emplaced backfill, waste containers which are failed at the time of disposal, *etc.*).

- **Events:** are generally of short duration and of natural or human origin. They could include earthquakes, faulting, flooding, intrusion by people into the repository, *etc.*
- **Processes:** normally includes the numerous natural processes which will dictate the evolution of the repository system after closure, such as container corrosion, water flow, radioactive decay, heat transport, radionuclide sorption, climate change, *etc.*

referred to as the *base case* and represents a continuation of the estimated present day regime. Other scenarios may be handled as perturbations or new flow pathways super-imposed on the base case.

The joint SKI/SKB scenario development project

The Swedish Nuclear Power Inspectorate (SKI) and Nuclear Waste Management Company (SKB) carried out a scenario development exercise, for a hypothetical repository for spent fuel and HLW, using the Sandia methodology as a starting point (Andersson *et al.*, 1989).

Two important features of this project were that: four groups of experts worked **independently** to identify relevant features, events and processes (FEPs) under different systems of classification; and **all** features, events and processes that might influence the long-term performance of the repository, *i.e.* not just disruptive phenomena, were listed.

The final merged list included about 150 FEPs. In order to assist in the organisation of this large number of FEPs remaining after screening, the concept of the Process System was introduced. This was defined as follows:

> The Process System is the organised assembly of all phenomena (FEPs) required for the description of barrier performance and radionuclide behaviour in a repository and its environment, and that can be predicted with at least some degree of determinism from a given set of external conditions.

Essentially, it includes all the phenomena that can be represented within the conventional models of barrier performance and radionuclide behaviour. The other phenomena are mainly external events. In the process of screening the initial list of FEPs, the majority are assigned to be part of the Process System. A relatively smaller number, called 'external FEPs' or 'EFEPs' were superimposed on the Process System, thus being used to form scenarios which can be represented by changing boundary conditions, or making additions, to the conventional models.

An attempt was also made to take a 'top-down' approach by defining three possible states of the main barriers and mapping those FEPs remaining after screening, and not already included in the Process System to these states, see Table 9.1.

The Canadian scenario analysis project

Atomic Energy of Canada Ltd. (AECL) have carried out a scenario analysis for the post-closure assessment of the disposal of spent fuel in deep crystalline rock (Stephens and Goodwin, 1990).

All relevant phenomena, termed *factors*, were recorded, including factors

Table 9.1 Illustration of consequences on the barrier states caused by individual FEPs, from the SKI/SKB study. Reproduced by permission of SKi.

Kept FEPs	Barrier States		
	Canister	near-field	Far field
Faulting	SC*	LE	SC*
near-field deficiencies	O	LE	O
Unsealed boreholes	O	O	SC*
Uplift/subsistence	O	O	LE
Glaciation	SC*	LE	SC*
Human actions on groundwater flow and composition	O	O	LE

Key to barrier states: O = ordinary
SC = short circuit
LE = less efficient
* = for part of repository only

affecting the performance of the biosphere. An initial massive list of over 1000 factors was reduced to about 250, of which 114 were identified as requiring quantitative treatment. To make the construction of scenarios more tractable the *central scenario* was introduced. This contains as many factors as possible and is similar to the Process System of the SKI/SKB exercise. Residual factors were then grouped, or screened out, and *alternative scenarios*, were formed by combining one or more of the residual factors, plus factors from the central scenario.

In view of the Canadian regulatory criteria, which include a risk guideline, the AECL group attempted to assign probabilities to scenarios. This was done by assigning relatively low values, 0.01 and 0.05, to the alternative scenarios that included one residual factor, hence 0.0005 to the alternative scenario that included both, hence the remaining 0.9395 to the central scenario.

SKI Project 90 and SITE-94

Following the joint SKI/SKB exercise to set in place the basis for scenario development, SKI continued to test the approach in their Project 90 assessment of a spent fuel repository at a hypothetical site in Sweden (SKI, 1991). The external FEPs that had been identified were lumped into groups having similar causes to form a set of 10 scenarios. These were analysed very simply by parameter variation in the central case used in the assessment. Thus, in Project 90, the difficult step of applying these lumped EFEPs to the Process System in order to define calculations of scenario consequence was not addressed.

Following Project 90, SKI have embarked on the SITE-94 project, which is designed to develop their assessment methodology further and build on the experience of Project 90. As part of this exercise the scenario approach is being developed in a more rigorous fashion than was possible in Project 90. A graphical model of the Process System has been constructed and carefully audited and screened for both content (of FEPs) and influences between FEPs to produce a Process Influence Diagram (PID) which forms the core of all subsequent scenario analysis. It is structured as a very large influence diagram with supporting text files on a computer database (see Figure 9.3).

Expert opinion has been used to attribute quantitative 'degrees of belief' to the significance of each influence ('influence levels'), thus allowing different versions of the PID to be constructed, reflecting different degrees of system uncertainty. At the time or writing, the project is exploring ways of incorporating conceptual and parameter uncertainty into the methodology. A Central Scenario has been defined which incorporates best estimate descriptions of the climatic evolution of Sweden over the next million years (King *et al.*, in press). A more rigorous approach is being made to evaluating the impacts of individual EFEPs on the Process System by determining the changes they cause in influence levels. Scenarios are defined both by consideration of an influence diagram containing all the EFEPs (see Figure 9.4), and by addressing extremes of parameter variation in the central scenario. The difficult task of translating scenarios into calculation cases for consequence analysis is also being studied. The concept of the 'Assessment Model Flowchart' (see Figure 9.4) has been developed to assist in making the transition from PID to calculation case. This approach ensures that all FEP's and influences are considered and their treatment recorded, thus providing an excellent performance assessment quality assurance tool. It is anticipated that, although SITE-94 will make further advances in scenario methodology, the system being developed will still require further testing in the future. This is in line with much current thought on scenarios which regards scenario development as a dynamic process which must grow and adapt to incorporate increased experience and increased understanding of a repository site and system.

Difficulties with probability assignment

A comprehensive scenario probability study has recently been made covering resource exploration, climatology, tectonics, seismicity and volcanology (Hunter and Mann, 1989). In principle, probabilistic information about

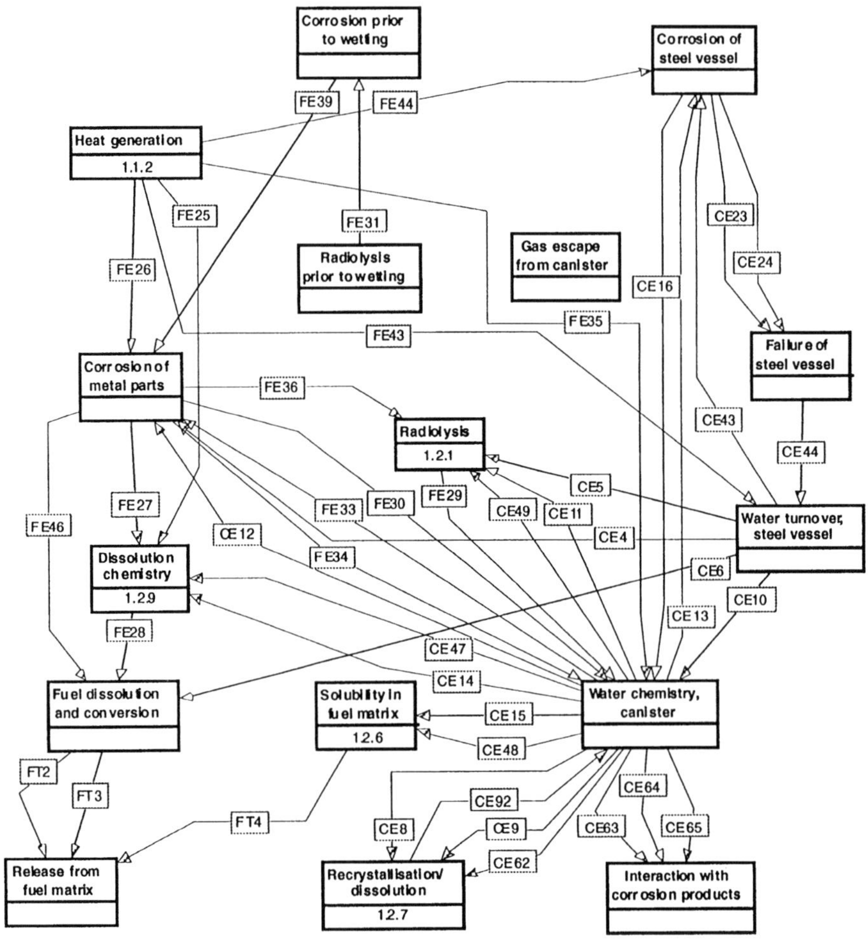

Figure 9.3 Part of the extensive Process System influence diagram being used by SKI as part of their SITE-94 project. Reproduced by permission of SKi.

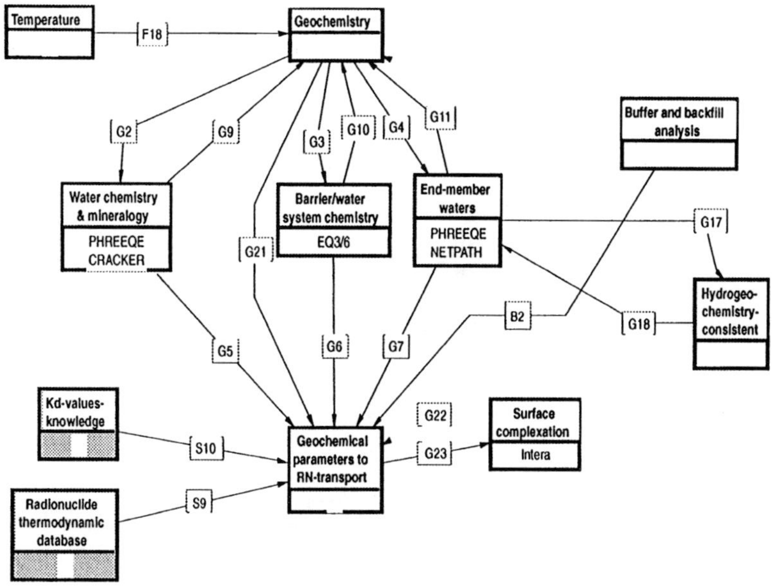

Figure 9.4 Part of the Assessment Model Flowchart being developed by SKI within SITE-94. Reproduced by permission of SKi.

component events and processes could be combined to give an overall scenario probability in a *bottom-up* approach. In practice, combining such quantities compounds the uncertainty in the overall scenario probability to such an extent that it can be less meaningful than if it had been elicited directly by expert judgement. Thus scenario probabilities must be assigned largely on the basis of expert judgement; these are essentially *degrees of belief*, as discussed in the SITE-94 work, rather than conventional probabilities.

It has been suggested that a *top-down* approach can be taken to scenario formation, analogous to the use of fault tree analysis for nuclear reactor safety. In this approach, consequences are postulated and the mechanism by which these states may be reached is then considered. In an early analysis by the UK DoE (Dalrymple *et al.*, 1986), a target event is defined, whose outcome is that of ultimate concern, in this case, whether risk is below, or exceeds, the risk target of 10^{-6} a^{-1}. The main factors that are considered to affect the outcome of the target event are the performance of the various barriers, *i.e.* vault (repository), geosphere and biosphere. A similar, but less probabilistically orientated analysis was touched on during the SKI/SKB Joint Scenario Development Project (see above).

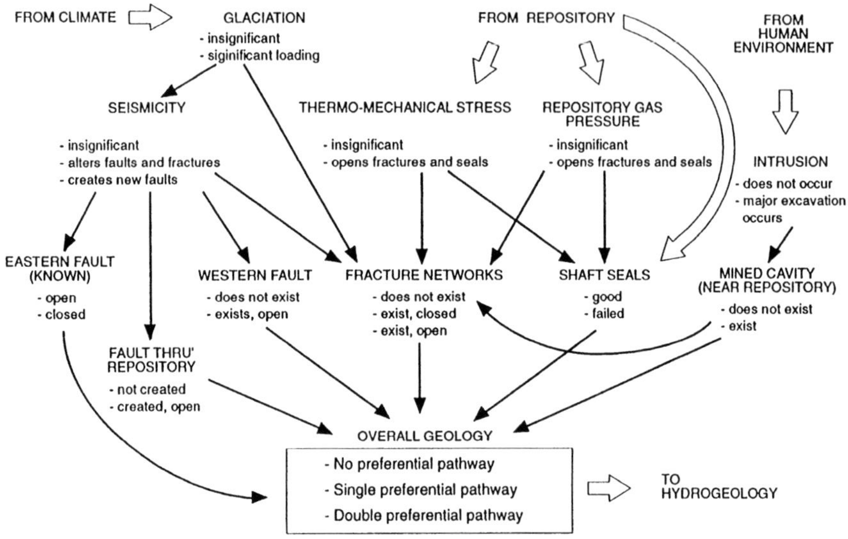

Figure 9.5 Part of an influence diagram relating scenario elements and states.

Scenario element approach

Working with a screened phenomenon list for guidance, this methodology divides the repository natural environment and radiological system into component parts, termed *scenario elements.* A comprehensive set of *alternative states* which each scenario element may adopt is then postulated. An influence diagram showing dependencies between scenario elements is drawn; this aids the definition of elements and states. Non-physical and unimportant combinations of scenario element states are screened to arrive at a set of scenarios for consequence analysis. Finally, degrees of belief are assigned to each scenario element state taking account of states of other elements via the influence diagram; hence scenario probabilities are derived. Figure 9.5 shows part of an influence diagram constructed in a trial application of the methodology.

The top-down approach allows a more justifiable assignment of probabilities, and hence enables calculation of risk for comparison with a risk target. The scenario set formed by all allowable combinations is clearly complete if the alternative states are complete. The difficulty is that the link between the extensive phenomenon lists and the alternative states must be made in order to demonstrate this.

9.2.3 Simulation approaches

Several models have been developed to simulate the long-term evolution of the environment, *e.g.* see (INTERA, 1983a, b; Courbouleix *et al.*, 1985; Ringrose *et al.*, 1990). To date, these have concentrated mainly on the effects of climate-driven change, and geological processes, although models for the simulation of human intrusive activities have also been proposed, *e.g.* Hunter and Mann (1989), Wittingham (1989). The value of such a model depends on the attention paid to initial selection of phenomena to be included; this should be explicitly documented, *e.g.* see Dames and Moore (1988).

In the main, these models are intended to explore the possible evolution of the environment in advance of calculations of radiological models, either deterministically, *e.g.* see Peaudecerf and Blanc (1990), or probabilistically, *e.g.* see INTERA (1983a). Used in this mode they can assist in the scenario development and may be of particular value in quantifying environmental parameters in otherwise qualitatively described scenarios.

U.K. DoE simulation approach

The U.K. DoE have developed an integrated systems based probabilistic safety assessment methodology (Thompson, 1989, 1993). The procedure is centred on the use of Monte-Carlo simulation to:

- generate samples of possible future evolutions of the natural environmental system, using the TIME2 and TIME4 codes (Ringrose *et al.*, 1990);
- analyse the effect of these temporal changes and their associated uncertainties on the repository nuclide release and groundwater transport system, using the VANDAL code (Laurens *et al.*, 1990), to estimate the risk incurred.

The approach is illustrated in Figure 9.6. Results from application of the methodology are presented by Thompson (1991).

The motivation for this course is two-fold. Firstly, to demonstrate compliance with a risk target it is necessary to investigate the full range of possible repository and environmental developments, and to assign probabilities coherently so that consequences can be combined. As argued by Thompson (1989), the use of traditional scenarios is not sufficiently rigorous for this purpose. Secondly, the developing disciplines of quantitative climatology and geomorphology do provide a scientific basis for the estimation of the variation with time of significant environmental parameters that affect the release and transport of radionuclides (Ringrose *et al.*, 1990; Boulton, 1990), and it would be remiss not to use the best current understanding.

A strength of the simulation approach is that the act of formulating a model forces the assumptions to be documented clearly. Also it provides a

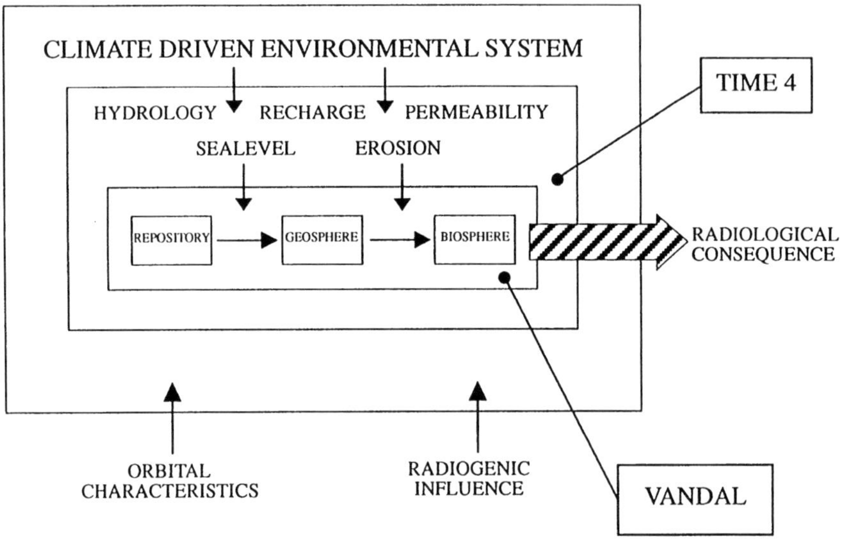

Figure 9.6 Conceptual view of a climate driven environmental system (Laurens *et al.*, 1990). Reproduced by permission of NEA/OECD.

rigorous way of dealing with interactions between phenomena, and creating coherent time sequences. Since the model is complex, there are considerable data requirements and the problems of understanding and presenting the results are challenging, see Sumerling and Thompson (1992).

9.2.4 Summary and conclusions

The *bottom-up* approach to scenarios analysis has been applied and further developed in major scenario analysis projects in the U.S.A., Sweden and Canada. The method has been particularly successful in identifying a broad range of relevant phenomena, providing a forum for experts to discuss these phenomena and hence broadening the knowledge and experience base on which quantitative assessments can proceed.

Phenomena which are continuous or readily predictable can be combined to define a conceptual process model (*e.g.* termed the Process System in the Swedish SKI/SKB analysis) that is used to evaluate consequences for a given scenario class. Event-tree analysis may be used to construct scenarios by combining phenomena that can be regarded as events leading to discrete consequences. These may provide boundary conditions for the process models.

A drawback of the bottom-up approach, especially significant for probabilistic safety assessment, is that it is not generally possible to derive a probability for a given scenario from the probabilities associated with its component phenomena. The *top-down* procedure allows scenario probabilities to be assigned coherently; therefore probability weighted results from individual scenarios can be combined to make an estimate of total risk. In both approaches, the scenarios generated are highly stylised and the estimates of probability or *degrees of belief* are subjective.

Environmental simulation provides a method of creating coherent time sequences that can be combined in a statistically meaningful way. However, a substantial investment is required to develop an integrated system model, and data requirements may be considerable.

Both scenario and environmental simulation approaches are limited by the lack of understanding and data concerning processes which may affect the environment into the far future. However, the probabilistic environmental simulation approach focuses attention on the elicitation of probability distribution functions for defined parameters, whereas the scenario approach poses questions about the likelihood of rather broadly defined classes of futures. In both cases, but especially the latter, reliance must be placed on subjective judgements; formalized elicitation techniques can be used to minimise the bias, and increase the defensibility, of subjective parameter estimates and of assignments of scenario degrees of belief (Bonano *et al.*, 1990).

A criticism that has been levelled at the scenario approach is that the terminology is rather vague (Thompson and Sagar, 1993). This stems partly from the existence of many programmes which have tended to follow their own course, each drawing partially from other programmes, and partly from the complexity of data management systems that need to be set up to define scenarios for analysis. Further developments should aim at rationalising terminology, as there is probably less difference in approach than may appear. A further criticism, which can only be answered by more rigorous work to define influences more quantitatively and to set up calculation cases with clarity, is that the eventual selection of scenarios often seems rather *ad hoc* in nature.

The choice of methodology may depend on the audience to which the assessment is addressed and also the safety criteria that are to be satisfied. A combination of scenario based and environmental simulation approaches may be necessary in order to develop a full safety case. Environmental simulations can aid scenario development (*e.g.* for helping define climatic evolution scenarios) and provide a quantitative base for parameters in scenario descriptions. The broader ranging qualitative discussions of scenario development can identify phenomena and interactions that are not considered in the environmental simulation model. The debate between scenario based and environmental simulation approaches has led scientists following both courses to examine their methodologies and provided an additional incentive to improve procedures on both sides.

9.3 Conceptual model development

For each significantly different scenario there is a need to develop a number of conceptual models which describe possible structures and behaviours of the system, and form the basis for mathematical modelling.

A conceptual model is taken here to be a set of hypotheses or assumptions describing the physical and chemical processes that affect the development of the repository system and the surrounding environment, together with geometry, structure, properties, initial and boundary conditions of the system. In fact just as with the word scenario, there is no universally agreed definition of the term 'conceptual model'.

Until recently, many assessments used a single conceptual model. Uncertainties in parameter values for a given conceptual model have been investigated extensively (OECD/NEA, 1990) but uncertainties arising from alternative conceptual models have until recently been largely ignored. And yet, these conceptual model uncertainties have the potential to give rise to the dominant uncertainties (Andersson *et al.*, 1991).

A case in point concerns the flow of water and the transport of radionuclides through a fractured rock mass. Figure 9.7 shows four alternative conceptual models for the rock mass permeability field. It is not appropriate to ask the generic question as to which of these, if any, is correct since the system is so complex and the extrapolations in time and space are so extensive that scientific validity is not achievable. Instead there is a need to ask whether the models are fit for purpose, such as for calculating the average flux of water through rock on a scale of tens of metres within order-of-magnitude accuracy. On limited time and space scales it is possible to test this fitness for purpose by defining surrogate performance measures and comparing independent predictions and measurements, as was done in the international Stripa Project (Hodgkinson, 1992).

In many ways, the status of methodologies for treating conceptual model uncertainty, resembles that of scenario development methodologies in 1987 before the NEA Scenarios Working Group began its state-of-the-art review (NEA, 1992). It is hoped that, as with scenarios, systematic procedures can be developed to elicit alternative conceptual models, ensure they are compatible with known facts, and propagate the results of alternative models through performance assessments.

One aspect of developing a systematic approach to conceptual model development, is the documentation of all modelling assumptions. For example, in the recent Uncertainty and Bias Audit project undertaken for Her Majesty's Inspectorate of Pollution (HMIP) this was achieved by eliciting

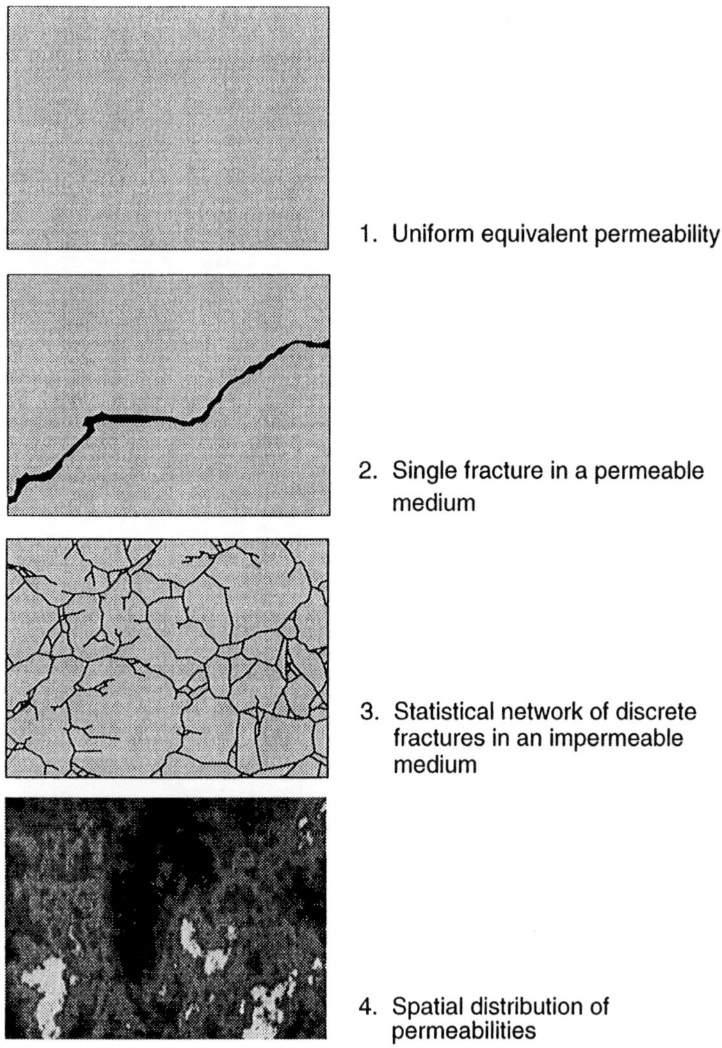

Figure 9.7 Alternative conceptual models for fractured rock permeability.

Decisions Assumptions and Omissions (DAOs) from sub-system Working Groups together with measures of the associated uncertainty and bias. This information is contained in a database which facilitates traceability and allows DAOs to be ranked in order of importance.

There are three approaches to propagating the effects of conceptual model uncertainty. Probabilists would argue that conceptual models should be sampled statistically as is currently done for parameters in probabilistic systems analysis. This in effect views alternative conceptual models as components of one super-model with uncertain parameters. The output from such an approach would thus be a weighted average of different conceptual models. On the other hand, the fuzzy logic perspective would be to take the weakest link in the chain, namely the conceptual model that gives the most pessimistic result. The argument here is that if the model is consistent with our understanding and is calibrated with known data then it cannot be ruled out. A third approach would be to state that it is not possible to combine results from different conceptual models; all one can do is document the possible sources of bias.

9.4 Mathematical model development

Mathematical models provide the primary tool of performance assessment. They encapsulate our general scientific understanding of processes and are made specific by information on a particular repository design and site. Mathematical models translate the assumptions of a conceptual model into the formalism of mathematics, usually sets of coupled algebraic, differential and/or integral equations with appropriate initial and boundary conditions in a specified domain. These equations are solved either analytically, semi-analytically or numerically to give the time and space dependence of the quantities of interest.

Mathematical models are required for two primary purposes. First, models are required to describe the evolution of the environment in the vicinity of a repository. Examples are the chemical evolution in the near-field, and the impact of climate change on the near-surface environment. Secondly, models are required to describe the transfer of radionuclides through the evolving environment.

In general, environmental changes affect radionuclide transfer but not *vice versa*, since radionuclides generally occur in trace quantities. Of course, there are exceptions, for example alpha radiolysis of water from actinides in a repository can affect the redox conditions and hence the mobility of radionuclides.

As discussed below, mathematical models come in all shapes and sizes. Some level of simplification is generally required in order to translate the concepts of a conceptual model into mathematical terms. This can take the form of simplifying the geometry or structure, for example considering transport in only one dimension or the medium to be homogeneous and isotropic. In addition, models often omit processes and interactions or simplify their description, for example neglecting kinetic terms in chemical reactions.

In general there is a hierarchy of models depending on the degree of simplification. At the most detailed level, research models are used to build an understanding of processes and structures. The aim of research modelling is to build an understanding of issues of importance to repository performance. This can be done by analysing the results of experiments or by using models to investigate the effects of interactions between various processes and structures.

Returning to the conceptual models for heterogeneous fractured rock shown in Figure 9.7, a typical research model would be the NAPSAC code (Herbert *et al.*, 1991) which generates statistical distributions of intersecting planar fractures, and calculates water flows and the transport of inert tracers (see Figure 9.8).

At the other end of the spectrum lie assessment models, for sub-systems (*e.g.* near-field) and the total system. Assessment models in use today usually have a simplified geometry, structure and representation of processes. This is due to computational constraints arising from the need to carry out a large number of calculations for sensitivity and uncertainty analysis, and to evaluate various design options. For example, assessments of radionuclide transport in fractured rock are usually carried out using one-dimensional models, or networks of such models, despite the fact that three-dimensional codes are available.

With the ever increasing power of computer hardware and the development of user-friendly and efficient software, it is probable that the distinction between research and assessment models will become more and more blurred. Moreover, it will become possible to simulate increasingly complex phenomena, for example coupled flow and kinetic chemical reactions in spatially heterogeneous media.

The use of detailed models in performance assessments is to be welcomed since the process of producing simplified models is far from rigorous and can introduce unquantified uncertainties and biases. Important aspects can be lost in the process of simplification, and for this reason the reduction is often done in such a way that it produces a pessimistic result.

A large number of computer codes have been developed for post-closure repository analysis, and the pace of development shows no sign of slowing down. For example, a recently updated directory (Ashton *et al.*, 1993) identified 320 codes on the following topics: nuclide inventory, corrosion, leaching, geochemistry, geomechanics, heat transfer, groundwater flow, radionuclide migration, biosphere modelling, safety assessment and site evolution.

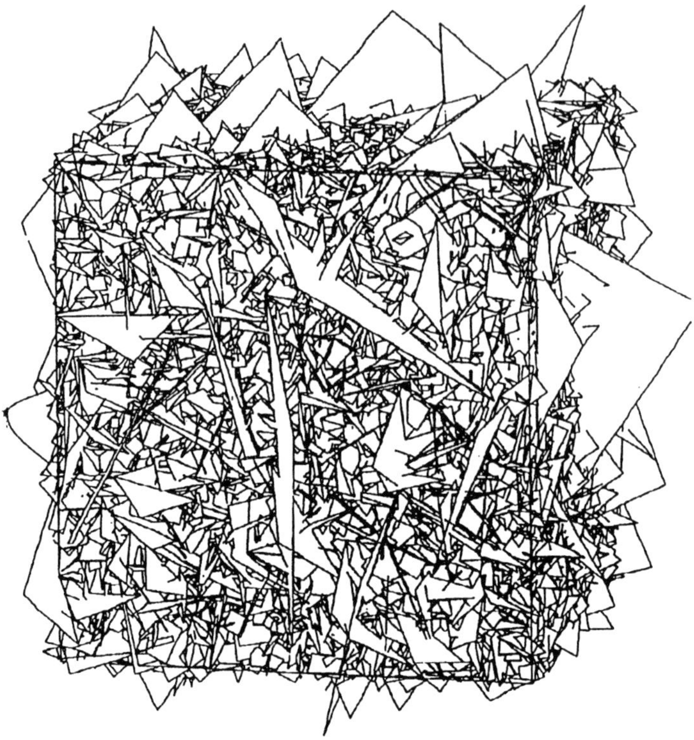

Figure 9.8 Fractures generated by the NAPSAC code (Herbert *et al.*, 1991). Reproduced by permission of SKB.

The following sections review the types of models that have been used for the important problems of radionuclide transport in groundwater through the near-field and geosphere.

9.4.1 Near-field radionuclide transport

The near-field of a repository is taken to encompass the engineered barrier system, comprising the waste packages and backfill material, together with that part of the surrounding geosphere that has been significantly affected by repository construction and waste emplacement. Models are required to calculate the repository source term, namely the time-dependent rate at which radionuclides enter the undisturbed geosphere.

A simple and robust approach to source term modelling is to imagine that following saturation of the repository (in groundwater-saturated hydrogeological systems) and the loss of physical containment, due to degradation of the containment materials such as corrosion of a steel canister, the radionuclides are uniformly dispersed throughout the near-field. The concentrations are determined according to the decay and ingrowth of radionuclides, linear sorption, elemental solubility limits and depletion due to advection to the geosphere (Robinson *et al.*, 1988). Such models are particularly useful for low and intermediate-level waste repositories where the engineered structures are generally more porous than for high-level waste repositories (NEA, 1988). A typical example is the STRAW code used by Nirex (Robinson and Williams, 1989).

The next level of source term model incorporates some degree of spatial dependence, and accounts for time-dependent transport processes throughout the near-field. For example, the STRENG code used by Nagra models dissolution of a glass waste matrix and diffusive transport of radionuclides through a clay buffer in addition to the phenomena included in the simpler homogeneous repository models (Grindrod *et al.*, 1990).

A wide range of additional phenomena have been incorporated into source term models, depending upon design features, rock properties and phenomena of particular concern. For example, the CALIBRE code used by the Swedish Nuclear Power Inspectorate includes the effect of oxidant production, from alpha-radiolysis of water, on the dissolution of spent nuclear fuel and on the redox conditions in a repository and its impact on the migration of redox sensitive radioelements (Robinson *et al.*, 1991). Also, CALIBRE includes the effect of solubility limits throughout the near-field, rather than just within the waste form, and this has been shown to lead to the formation of precipitation fronts in some situations (Worgan and Apted, 1992). Finally, CALIBRE includes both the transport of radionuclides along a fracture crossing the deposition hole and diffusion in the rock matrix throughout the near-field.

Further processes that have been incorporated into source term models include non-linear sorption (Robinson *et al.*, 1992) and temporal variation of physical and chemical properties (Maul *et al.*, 1993). A comprehensive review of near-field models has recently been produced (Andersson *et al.*, 1993), and Chapter 4 surveys the modelling of key near-field processes.

9.4.2 Geosphere radionuclide transport

There is an extensive literature on fluid flow in the geosphere for radioactive waste disposal applications and for related fields such as contaminant hydrogeology and petroleum reservoir engineering (de Marsily, 1986). The modelling of key geosphere processes is considered in Chapter 5.

The first and most straightforward task involves the calculation of groundwater flow for specified boundary conditions and permeability fields. For rocks that can be approximated by a collection of blocks each with a uniform permeability, there are numerous codes which can be used to compute the groundwater flux and head distribution (Ashton *et al.*, 1993). These usually make use of finite-element or finite-difference techniques, and have been widely tested, for example in the international HYDROCOIN project (NEA, 1992). Particular examples of codes which have been widely used to compute three-dimensional groundwater flows for radioactive waste applications, are CFEST (Gupta *et al.*, 1987) NAMMU (Atkinson *et al.*, 1985) and SWIFT (Reeves *et al.*, 1986).

In recent years, it has been increasingly realised that many rock masses are highly heterogeneous in their hydrogeological properties. This has an important impact on groundwater flow, and an even more profound impact on solute transport. Essentially, solutes are channelled along localised paths of least resistance, rather than being dispersed in a relatively uniform plume. Various geostatistical approaches have been developed to generate permeability fields which account for heterogeneity in a statistical sense. These methods discretise the permeability field to a much higher resolution than the traditional permeable medium approach, and assign patterns of permeabilities to volume elements which reflect the measured variability. Recent tests of these methods have been carried out in relation to flow in the Culebra Dolomite aquifer overlying the WIPP facility in New Mexico (Zimmerman and Gallegos, 1993).

For fractured rock masses, where the major flows are through a network of intersecting fractures (see Figure 9.8), a related statistical approach has been developed and used for example in the Stripa Project (Hodgkinson, 1992). The discrete fracture approach is clearly more realistic for modelling on the scale of individual fractures, and has proved useful for integrating hydraulic test data with information on fracture geometries. One application is to deduce permeability distributions for input to permeable medium models which average over discrete fracture effects. The major discrete fracture codes are FRACMAN/MAFIC (Dershowitz *et al.*, 1991; Miller, 1990) and NAPSAC (Grindrod *et al.*, 1991).

Using the various codes discussed above it is possible to model fluxes of water through a rock mass in three dimensions, although for reasons of time, cost and the availability of data, two-dimensional simulations are often used. The modelling of solute and colloid transport in such flow fields is considerably more complex from conceptual and numerical perspectives, as well as requiring a wide range of additional information.

The traditional approach to modelling solute transport in the geosphere is to assume a distribution of kinematic porosity (often taken to be uniform within rock units) to convert groundwater flux into groundwater velocity, and to include this velocity field within a transport equation accounting for advection, linear equilibrium sorption and hydrodynamic dispersion, as described in Chapter 5. Concerns have been expressed over the adequacy of this formulation, for example its simple treatment of chemical effects and dispersion, and lack of matrix diffusion. However, even for this relatively simple formulation, there have been rather few applications in two dimensions and even less in three dimensions, due to computational constraints with finite-element and finite difference codes such as CFEST, NAMMU and SWIFT. Fortunately, more computationally efficient, accurate and more user-friendly radionuclide transport codes are under development, such as LAPLACE-2D (Robinson *et al.*, 1994).

To date, most radionuclide migration calculations have been performed using one-dimensional models, or networks of such models. These are computationally efficient and can include a variety of transport processes such as matrix diffusion and colloid transport. The CRYSTAL code is a typical example of a one-dimensional code with matrix diffusion (Worgan and Robinson, 1992), and the COLLAGE code can be used for transport in the presence of colloids (Grindrod *et al.*, 1992). A simple one-dimensional advection-dispersion-sorption model is included in the PAGODA code (Robinson, 1991).

9.5 Consequence analysis

For each scenario and conceptual model considered in an assessment, there is a need to analyse the consequences using a suitable mathematical model with appropriate parameter values and boundary conditions. This section surveys the various types of consequence analyses that have been used in the radioactive waste disposal context.

9.5.1 Scoping calculations

Scoping calculations, often based on simple analytical formulae, are useful in a number of ways. The first is to place the hazards of radioactive waste disposal in perspective with other environmental concerns. Lord Marshall's Garden Lecture is one example of this. Another is the calculation of time-dependent hazard indices for radioactive waste.

Lord Marshall as Chairman of the UK Central Electricity Generating Board (CEGB), which until 1989 generated and supplied electricity throughout England and Wales, used to try to put the risks of radioactive waste into context by considering the natural radioactivity in a 'garden unit' of soil, illustrated in Figure 9.9 (Lord Marshall of Goring, 1988). Calculating the amount of uranium, potassium and thorium in the soil he compared the risks from ingesting or inhaling the radioactivity associated with a 'lifetime's supply' of electricity from nuclear power, amounting to 16 litres of low level waste, 4 litres of intermediate level waste, and 0.1 litres of high level waste.

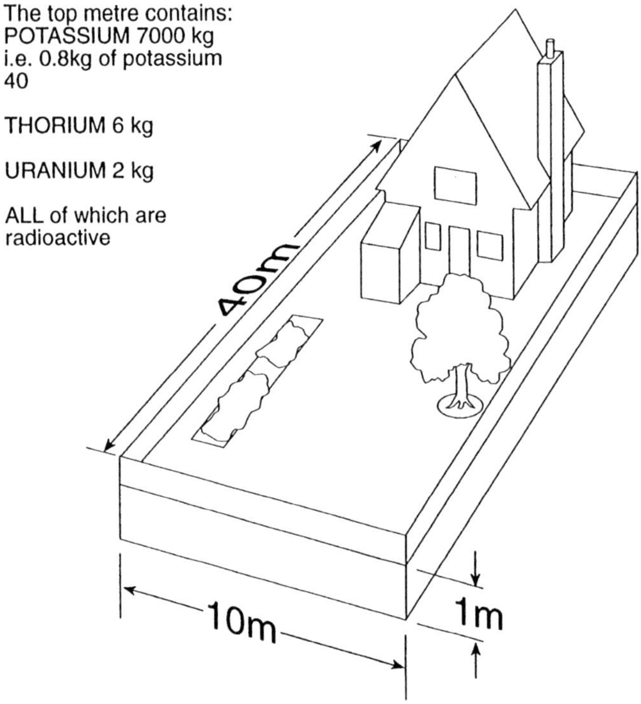

Figure 9.9 A 'garden unit' of radioactivity (Lord Marshall of Goring, 1988). Reproduced by permission of Nuclear Electric plc.

He concluded, for example, that for low level waste the risks were a factor of 10,000 below those for the garden unit!

Most people would not recognise these arguments as constituting a performance assessment, but they certainly constitute a simplified assessment for a particular purpose. The assessment was carried out by a waste producer in order to put the risks of radioactive waste into perspective, and in terms of the determining factors considered above, it could be regarded as a strategic assessment attempting to influence the public. In this case the constraints of regulatory regimes did not apply; the originator of the assessment could determine his own end points to illustrate the case he was making.

Another example of this is shown in Figure 9.10, which uses a generic estimate of the transfer rate for an atom at 600 m in rock to the surface, in order to calculate the expected number of deaths per year for each GWe-year of electricity produced (Cohen, 1986).

Scoping calculations have an important role to play in scenario development, where it is required to screen out processes and interactions having an insignificant effect on the system performance. For example, in Project-90 (Andersson *et al.*, 1991) while the major transport pathway for corrodants and radionuclides was taken to be via fractures, scoping studies were undertaken to estimate transport to a tunnel through the buffer and backfill material (Shaw and Worth, 1992).

Last but by no means least, scoping calculations are an intrinsic part of checking the order-of-magnitude of more detailed calculations, which is a standard procedure in engineering analysis.

9.5.2 Bounding calculations

One approach to dealing with uncertainty (Chapter 10) is to choose scenarios, conceptual models, mathematical models and parameter values which overestimate the consequences. Such bounding calculations have the advantage that they are generally rather straightforward to explain and defend to a variety of audiences. For this reason they are particularly suited to radioactive waste disposal decisions which straddle the tightrope between science and public policy.

However, there are a number of problems with this approach which have become apparent. The first is how to establish that an assessment is pessimistic, without performing more detailed calculations. Recourse is usually made to arguments based on mental models, but these can be misleading when dealing with complex systems. For example, it could be argued that it is pessimistic to omit dispersion in the geosphere as this only spreads solutes out in space. However, this is not always the case, for example when radionuclides dispersed ahead of the central plume arrive before significant radioactive decay has occurred (Hadermann, 1980). In general, it is difficult to argue that particular scenarios or models are intrinsically pessimistic; it depends on the parameter values and detailed assumptions.

Another problem with presenting bounding calculations is that critics might go on to make even more pessimistic assumptions. My worst case is worse than yours! Also, while this approach may be sensible for the intrinsically safe systems used in radioactive waste disposal, if adopted for more active systems such as chemical plants, it could impede economic progress.

A balanced view of the use of bounding calculations is that they have a useful role to play in presenting safety cases to wide audiences, but they need to be backed up by more detailed models and uncertainty analysis.

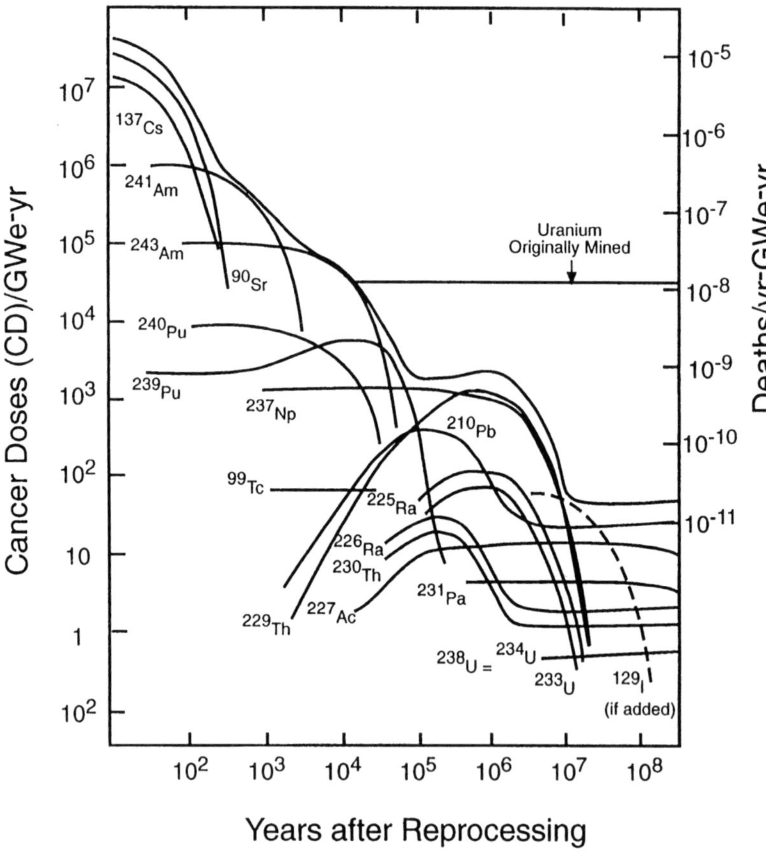

Figure 9.10 Number of cancers expected if all reprocessed waste produced by generation of 1 GWe-yr of electricity were eaten by people, versus time. Individual curves show contributions of individual radionuclides, and the thick curve is their sum. Scale on right gives deaths/yr using an estimate that the transfer probability from buried waste into human stomachs is $4 \times 10^{13}/yr$. The dashed line shows the number of cancers expected due to radiation if the 160 te of uranium (+ daughters) needed to produce 1 GWe-yr of electricity were eaten by people. From Cohen, B.L. (1986) Risk analysis of buried wastes from electricity generation, *American Journal of Physics* **54(1)**. Reproduced by permission of American Journal of Physics.

9.5.3 Deterministic calculations

The deterministic approach to consequence analysis makes use of models with fixed values of the parameters. The major advantage of this approach is that the most detailed models which encapsulate an in-depth understanding of processes and interactions can be used. Typically, a best estimate set of parameters would be used for a base case calculation. This provides a point of reference against which results from other calculations can be measured. A set of calculational cases whose parameters span the range of interest would usually be evaluated in order to build up an appreciation of possible consequences, and to develop an understanding of the system.

An example of the deterministic approach to consequence analysis is to be found in Project 90 (Andersson *et al.*, 1991). Here, a set of near-field and far-field cases were defined for use with the CALIBRE and CRYSTAL codes (Robinson *et al.*, 1991), in order to develop an understanding of uncertainties arising from an incomplete knowledge of scenarios, conceptual models and parameters.

A drawback to the deterministic approach is that there is usually no systematic or complete coverage of the uncertainty space. Also, it is difficult to ascribe probabilities in a coherent way to an *ad hoc* set of cases, and thus to estimate the risk, if this is a required end-point.

9.5.4 Probabilistic calculations

Over the past decade, there has been an increasing use of probabilistic system assessment (psa) in radioactive waste disposal assessments, in part stimulated by the NEA Probabilistic System Assessment Group (NEA, 1990).

This approach focusses on integrated models of the total system, sometimes including the time-dependent environment (Thompson and Sagar, 1993). Uncertainties are quantified in terms of probability density functions for the model parameters, and these are propagated through the system model to give a distribution of consequences. This propagation is usually carried out by statistical sampling of the input distributions using *e.g.* Monte Carlo or Latin Hypercube techniques. However, for relatively simple models, more exact techniques are available (Robinson and Hodgkinson, 1987; Shaw and Grindrod, 1989).

Figure 9.11 shows an example of the output from a number of psa codes for the PSACOIN Level E test case in which the mean dose from the chain Np-237 – U-233 – Th-229 was estimated and compared to an exact solution (NEA, 1989).

The ensemble of results produced by a probabilistic system assessment provides a rich seam for subsequent analysis. First, various means, moments, percentiles and other statistics can be derived. Secondly, the distribution

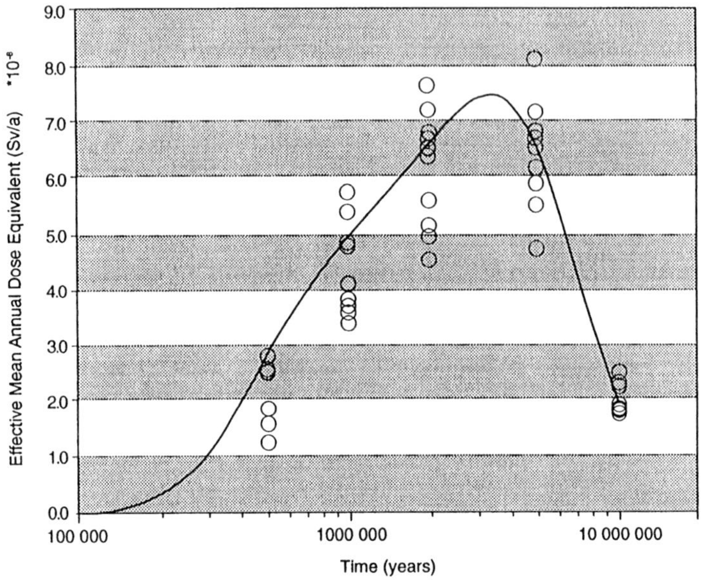

Figure 9.11 Example comparison of results from the PSACOIN Level E exercise. Participants' estimates of the mean dose from the chain $^{237}Np -^{233} U -^{229} Th$ at 5 times are plotted together with the exact solution (continuous line). Reproduced by permission of NEA/OECD.

of consequences gives a direct measure of uncertainty. Thirdly, sensitivity analysis techniques can be applied in order to determine how the results depend on the input assumptions and parameters, and their relative importance. The ranking of input parameters can provide useful feedback to experimental programmes and repository design studies.

The strength of psa lies in its ability to be comprehensive in exploring the space of the phenomena considered, and their associated model parameters. Its weakness is the need to make use of simplified models and the possibility that the statistical sampling may choose parameter combinations outside their range of validity. For this reason, a combination of probabilistic and deterministic approaches is often used, with detailed calculations being performed as a check in high consequence regions of parameter space. However, with the inexorable increase in the power of computers and the development of highly efficient software, it may soon be possible to run today's detailed models in probabilistic mode.

Of course, probability theory is not the only approach to dealing with our lack of knowledge of repository systems. Fuzzy logic offers an alternative paradigm, as discussed in Chapter 10, and software has been developed to assess radionuclide migration models using this technique (Brown and Robinson, 1993).

9.6 Uncertainty and sensitivity analysis

Uncertainty analysis is central to the assessment of radioactive waste disposal performance because of the large extrapolations in time and space that must be made. This topic is discussed in Chapter 10 together with sensitivity analysis techniques for ascertaining the most significant assumptions and parameters.

In the past, these techniques have been applied extensively to parameter uncertainty in consequence calculations. In future, they need to be applied to the total system including the choice of scenarios and conceptual models.

A major problem with uncertainty analysis is its unfamiliarity outside certain technical disciplines. It is thus difficult to communicate to some groups involved in decision making. This has led some people to revert to the use of pessimistic bounding models which can be defended more easily in public. However, it should be possible to develop improved techniques for presenting uncertainties, especially as the concepts become more familiar, for example in weather forecasting.

9.7 Confidence building

How are decision-makers to be convinced of the safety of disposing of radioactive waste underground? The first step is to decide what is meant by safety, and this has been done by national and international agencies and codified into regulations as discussed in Chapter 1. Of course, even this is a contentious subject. While there are common threads running through the various national regulations, they differ in detail and are subject to change, as is currently happening to the high-level waste regulations in the U.S.A. At the end of the day, our best hope of building confidence in regulations is peer review by international agencies such as the NEA, IAEA and ICRP.

Having decided upon the regulations, how do we ensure that they are met? For periods of tens to hundreds of years following disposal it is possible to monitor a repository, especially for shallow land burial, thus giving confidence in its performance or evidence that remediation measures are required. For deep disposal, we are generally concerned with what might happen in the distant future and thus there is no hope of monitoring the repository and comparing this directly with regulatory limits. Consequently, we need to compare forecasts based on mathematical models with regulatory criteria.

What confidence can we have in these forecasts? On the face of it, they are based on complicated science and mathematics embodied in equally complex computer codes. To some extent it is possible to test models. For example, over limited time and length scales, models can be compared with the results of independent experiments and a judgement made as to whether they are fit for purpose, as was done for example in the Stripa (Hodgkinson, 1992) and INTRAVAL projects (SKI/OECD/NEA, 1993; Chapman *et al.*, 1994). However, the purpose of such models is inevitably limited to time and length scales much smaller than those required for repository assessments.

In addition, validation exercises based on laboratory and field experiments can be complemented by the study of natural evidence on appropriate time and length scales. However, the initial conditions and evolutionary history of analogue sites are inevitably poorly understood, so that this evidence can only be supportive, not definitive. The use of natural systems data in confidence-building is discussed further in Chapter 10.

Computer codes themselves can be tested fairly rigorously by comparison of test case results with known solutions, and by making use of sophisticated logical analysis tools that have been developed in recent years. Thus, verification that a computer code performs as required is perhaps the most objective aspect of the problem of building confidence.

There needs to be confidence that the chain of activities from data acquisition and interpretation through to model development and application, has been carried out to sufficiently high standards. Quality Assurance has a role to play here, in laying down agreed procedures for each type of activity, and providing documentary evidence that they have been followed. Of course, it can be tedious to produce a documentation trail but fortunately computer systems which keep a log of activities are likely to come to our aid in this respect. At the very least, QA should provide a paper trail which allows the train of argument to be followed: right or wrong.

It cannot be emphasised too strongly that repository performance assessments are not just based on scientific facts. A very large element of expert judgement enters into virtually all aspects of the chain of reasoning leading to a forecast of repository performance. This ranges from the assumed behaviour of human beings living far into the future to assumptions about the range of values for unmeasured, and sometimes unmeasurable, parameters appearing in environmental evolution and radionuclide transport models. How can we have confidence in the judgements of these experts? All that can be done here is to follow a structured approach to expert elicitation, and ensure that the whole processes is documented to allow for review.

In all of the above areas, confidence is enhanced by subjecting the work to peer review. At the top level, regulatory agencies peer review the work of implementing agencies both by the conventional approach of commenting on reports, and also by carrying out independent research and assessments. Also, there are many places where the work of radioactive waste assessments touches established academic disciplines. Here the normal professional process of peer review comes into play to build confidence by engaging a wider circle of people into the debate.

In summary, how do we ensure confidence in a radioactive waste performance assessment? Fundamentally, an assessment is a chain of argument, and confidence is built by examining each link of that argument using scientific reasoning where possible but relying on peer review in virtually all areas. Peer review requires that all aspects of the case are documented and thereby open to scrutiny. The most rigorous form of scrutiny is to carry out independent assessments and especially to use a variety of approaches and models. At the end of the day, all that can be done is to put forward an argument based on evidence and allow these to be scrutinised by peers, as in a court of law.

9.8 References

Andersson, J., Andersson, K., Carlsson, T., Kautsky, F., Norrby, S., Sundstrom, B. and Wingefors, S. (1991) The integrated performance assessment in SKI Project-90. In *High Level Radioactive Waste Management, Proceedings of the Second Annual International Conference*, Las Vegas, Nevada, 1166-1173.

Andersson, J. (ed.) (1989) The Joint SKI/SKB Scenario Development Project. *SKI Report* TR 89:14. Swedish Nuclear Inspectorate, Stockholm, Sweden.

Andersson, K., Apted, M.J., Arthur, R.C., Conca, J.L., Salter, P.F., Worgan, K. and Zhou, W. (1993) Survey and review of near-field performance assessment. *NEA Report*, to be published. NEA/OECD, Paris, France.

Ashton, J., Broyd, T.W., Jones, M.A., Knowles, N.C., Liew, S.K., Mawbey, C.S., Read, D. and Smith, S.L. (1993) A directory of computer programs for assessment of radioactive waste disposal in geological formations, second edition. *CEC Report* EUR 14201/1&2 EN. Commission of the European Communities, Luxembourg.

Atkinson, R., Herbert, A.W., Jackson, C.P. and Robinson, P.C. (1985) NAMMU user guide. *UKAEA Report* AERE-R11364. United Kingdom Atomic Energy Research Establishment, Harwell, U.K.

Bonano, E.J. *et al.* (1990) Elicitation and Use of Expert Judgement in Performance Assessment for High-level Radioactive Waste Repositories. *US Nuclear Regulatory Commission Report* NUREG/CR-5411; *Sandia Laboratory Report* SAND89-1821. Sandia National Laboratory, Albuquerque, New Mexico, U.S.A.

Boulton, G.S. (1990) Time-dependent modelling of environmental change: the effect of Quaternary glaciations. In *Safety Assessment of Radioactive Waste Repositories*, NEA/IAEA/CEC, Paris 1990, pp 363-375.

Brown, R.C. and Robinson, P.C. (1993) QUEST: technical overview and user guide. *Intera Report* IM2617 Version 2. Intera Information Technologies Ltd, Henley-on-Thames, U.K.

Chapman, N.A., Andersson, J., Skagius, K., Wene, C-O. and Wiborgh, M. (1994) Scenario development for SITE-94. *SKI Technical Report*. Swedish Nuclear Inspectorate, Stockholm, Sweden.

Chapman, N.A., Andersson, J., Bogorinski, P., Carrera, J., Hadermann, J., Hodgkinson, D.P.H., Jackson, C.P., Neretnieks, I., Neuman, S., Nicholson, T., Tsang, C-F. and Voss, C. (in press) Developing groundwater flow and transport models for radioactive waste disposal – six years of experience for the INTRAVAL project. In *GEOVAL '94*, OECD/NEA, October 1994. OECD/NEA, Paris, France.

Cohen, B.L. (1986) Risk analysis of buried wastes from electricity generation, *American Journal of Physics* **54(1)**.

Courbouleix, S. *et al.* (1985) Étude Géoprospective d'un Site de Stockage, Vols 1-9. Bureau de Recherches Géologiques Minières, *CEC Report* EUR 9866 FR. Commission of the European Communities, Luxembourg.

Cranwell, R.M., Guzowski, R.W., Campbell, J.E. and Ortiz, N.R. (1982) Risk Methodology for Geologic Disposal of Radioactive Waste: Scenario Selection Procedure. *US Nuclear Regulatory Commission Report* NUREG/CR-1667; *Sandia Report* SAND80-1429. Sandia National Laboratory, Albuquerque, New Mexico, U.S.A.

Dalrymple, G.D. *et al.* (1986) First Report on the Acquisition of Data for Use in the Probabilistic Risk Assessment of Underground Disposal of Radioactive Wastes. *UK DoE Report* DoE/RW/86.071. Department of the Environment, London, U.K.

Dames and Moore (1988) Background studies: climatic and geomorphological aspects of the evolution of shallow-land burial sites for radioactive waste disposal. *UK DoE Report* DoE/RW/88.055. Department of the Environment, London, U.K.

Dershowitz, W., Lee, G. and Geier, J. (1991) FRACMAN version beta 2.3: interactive feature data analysis, geometric modelling and exploration simulation: user documentation. *Project Report* 913-1058. Golder Associates Inc., Redmond, Washington, U.S.A.

Grindrod, P., Brown, R.C. and Gealy, N.D. (1992) COLLAGE: a numerical code for radionuclide migration through a fractured geosphere in aqueous and colloidal phases. *SKI Technical Report* 92:19. Swedish Nuclear Inspectorate, Stockholm, Sweden.

Grindrod, P., Herbert, A., Roberts, D. and Robinson, P. NAPSAC technical document. *Stripa Project Technical Report* 91-31. Swedish Nuclear Fuel and Waste Management Company, Stockholm, 1991.

Grindrod, P., Williams, M., Grogan, H. and Impey, M. (1990) STRENG: a source term model for vitrified high level waste. *Nagra Technical Report* NTB 90-48. Nagra, Wettingen, Swizerland.

Gupta, S.K., Cole, C.R., Kincaid, C.T. and Monti, A.M. (1987) Coupled fluid, energy and solute transport (CFEST) model: formulation and user's manual. *Office of Nuclear Waste Isolation Report* BMI/ONWI-660.

Guzowski, R.V. *et al.* Potential scenarios for use in performance assessment of high-level radioactive waste repositories in unsaturated tuff. *US Nuclear Regulatory Commission Report* NUREG/CR-4770; *Sandia Laboratory Report* SAND86-7170. Sandia National Laboratory, Albuquerque, New Mexico, U.S.A.

Guzowski, R.V. (1990) Preliminary identification of scenarios that may affect the escape and transport of radionuclides from the Waste Isolation Pilot Plant,

Southeastern New Mexico. *Sandia Laboratory Report* SAND89-7149. Sandia National Laboratory, Albuquerque, New Mexico, U.S.A.

Hadermann, J. (1980) Radionuclide transport through heterogeneous media. *Nuclear Technology* **47:** 312-323.

Herbert, A., Gale, J., Lanyon, G. and MacLeod, R. (1991) Modelling for the Stripa site characterization and validation drift inflow: prediction of flow through fractured rock. *Stripa Project Technical Report* 91-35. Swedish Nuclear Fuel and Waste Management Company, Stockholm, Sweden.

Hodgkinson, D.P. (1992) An overview of fracture flow modelling at Stripa. *Proceedings of the Fourth International Symposium of the OECD/ NEA Stripa Project,* October 1992. Swedish Nuclear Fuel and Waste Management Company, Stockholm, Sweden.

Hodgkinson, D.P. and Sumerling, T.J. (1990) A review of approaches to scenario analysis for repository safety assessment. In *Safety Assessment of Radioactive Waste Repositories,* NEA/IAEA/CEC, Paris 1990, pp 333-350.

Hunter, R.L. (1983) Preliminary scenarios for the release of radioactive waste from a hypothetical repository in basalt of the Columbia Plateau. *US Nuclear Regulatory Commission Report* NUREG/CR-3353; *Sandia Laboratory Report* SAND83-1342. Sandia National Laboratory, Albuquerque, New Mexico, U.S.A.

Hunter, R.L. and Mann, C. (eds.) (1989) Techniques for determining probabilities of events and processes affecting the performance of geologic repositories: Volume 1 - Literature Review. *US Nuclear Regulatory Commission Report* NUREG/CR-3964; *Sandia Laboratory Report* SAND86-0196. Sandia National Laboratory, Albuquerque, New Mexico, U.S.A.

INTERA (1983a) GSM: geologic simulation model for a hypothetical site in the Columbia Plateau: large computer version. Nuclear Waste Isolation Report, ONWI-447. INTERA Environmental Consultants Inc.

INTERA (1983b) FFSM: far-field state model. *Office of Nuclear Waste Isolation Report,* ONWI-436. INTERA Environmental Consultants Inc.

Kahn, H. *et al.*. (1967) *The Year 2000, a Framework for Speculation.* Macmillan.

King, L.M., Chapman, N.A. and Kautsky, F. (in press) Ice-front scenarios: SKI's SITE-94 Project to evaluate the future behaviour of a deep repository for spent fuel. In *Proceedings of Géoprospective Colloquium.*

Laurens, J-M, Thompson, B.G.J. and Sumerling, T.J. (1990) The development and application of an integrated radiological risk assessment procedure using time-dependent probabilistic risk analysis. In *Safety Assessment of Radioactive Waste Repositories,* NEA/IAEA/CEC, Paris 1990, pp 627-635.

Lord Marshall of Goring (1988) Nuclear Waste and NIMBY. UK Central Electricity Generating Board.

de Marsily, G. (1986) *Quantitative Hydrogeology: Groundwater Hydrology for Engineers.* Academic Press.

Maul, P.R., Cooper, N.S. and Robinson, P.C. (1993) MENTOR: a prototype computer code for assessing disposal options for TRU wastes. *Intera Report* IS3576-1. Intera Information Technologies Ltd, Henley-on-Thames, U.K.

McCombie, C. and Zuidema, P. (1992) Application of performance assessment as a tool for guiding project work. *Proceedings of the Third International Conference on High Level Radioactive Waste Management,* Las Vegas, April 12-16.

Miller, I. (1990) MAFIC version beta 1.2 matrix/fracture interaction code with solute transport – user documentation. Golder Associates Inc., Report prepared for Battelle Memorial Institute, Office of Waste Technology Development, Willowbrook, Illinois, U.S.A.

NEA (1988) Near-field assessment of repositories for low and medium level radioactive waste, Proceedings of an NEA Workshop, Paris. NEA/OECD, Paris, France.

NEA (1989) PSACOIN Level E Intercomparison. OECD/NEA, June 1989. OECD/NEA, Paris, France.

NEA (1990) Disposal of Radioactive Waste: The International Probabilistic System Assessment Group: Background and Results. NEA/OECD, Paris France.

NEA (1992) Systematic Approaches to Scenario Development: A report of the NEA Working Group on the Identification and Selection of Scenarios for Performance Assessment of Radioactive Waste Disposal. NEA/OECD, Paris, France.

Peaudecerf, F.P., and Blanc, P.L. (1990) L'evolution a long terme de l'environment des stockages: choix des scenarios et methodes. In NEA/IAEA/CEC, Paris 1990, pp 351-362.

Reeves, M., Ward, D.S., Davis, P.A. and Bonano, E.J. (1986) SWIFT II self-teaching curriculum: illustrative problems for the Sandia Waste - isolation flow and transport model for fractured media. *Sandia Report* SAND84–1586, NUREG/CR-3925. Sandia National Laboratory, New Mexico, U.S.A.

Ringrose, P.S., Chadwick, A.F., Kleissen, F.A.T., Larkin, J.P.A., Polloket, D.T. and Wilmot, R.D. (1990) Probabilistic simulation of the long-term evolution of radioactive waste disposal sites. In *Safety Assessment of Radioactive Waste Repositories* NEA/IAEA/CEC, Paris1990, pp 427-436.

Robinson, P.C. (1991) PAGODA guide. *Intera Report* IM2795-1. Intera Information Technologies Ltd, Henley-on-Thames, U.K.

Robinson, P.C. and Hodgkinson, D.P. (1987) Exact solutions for radionuclide transport in the presence of parameter uncertainty. *Radioactive Waste Management and the Nuclear Fuel Cycle* **8(4)**: 283-311.

Robinson, P.C., Hodgkinson, D.P., Tasker, P.W., Lever, D.A., Windsor, M.E., Grime, P.W. and Herbert, A.W. (1988) A solubility-limited source-term model for the geological disposal of cemented intermediate-level waste. *UK Department of the Environment Report* DOE/RW/ 88.015. UK Department of the Environment, London, U.K.

Robinson, P.C. and Williams, M.J. (1989) STRAW-PC : description and user guide. *UK Nirex Ltd. Report* NSS/R156. UK Nirex Ltd, Harwell, U.K.

Robinson, P.C., Williams, M.J., Impey, M.J. and Brown, R.C. (1994) LAPLACE-2D technical details. *Intera Report* IM3841-1 Version 1. Intera Information Technologies Ltd, Henley-on-Thames, U.K.

Robinson, P.C., Williams, M.J. and Worgan, K.J. (1992) A description of MESHNOTE. *Intera Report* IM3479-1. Intera Information Technologies Ltd, Henley-on-Thames, U.K.

Robinson, P.C., Worgan, K.J., Shaw, W.T. and Wingefors, S. (1991) CALIBRE and CRYSTAL: near-field and geosphere models for Project-90. In *High Level Radioactive Waste Management: Proceedings of the Second Annual International*

Conference, Las Vegas, Nevada, 1174-1180.

Shaw, W. and Grindrod, P. (1989) Investigation of the potential of fuzzy sets and related approaches for treating uncertainties in radionuclide transfer predictions. *CEC Report* EUR 12499EN. Commission of the European Communities, Luxembourg.

Shaw, W. and Worth, D. (1992) Scoping calculations for canister-tunnel migration of corrodants, Oxidants and Radionuclides. *SKI Technical Report* 91:19. Swedish Nuclear Inspectorate, Stockholm, Sweden.

SKI (1991) SKI Project 90 Volume I. *SKI Technical Report* 91-23. Swedish Nuclear Inspectorate, Stockholm, Sweden.

SKI/OECD/NEA (1992) The International HYDROCOIN Project: Summary Report. Swedish Nuclear Power Inspectorate (SKI) and OECD Nuclear Energy Agency (NEA).

SKI/OECD/NEA (1993) The International INTRAVAL Project: Phase 1 Summary Report. SKI and OECD-NEA. Swedish Nuclear Inspectorate, Stockholm, Sweden – NEA/OECD, Paris, France.

Stephens, M.E. and Goodwin, B.W. (1990) Scenario analysis for the post closure assessment of the Canadian concept for nuclear fuel waste disposal. In *Safety Assessment of Radioactive Waste Repositories* NEA/IAEA/ CEC, Paris 1990, pp 405-416.

Sumerling, T.J. and Hodgkinson, D.P. (1992) The treatment of uncertainty in future states for radioactive waste disposal. In *Probabilistic Safety Assessment and Management, Volume 2*, Elsevier, pp 641-648.

Sumerling, T.J. and Thompson, B.G.J. (1992) Application of a probabilistic system-model based methodology for the performance assessment of deep underground disposal of nuclear wastes. In *High Level Radioactive Waste Management, Proceedings of the Third International Conference*, Las Vegas, April 1992, Volume 2, pp 1647-1657.

Thompson, B.G.J. (1989) The Time Dimension in Risk Analysis : Examples from Recent Work in the United Kingdom. In *Risk Analysis in Nuclear Waste Management*, (eds. A. Saltelli *et al.*), pp 231-262.

Thompson, B.G.J. (1993) Application of probabilistic risk analysis to the post-closure radiological assessment of underground disposal of radioactive wastes. Presented at PSAM 1991, Beverly Hills.

Thompson, B.G.J. and Sagar, B. (1993) The development and application of integrated procedures for post closure assessment, based upon Monte Carlo simulation: the probabilistic systems assessment (psa) approach. *Journal of Reliability Engineering and System Safety* **42:** 125-160.

Worgan, K.J. and Apted, M.J. (1992) The effect of precipitation fronts induced by radionuclide chain decay and elemental solubility limits on near-field mass transport. In *Scientific Basis for Nuclear Waste Management*, (ed. C. Sombret). Materials Research Society, Pittsburgh, Pennsylvania, U.S.A., pp. 705-712.

Worgan, K. and Robinson, P. (1992) CRYSTAL: a model of a fractured rock geosphere for performance assessment within SKI Project-90. *SKI Technical Report* 91:13. Swedish Nuclear Inspectorate, Stockholm, Sweden.

Zimmerman, D.A. and Gallegos, D.P. (1993) A comparison of geostatistically-based inverse techniques for use in performance assessment analysis at the WIPP

Site: results from test case No. 1. In *High Level Radioactive Waste Management, Proceedings of the Fourth Annual International Conference*, Las Vegas, Nevada, 1426-1436.

10

Uncertainty and confidence building

This chapter deals with issues such as uncertainty, model validation, building confidence in performance assessments through information from natural systems, and radioactive waste in perspective with other environmental hazards.

10.1 Uncertainty

10.1.1 General issues in uncertainty

In any performance assessment exercise there are numerous sources of uncertainty, variability and possible inconsistency (NEA, 1987a). In order to manage and quantify such uncertainty, a variety of tools and techniques can be employed. The objective of this chapter is to examine issues and methodologies from the standpoint of performance assessment management and the compilation of a defensible technical safety case. The section is split into three:

- a general discussion on types and sources of uncertainty within the whole process of performance assessment;
- a review of current practices and a discussion of the issues arising from their implementation;
- a discussion of novel methodologies for dealing with uncertainty, which could be useful for performance assessment.

10.1.2 Types of uncertainty

There are numerous sources of uncertainty within the assessment process, which overlap and interfere in a complex manner. Such inter-relationships make a formal classification rather subjective. However, a number of types of uncertainty can be singled out and are discussed briefly as follows.

Regulatory Uncertainty. How is performance measured and how can a technical safety case be justified within the regulatory framework? Safety criteria are themselves uncertain although their technical interpretation may often establish the framework and methodology to be used within an assessment.

Fundamental Processes, such as time-dependent sequences of events may be combined into discrete sets of scenarios. However the ranking of such situations in terms of probability, possibility, level of belief, *etc.* must inevitably involve some subjective decisions, and thus introduce a selective uncertainty.

Having identified particular **scenarios** and processes for analysis, a conceptual model must be specified. Which sub-processes are important and what assumptions will be made? All models will encompass some simplification of reality, and there will be choices concerning model types (determinism or simulation), processes to be included, and their sophistication, and state variables (block averages or point values). The range of modelling approaches gives rise to so called **conceptual model uncertainty** since it is not clear *a priori* or *a posteriori* how any choice may bias or affect results. This is an area which has historically been under-studied but is now receiving more full attention. It is discussed in detail in Section 10.1.9.

Any model will require site information in order to condition or calibrate it. However, there is likely to be considerable uncertainty in characterising specific sites, particularly with reference to the data input requirements of the model. The quantity, quality and type of **site characterisation data** available are all issues. In addition, the way in which geological data are represented in model calculations is also of interest. Measured permeabilities and porosities may perhaps be straightforward to incorporate (including their spatial distribution), but fracture distributions or intersection frequencies may be difficult to determine from the available data. At the same time much detailed information on the geological structures may be utilised in only a much simplified form.

Both site characterisation and conceptual model uncertainties are related to model **parameter uncertainty**. Model parameters may be uncertain because:

- values are unknown (ignorance), yet could, in principle, be obtained;
- measured values are highly variable (inconsistent data);
- they represent averaged or derived quantities which cannot be measured directly (that is, can only be calibrated);
- external effects, which give rise to parameter variation, are uncertain (for example, ambient chemistry, temperature, or stress) so a variety of situations must be dealt with.

There may also be parameters regarded as genuinely stochastic. For example, **spatially variable** fields may be represented in a number of ways, and these provide further sources of conceptual uncertainty (how should such fields be devised, represented and manipulated?) and parametric uncertainty (how are variograms, scaling laws or random functions to be calibrated?).

Once models are identified and calibrated there is a need to discuss the validity of their results in predicting consequences on the basis of the available data. **Validation** is usually addressed with reference to particular sites or experimental data sets.

There may be numerical inaccuracies introduced in solving mathematical models, particularly when fast methods must be adopted within a Monte Carlo ensemble.

In addition to all model selection, solution, calibration and validation issues, there remains an overall uncertainty introduced by the adoption of particular techniques used to quantify and manage the uncertainties themselves. Undoubtedly the probabilistic approach is the best developed and most favoured approach. However, there are issues which do not easily fit in with such methods and it is valuable to discuss alternative approaches

10.1.3 Management of uncertainty

As has already been stated, uncertainty always occurs in modelling any physical system. The aim of any performance assessment is not to remove such uncertainties but rather to manage them and to reduce their effects. The management of uncertainty has four main components:

1. **Awareness**. Uncertainties cannot be managed if they are not known about. A performance assessment should identify all major potential locations of uncertainty. This might be done through a Bias Audit.
2. **Importance**. Some uncertainties have significant effects on the derived safety case, while many others are unimportant. Before attempting to reduce uncertainties it is first necessary to determine whether the uncertainty has a significant effect on the overall outcome. This will involve a sensitivity analysis of some kind.
3. **Reduction**. Having ascertained the importance of particular uncertainties, measures can then be undertaken to reduce them. For parameter uncertainty, observational studies can limit the known range. Conceptual model uncertainty should be reduced by further research. Stochastic uncertainties cannot be easily reduced, though design changes may help to reduce their effect (for instance, by moving the location of the repository away from a particular rock formation).
4. **Quantification**. The effect of uncertainties on the final safety assessment needs to be quantified. Some uncertainties are more difficult to quantify

than others, but all important uncertainties need to be quantified, even if there are uncertainties in the uncertainties.

10.1.4 Current practice

Current approaches to handling parameter uncertainty can be classified as:

- conservative, or worst case;
- best estimate, often with '*What if ?*';
- sensitivity analysis;
- probabilistic.

In the conservative approach, an attempt is made to be pessimistic about each parameter value. This approach can extend to conceptual and scenario uncertainties in a natural way. The major weakness of the approach is that it is hard to set a single worst value – it is always tempting to go a little further. Also, there are technical difficulties in some cases, since it is not always obvious what is conservative for a particular parameter.

In the best esimate approach, a central set of parameter values is taken to be representative. Often this approach is coupled with the use of what-if calculations to explore other (more extreme) possibilities. Again, the same approach can be applied to scenarios and to conceptual uncertainties. This approach has the problem that the status of what-if calculations is not clear – if they give unacceptable consequences then what does this imply? Also, it invites a view that each what-if calculation represents a situation that could happen.

In many cases a sensitivity analysis may be performed. This identifies the most influential parameters for the result(s) of interest. In a local sensitivity analysis, the derivatives of the result with respect to each parameter, for a particular set of parameter values, are calculated. In a global sensitivity analysis the behaviour of the result over a full range of parameter values is explored, often by using regression analysis methods. In the context of a probabilistic assessment the influence of parameter distribution characteristics on result distribution characteristics can be considered.

In the probabilistic (PSA) approach (Apostalakis, 1991), all uncertain parameters are given probability distribution functions to characterise their uncertainties. Similarly, scenarios can be given probabilities (although this has rarely been attempted in full), and conceptual models could be given probabilities of correctness. For regulations requiring a risk to be calculated, the mean consequence is taken. For dose-based regulations other properties of the distribution of consequences are more appropriate, *e.g.* a 95th percentile. A difficulty with the probabilistic approach is that it requires more computational effort and more man-time to set up and interpret the calculations than a deterministic approach.

More fundamentally, the use of probability theory may not be appropriate for the types of uncertainty that are to be addressed. Basically, it is not the parameters that are uncertain – it is the experts! Thus, the uncertainties are largely subjective and not well characterised by probabilities. Following a description of the way the PSA approach is implemented, two examples of concern with respect to the PSA approach are presented: risk dilution; and the handling of sparse information. An alternative novel approach to handling uncertainties is then presented.

10.1.5 Implementing a PSA approach

The basic approach used to implement the PSA methodology is simply stated: a (large) set of samples is taken covering the parameter ranges and these sample results are used to estimate statistical properties of the system (*e.g.* means or percentiles). Different sampling strategies are available – in particular pure Monte Carlo, Latin Hypercube sampling and Importance sampling. The basic approach to these is common and is presented below. For simplicity we focus only on the mean of some result.

If $R(\mathbf{p})$ is a result depending on a vector of parameters $\mathbf{p}$ $[= (p_1, p_2, \ldots, p_m)]$ and the parameters have probability density functions $f_i(p_i)$ then the mean is defined by:

$$\bar{R} = \int_{p_1}\int_{p_2}\cdots\int_{p_m} F(\mathbf{p})R(\mathbf{p})dp_1\,dp_2\ldots dp_m$$

where $F(\mathbf{p}) = \prod_{i=1}^{m} f_i(p_i)$ is the joint density function. (In fact this need not be a product but is generally taken to be so).

Now if a set of samples $\mathbf{p_j}$, $j = 1, \ldots N$ is generated using joint density function G (again usually a product of individual g_i). The mean can be estimated as:

$$\hat{R} = \frac{1}{N}\sum_{j=1}^{N}\frac{F(\mathbf{p_j})}{G(\mathbf{p_j})}R(\mathbf{p_j})$$

This estimate is minimum variance and unbiased (ie it is the best from the sampled points and on average gives $\bar{R}$). There are a few things to note.

Just because $\hat{R}$ is unbiased does not imply that half of the time is too big and half too small – for radioactive waste assessments the distributions are often skewed severely and $\hat{R}$ will be usually a little to small (perhaps zero) and rarely a lot too big.

If $G = F$ this is pure Monte Carlo sampling. If G is chosen to be proportional to $F.R$ then a perfect scheme is obtained (it is always correct). The catch is we need to know the answer to do this, but it is the basis of Importance sampling when an approximation to $F.R$ is used to give a good choice of G.

It is possible to derive confidence limits around $\hat{R}$ but these are only reliable for large enough sample sets. Unfortunately, because of skewed distributions, the number of samples required can be very large.

10.1.6 Risk dilution in PSA

Risk dilution is said to occur in a situation where an **increase in uncertainty** leads to a **decrease in risk** and *vice versa*. It does not arise in all circumstances, and is largely a product of the use of the mean dose at a particular time as the risk measure.

Given a dose model $D(\mathbf{p}, t)$, we calculate the risk, R as a function of time, t, using:

$$R(t) = \gamma \int_{\mathbf{p}} D(\mathbf{p}, t) f(\mathbf{p}) d\mathbf{p}$$

where $f(\mathbf{p})$ is the joint probability density function for all the parameters.

Now consider a very simple case. Imagine a constant release from the repository lasting 1000 years from the time of closure. The released contaminants move slowly to the surface without significant decay, spreading or dilution. At the surface they enter a river and give rise to a dose through the drinking water pathway. Imagine that the only uncertain parameter is the travel time from the repository to the river.

If the travel time is known to be, say, 5000 years then the risk is simply calculated to be uniform between 5000 and 6000 years after closure and zero outside that band (Figure 10.1).

Now suppose that the travel time is not so well known, and expert judgement indicates that anything between 3000 and 7000 years is equally likely. In this case a dose could occur at any time between 3000 and 8000 years after closure. The risk is therefore spread, and the peak risk is **smaller,** because there is a large probability against the dose actually occuring at any particular time (Figure 10.2).

The phenomenon of risk dilution undermines the philosophical basis for the probabilistic approach. It is certainly important to be aware of it if such an approach is used. The extent of risk dilution can be estimated by comparing the maximum risk as defined above with the risk from the mean maximum dose, *i.e.* with:

$$R = \gamma \int_{\mathbf{p}} \max_t D(\mathbf{p}, t) f(\mathbf{p}) d\mathbf{p}$$

10.1.7 Handling sparse information in PSA

This second example is generic to all probabilistic calculations. Imagine a situation where the performance assessment model has two important

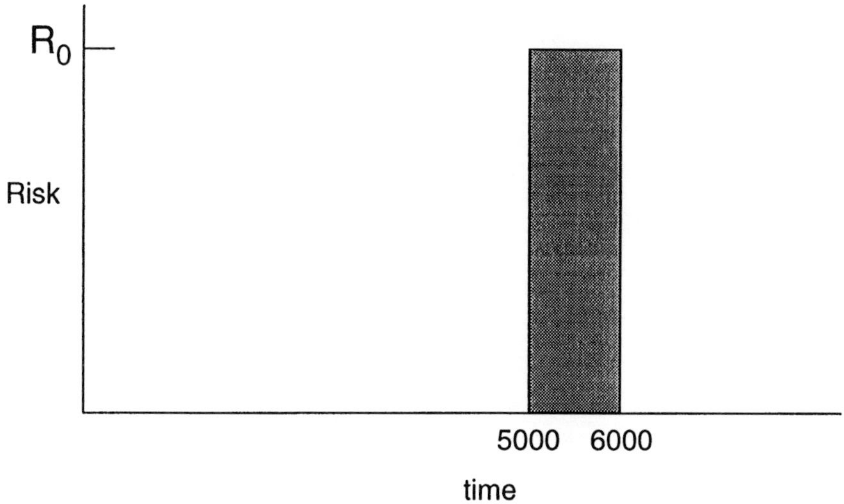

Figure 10.1 Narrow risk versus time, peaking at R_0.

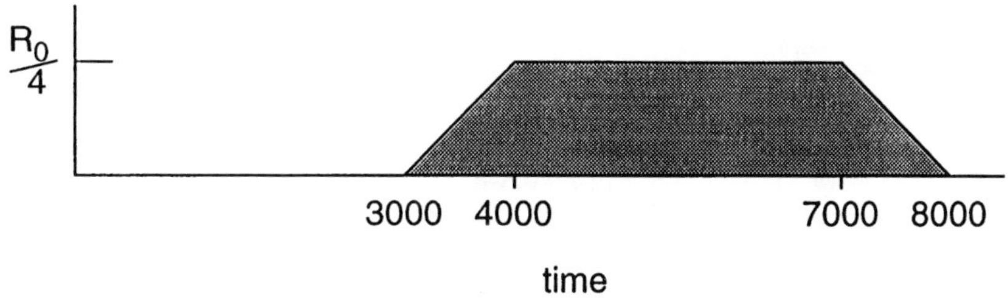

Figure 10.2 Wide risk versus time, peaking at $R_0/4$.

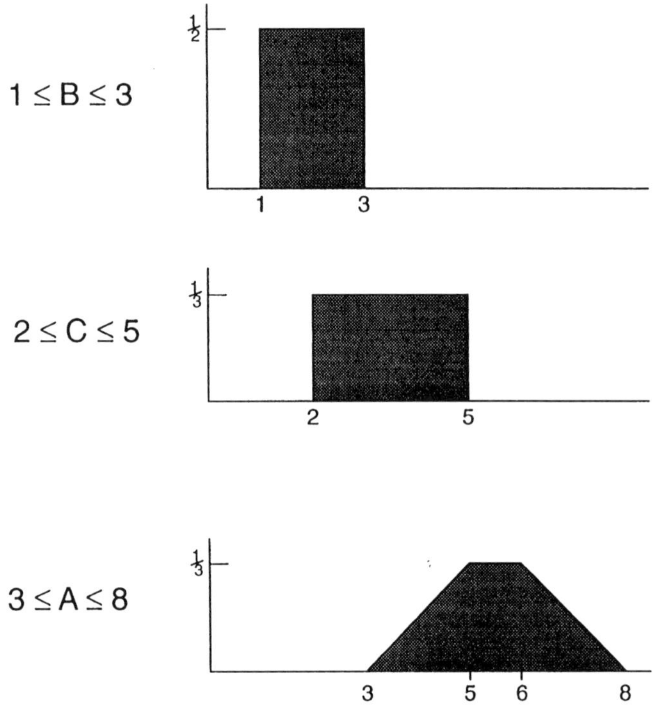

Figure 10.3 $A = B + C$.

parameters which appear as a simple sum. Ignoring the other parameters we just consider:

$$A = B + C$$

Now suppose that expert judgement for B and C is based on very little evidence or measurement, so that all that can be said is that B is between 1 and 3 while C is between 2 and 5. The probabilistic approach takes uniform distributions for each parameter (sometimes justified on minimum entropy grounds) and calculates the distribution for A.

The calculated distribution shows that A is more likely to be near the centre of the range than at the ends (Figure 10.3).

Obviously, if this were a dice game then all would be well, but does ignorance work like this? In any case, if we had chosen $\log B$ as a basic parameter, we would have gained a different answer.

10.1.8 Nested set analysis

The following section discusses a new approach to handling parameter uncertainty. It is readily extended to handle all types of uncertainty in a consistent framework. The aim is to combine the best aspects of each of the other approaches (best estimate, conservative and probabilistic).

The approach, called Nested Set Analysis here, arose from consideration of the use of fuzzy set theory. It is not necessary to base it on fuzzy sets, but this may be useful from a philosophical viewpoint. It is emphasised that this approach is novel and is still being developed; it is presented here to provoke discussion of the basic issues involved. The approach aims to give a robust statement of safety, but allows for degrees of robustness. There are many ways of viewing this, just one of these is given below.

In calculating the consequences of a radioactive waste disposal practice a series of decisions (choices) will have to be made regarding features, models, data *etc.* Each decision gives alternative choices, and it is the existence of these alternatives which leads to uncertainty. A complete set of choices corresponds to a single statement of the consequences of the disposal practice (*e.g.* a dose or a series of doses against time).

For any particular decision, we can imagine cataloguing the full set of choices. Now each choice can be classified according to how possible it is that this is the correct choice, or conversely according to how surprising it would be if this choice was found to be the true one. The classification can be labelled in various ways, for example as {completely possible; quite possible;...; almost impossible; impossible} or {unsurprising; slightly surprising; surprising; very surprising; ...}. A number between 0 and 1 could be assigned to the classification. Alternatively, the number could be arrived at directly by considering the proportion of experts who would consider that each particular choice is possibly correct; this corresponds to a membership in fuzzy set theory.

The classifications define a collection of nested sets, or nested intervals for parameter values. The innermost set or interval contains members with the highest membership (unity) and each subsequent set contains all members with memberships higher than some value (Figure 10.4).

Now the idea is to perform an assessment for each level of the nested sets in a consistent fashion. That is, the classification of the results is arrived at by taking the same level of membership for all decisions. This gives the nested intervals for the results. Any result in the innermost interval can be achieved by making choices all of which are in the innermost intervals. In general, each interval consists of results which can be obtained using choices which have membership equal to or higher than a given level. To look at it from the other side, results outside any particular interval require at least one choice with membership lower than the given level.

In practice, an analysis of all performance assessment choices can be made

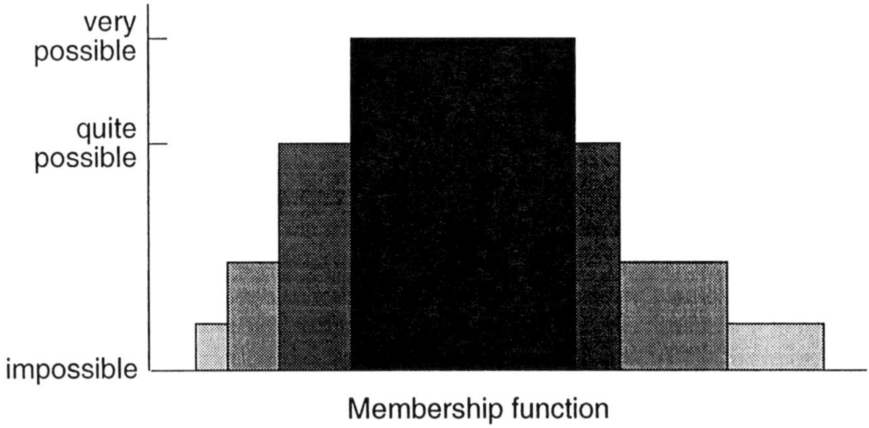

Figure 10.4 Membership and nested intervals.

together using a Monte Carlo approach, with use of minimum-maximum rules for combination of membership levels.

The fundamental philosophy behind this approach is that a consequence is less possible than x if it can not arise using a set of component choices which are as possible as or more possible than x.

10.1.9 Conceptual model uncertainty

What is conceptual model uncertainty?

At the first stage of modelling within any assessment there will be a range of features and processes which have been identified as potentially important in determining the overall system performance of the near-field, the geosphere, or the biosphere. These may be thought of as a story board, or inference diagram, relating the chemical and physical processes involved, and their relative time scales. However the translation of such information from this qualitative conceptual outline, into a quantitative mathematical model necessitates some choices between alternative suitable models, the specification of their level of resolution, their data requirements, and explicit interpretation of their results in terms of suitable performance measures. At the very start of this process there may be a number of conceptual alternatives, each of which is designed to approach similar problems, from slightly different (mathematical) starting points (Figure 10.5). Each provides a description of the (sub)system in mathematical terms, though there may be differences of scale, detail, or emphasis on the various fundamental processes under consideration. This range of possible conceptual (and hence mathematical) models which are

Figure 10.5 An example of different conceptual models of the same problem.

available and the need to make a choice give rise to **conceptual model uncertainty**.

For example, suppose we wish to model fluid flow through a porous medium composed of well characterised solid grains packed together surrounded by saturated pore space. One modelling approach might be to generate a suitable distribution of grains and solve the Navier-Stokes equations in the surrounding pore space, allowing for inertial and viscous effects to play a role as the flow responds to imposed pressure differences and the solid boundaries. Alternatively it could be decided that the flow will be relatively slow, so that a Stokes flow approximation can be made, resulting in a linear model relating the fluid velocity to the pressure distribution, a substantial simplification. However this would still require precise information concerning the distribution and geometry of the pore space: is such resolution, accessible, practicable, or desirable? A much more simple alternative would be to assume an empirical law such as Darcy's law, relating the (averaged) flow though the rock to the pressure gradient, by defining suitable *effective* parameters such as permeabilities. This provides a substantial model simplification, but incurs the uncertainty as to whether Darcy's law is applicable to the particular medium we have in mind; the length scales over which parameters should be allowed to vary; the resolution at which calculations should be made; and many other issues. What kind of boundary conditions would be appropriate? Should variable density effects be considered? How should isotropy or heterogeneity be represented/generated? Can the problem be reduced by assuming planar, or cylindrical symmetry?

Clearly, once the modellers are asked to specify conceptual and mathematical models appropriate to the conceptual plan, or storyboard, there is a long list of decisions to be made, incorporating assumptions each of which may or may not be valid.

It is helpful to consider that a particular conceptual model, proposed or adopted, is really just a list of assumptions. If made explicit this list provides all the information regarding the model, its limitations, and advantages. Such lists would certainly be useful in facilitating comparisons between different models.

However model assumptions are all too often hidden within the relevant documentation, or implicitly accepted by those close to a particular speciality. Worse still, sometimes particular assumptions, or practices, take on a life of their own without ever being critically examined (the willingness of modellers to allow for dispersion by imposing a fixed Peclet number, for example).

It might be argued that the uncertainty attached to, or induced (accepted) by, the adoption of a particular conceptual model is some kind of aggregation of the uncertainties attached to each one of the assumptions required. However extreme care should be taken here since the conditional nature of the separate assumptions is often hidden.

Being aware of conceptual uncertainty

Once a particular sub-system concept has been specified it is often possible to make scoping calculations to try to assess performance, or to see if any substantial issues arise. For example, a simple calculation assuming completely anaerobic conditions within a repository, and a constant iron corrosion rate may give a rough estimate as to the amount of hydrogen released within the near-field; will it simply react and diffuse away, or should we consider the migration of a gaseous phase? Such calculations provide initial estimates which focus the modelling priorities. Having identified a relevant issue, conceptual model alternatives should be sought and considered.

For the example of near-field gas migration, one might have three or four model alternatives:

- ignore gas migration and assume gas is held sealed within the near-field, with transport away only in the aqueous phase;
- use a classical two-phase flow approach and allow the near-field to be

partially saturated, monitoring both aqueous and gaseous flows;
- adopt a moving interface model allowing the gas to displace water and possibly exit to the geosphere;
- assume all gas escapes rapidly along a preferential path left during excavation, so that its effect is limited to temporary chemical changes within the near-field.

Clearly these alternatives represent vastly different commitments in terms of effort, and data requirements. Thus practical constraints come into play and some conceptual models may be ruled out of mainline assessments on the grounds that their application is too onerous and would significantly unbalance the focus of the programme.

At the very least, alternatives should be sought and documented. This may be achieved in part by reviewing the concepts employed in other assessments or related disciplines, and represents an objective information collection exercise which should not, as far as possible, be compromised by the expertise, or code availability, expected to reside within the programme, or knowledge of a specific most favoured model. On the contrary, where there is an advantage in adopting one particular model in terms of progress which could be made, this should be clearly understood. Alternative concepts should then be identified and employed to scope the uncertainty in mainline performance in various scenarios.

Is conceptual uncertainty treatable as parametric uncertainty?

It is sometimes argued that a viable approach to conceptual model uncertainty is to parameterise the alternatives by some (discrete) parameter and then employ the methodology of parameter sampling, as used in say Probabilistic Safety Analysis (PSA). Thus if there are two conceptual approaches, and model one is deemed twice as likely as model two to be representative of the system, results from model two are weighted in some way to be half as important as those from model one, or sampled half as often.

Parametric uncertainty allows for uncertainties to be expressed by distribution functions over known ranges. This implicitly requires that all possible alternatives are known. However we may never have access to the full range of conceptual models which could be applied. Thus the 'Universal Set' of possible alternatives for conceptual uncertainty remains unknown and possibly unbounded. How, then, could it be sampled in, say, Monte Carlo analysis? If a PSA approach is adopted it also requires that we accept that alternative concepts (possibly directly opposed to one another) may contribute together in producing an aggregated performance measure. This last issue touches on the whole philosophy of how uncertainties should be quantified and is discussed elsewhere in these notes with particular reference to alternatives to PSA; worst case (weakest link), interval analysis, nested sets, *etc.*

From a practical point of view, adopting a single probabilistic paradigm to cope with all uncertainties is mostly unhelpful, since it requires selective judgements to be made regarding subjective modelling choices, and the probabilities of their correctness, besides the inexplicit conditional nature of the underlying assumptions. This would result in experts being encouraged to produce more and more fictitious detail.

Thus although a sampling approach may be most appropriate for distributed model parameters (fracture apertures, flow rates, *etc.*) where we may (objectively) measure actual ranges of values, this is not appropriate in dealing with conceptual alternatives from either a theoretical or practical standpoint.

What about site specific information?

As specific information regarding site data or tested experimental performance of barriers becomes available, conceptual models may be ruled out by demonstrating that parameters or phenomena assumed to be relevant are not. However care must be taken where scaling issues (in space and time) are important. For example a geological site characterisation should decree whether the major water bearing paths are occasional fracture zones or highly transmissive porous layers (*e.g.* conglomerates or highly fractured rock). In the first instance this sets out the kind of concepts which should be considered for the geosphere. Similarly a hydrogeological characterisation should provide some discussion of the magnitudes of the fluid flows, pressure heads, recharge *etc.*

However even for a simple system there will still remain a number of alternative competing concepts. For example if the major conduit for groundwater flow is a single (near planar) thin layer of highly permeable porous rock with no significant discharge or recharge within the model region, models may adopt a two dimensional geometry; a three dimensional geometry; a large scale block numerical approach to the distribution of permeabilities; a high resolution numerical approach, allowing for a statistical distribution of permeabilities (fractals, kriging); or a desk-based analysis with representative values, and closed form solutions, examining sensitivity; or some other alternatives. All this before attention is given to the dispersive migration of reactive solutes, about which there is uncertainty even in the absence of geochemistry.

Flow and tracer tests in similar rocks, preferably with well-characterised flow paths are invaluable. This allows different conceptual models to be evaluated. However various models may upscale in a different way, so there is a danger in building false confidence at 10-50 m scales when the assessment requires calculations over 500-5000 m scales.

Validation

The term validation refers to the process by which (sub)models, or detailed sub-system models, or even the conceptualisation of key processes, are shown to be realistic representations of the natural system by direct comparison with the observed behaviour within the field or laboratory.

A number of international exercises have been carried out: PSACOIN (NEA, 1987b; 1989; 1990a; 1991; 1992a; 1993); INTRACOIN (INTRACOIN, 1984; 1986); HYDRACOIN (NEA, 1992b) INTRAVAL (NEA, 1990b); GEOVAL (NEA, 1990c); CHEMVAL (Read and Broyd, 1989; Read, 1990, 1991); the Stripa Project (NEA, 1992c); the CoCo Club; PEGASUS; INTERCLAY *etc.*, aimed at specific fields of underlying research supporting performance assessment. In addition, international projects considering natural analogues are often used in validating concepts in a more general sense.

The difficulties with such programmes are underlined by the nature of their conclusions. In general models are calibrated first and then used to predict further behaviour of experimental or natural systems. Thus, proposed approaches are sometimes shown to be inapplicable, but a 'good' result cannot necessarily be extrapolated to a positive statement regarding validity. Hence validation can only be inferred as the negative of invalidation. More importantly, since field and laboratory systems are often simplified within the models, validation studies tend to result in

- open-ended discussions of the issues including the range of possible models that could be applied,
- calls for further (detailed) processes to be considered (particularly where a known process has been omitted),
- no universally applicable, detailed conclusions.

All this takes place with references to the underlying research programmes, rather than the needs of performance assessment. Thus an enormous effort might be expended upon issues considered to be peripheral to the needs of the mainline assessment.

A simple example is the influence of scaling. Since we can only build confidence at limited time and spatial scales, there is a need to consider the up-scaled impact of such processes and their relevance and representation. Field experiments may be available but many of the issues encountered are site-specific and may involve considerable characterisation. Elsewhere, and particularly at repository sites, such information may be less complete. Thus generic know-how may be extrapolated to other sites, whereas specific details of the model calibration are of less general utility.

Any validation study has a cyclic structure, which can be iterated, as concepts and models become refined. The situation is depicted schematically in Figure 10.6. On the basis of previous knowledge and experiments models are formulated. These are used in calibration and prediction exercises. Poor

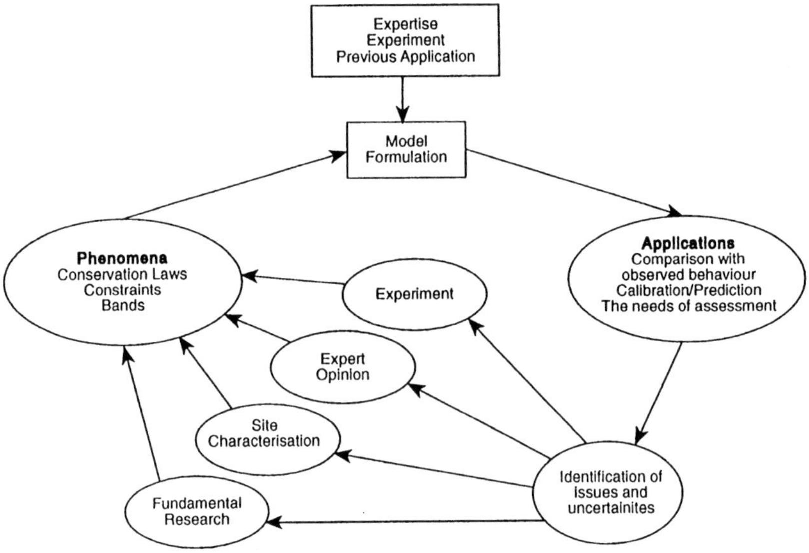

Figure 10.6 Cyclic structure of the validation process.

performance must result in doubts about validity. Key uncertainties are identified and further processes sought to clarify issues (expert opinion, site characterisation, fundamental research, experimental programmes). This leads to further constraints and fundamental knowledge of behaviour which should be reflected by the model solutions. Accordingly, the process descriptions can be modified, and refined models can be reconsidered in the calibration prediction phase. Hence the modelling and validation process should drive the experimental and site characterisation programmes and vice-versa.

Validation is an iterative process, not a product, and should consider the needs of PA as well as the requirement for, and availability of, site specific data. Above all it should be well-documented and a defensible trail produced in support of the final models to be employed within the assessment.

A number of sub-issues is raised by validation programmes. Some of these are briefly reviewed.

Expert judgement. Where models or methods depend upon expert subjective judgement there may be no way in which such actions can be validated, and the cycle is broken. For example, in designing a procedure for specifying input probability density functions to a PSA, there is no absolute truth or experiment that can be performed: the calibration/prediction stage cannot be carried out. Thus such processes cannot be validated, instead they should be defended, by setting out the methodology and following it in a traceable, audited fashion.

Purpose. Considering various models, and their mathematical formulation, they should be 'fit for their purpose'. Hence it is not necessary to validate and defend detailed process models which can have no impact within the assessment system, and the evaluation of model performance should be oriented towards its future application.

Peer review. This plays an essential role in validation. By exposing (sub)system models to the international community, best practice and experience can be shared, so that the views of experts outside of one's own team can have an impact.

Multiple model results. Why should our conceptual uncertainties be small? If different, competing approaches can perform similarly on calibration/prediction procedures within a validation programme, they are equally valid (or invalid). Nevertheless, when upscaled they may diverge. The resulting uncertainty is a result of our conceptual uncertainty in the site-specific 'inverse' problem: trying to condition our models on site data. Future iterates of the validation cycle should seek to devise tests to exploit and test the differences in model approaches. If this cannot be done, or the divergence is due to site-specific considerations then models cannot be ruled out on the basis of their validity.

Site-specific data. The validation of models is clearly site-specific. It is difficult to point to calibrated models which can be extrapolated from site to site. It is rather the fundamental process (diffusion, advection) which can have a wide application, though once a conceptual site model is imposed (ambient conditions, spatial structure/variability) the resulting models require validating afresh. Hence the reliance on site data not specific to the assessment should be avoided.

In summary, validation or rather our confidence in detailed models and assessment models is something which must be continuously updated during, and driving, the data collection within an assessment. The data and experiments should sharpen the questions, (and the models), rather than providing definitive answers.

Up-scaling models

A number of alternative conceptual models may be satisfactory where judged against:

- limited experimental information;

- results at small time or spatial scales.

As an illustration, consider the following example for up-scaling geosphere transport.

A non-reactive tracer migrating within a saturated porous medium is evaluated using three conceptual models:

- a one dimensional advection-dispersion model incorporating matrix diffusion;
- a pure channelling model, assuming the tracer is distributed across a collection of independent channels with various apertures (and hence axial velocities);
- a channel network model allowing tracer particles to 'jump' from fast channels to slow channels and vice versa (a network effect).

The behaviour of these models all scale up in time (and space) differently, since the concepts underlying them rely on fundamentally different assumptions concerning the geometry of the flow path network. They are described elsewhere in this book (Chapter 5) as examples of tracer transport models for the geosphere. For our current purposes we note that in their simplest forms each has three free parameters that may be used to calibrate them to some given data.

Suppose there are some tracer breakthrough data available, from a hydrogeological field test, in the form of a cumulative breakthrough as a function of time, at a distance L from a point source injection. Figure 10.7 shows such data along with three curves obtained by calibrating each model to those known data. Each fits the data quite well; various measures of 'goodness' of fit might be considered. How reliable are these calibrated models when they are used as predictors of similar behaviour over other, longer, length scales? This is at the heart of conceptual model uncertainty; many models may perform well when calibrated by a limited amount of data valid over limited ranges. However the conceptual model uncertainty which remains after parameters are fitted may be illustrated directly by a cross-comparison of the models' behaviour at new length scales.

In Figure 10.8 the calibrated models have been taken and used to predict the subsequent tracer breakthrough at lengths $2L$ and $4L$ (assuming the medium remains similar to that encountered over the first length, L).

It can be noted that the scaling behaviour of the tracer plumes is quite different in each case despite the relatively small increase in scale. In fact, for Fickian-like (dispersion) processes, the spatial variance of the tracer distribution always grows linearly in time, whereas it grows quadratically for the pure channelling model. The channel network model is dominated by channelling effects for small time, but is asymptotic to an equivalent dispersive process for large times. Thus the disparity between the models gets worse as

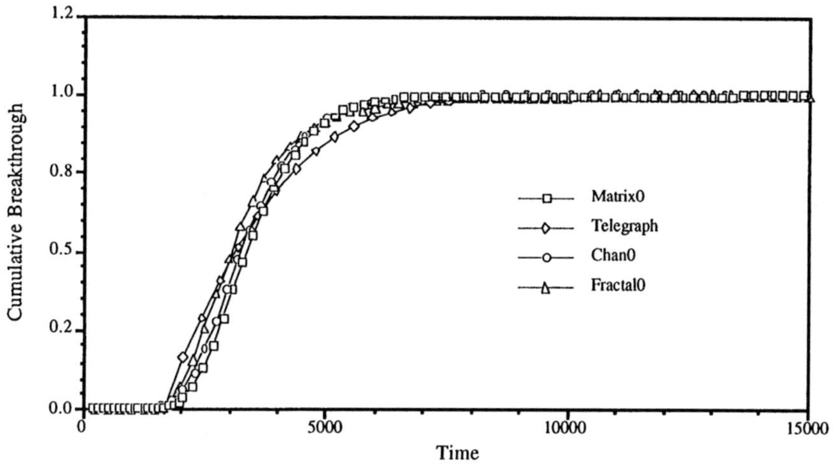

Figure 10.7 Data + 3 calibrated curves.

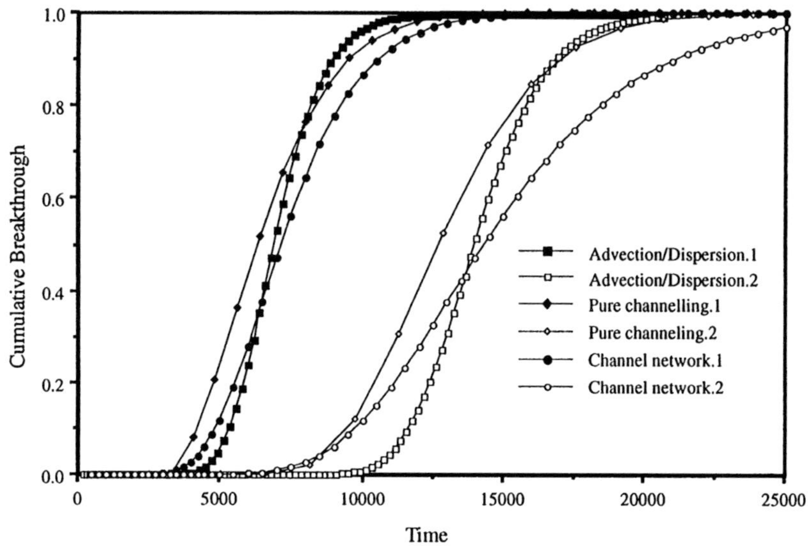

Figure 10.8 3 curves each at $2L$ and $4L$.

behaviour is extrapolated to larger and larger spatial scales; the calibration becomes unreliable and the performance of the models more uncertain.

This behaviour could not be 'got around' by distributing the 'calibrated' parameters of one of the models probabilistically, since all Fickian processes scale in a different manner to channelling processes, and both to channel network processes. What is required is new information concerning the real system performance at these new length scales in order to validate, or invalidate, the upscaled models. Where the performance of alternative conceptual models is sufficiently divergent that a single choice cannot be made with certainty, and a subjective (pragmatic?) decision must be made, the subsequent assessment results may well be biased, and consequent scoping calculations should be made employing alternative concepts to assess the extent of the residual uncertainty.

Conservative and pragmatic assumptions

In practice, conceptual decisions are often justified by their conservatism, there being no wish to obtain a more desirable overall performance by the introduction of uncertainty. Furthermore, in order to make progress, simplified assessment models may be adopted as a means of making robust calculations within acceptable run times. Thus conceptual model uncertainty may well be introduced by the need to adopt simplifications of high resolution or complex phenomena, which in turn require calibration and verification based on the known behaviour of more sophisticated (sub) model calculations.

For example, many models considered are linear, even when there is a large amount of evidence that the processes under consideration are truly non-linear. The reason for their adoption include:

1. Historical model development, where assumptions within the literature are rarely questioned (for example, the superposition of the random motion of particles in deriving diffusion processes).
2. Local linearity, where non-linear effects are thought to be extreme (for example, conservative k_Ds used instead of isotherms).
3. Empirical laws, such as simple fits to variable data (Darcy's law, for example) which might include non-physical variables (for example, permeabilities).
4. Ignorance – the need to allow for a variable dependence with little or no data (the apparent scale dependence of dispersivities and the adoption of constant Peclet numbers).
5. The need to obtain analytical and/or numerical solutions.

Of these, the last item is perhaps the most compelling (and the *raison d'être* for the others), since the theory of linear systems is highly developed, and exact solutions (for verification) may be available, as well as transform-based techniques which avoid the need to time-step. This may be particularly important within Monte Carlo analyses where thousands of calculations must be achieved within an acceptable time.

The key to the acceptance of such simplifications is:

- the demonstration of the qualitative understanding of the processes and their interdependences;
- the demonstration of quantitative model performance through the comparison of models with experimental or analogous behaviour;
- the use of conservative assumptions in order to avoid underestimating risks as a consequence of model choice.

Similar considerations to those above apply to the reduction of dimensionality in geometric descriptions of flow and transport problems. In such cases, three dimensional problems are replaced by their two or one dimensional analogues. This clearly has calculational advantages, but is not simply a pragmatic assumption. For example, since the model may be linear, it is possible to average (or sum) behaviour geometrically to yield a system dependent upon few spatial directions. In addition, the heterogeneous nature of natural systems may tend to channel flow and activity to a few, least resistive paths. Thus the solution of a full three dimensional problem may well add nothing to the results of the geometrically simplified problem.

Finally, it should be noted that some attempts can be made to add back some realism to simplified systems. For example, representative one dimensional flow paths may model a fracture zone by the imposition of extra non-realisable (impossible!) properties, such as effective surface areas for sorption, *etc.* All such assumptions must be first posited at the conceptual level; their validity is open to scrutiny and question.

A proposed framework

Some thought on the issues outlined above soon leads to the conclusion that there is no best way to deal with conceptual model uncertainty. Indeed it also seems evident that there is no point in placing all one's eggs in one basket, and imposing one particular structure to the exclusion of all others. However some corollaries to our discussion are evident. Recall that if models are elicited explicitly, with lists of assumptions required to be made, then their comparison at an executive level may at least provide a qualitative understanding as to why various assessment results differ, and which were the critical modelling decisions. Following on from this idea we suggest that conceptual model uncertainty could be treated via an audit of all possible conceptual decisions made; all alternatives discussed, employed, or ruled out (for what ever reason); and estimates for potential shifts, or biases, in the result which would follow from the adoption of alternatives.

This is called a **bias audit**. Its central objective is to follow through the modelling assumptions made within an assessment and give a quantitative estimate as to how representative are the given results, the alternative conceptual models and hence other assessments which might have been realised.

For each decision which is made in specifying conceptual models to be used in an assessment there are at least two issues which should be considered: how correct are the assumptions that are being made? and how much impact would these have on the assessment results if they were shown/found to be incorrect?

The task of answering these questions necessarily falls to experts or expert groups, and the information elicited will be subjective (that is; not verifiable by independent means, except perhaps checking and re-checking the experts for consistency). It is evident that the range of modelling assumptions that could ever be made is a discrete (countable) set; yet it may not be finite. However within any individual assessment only finitely many conceptual models will be considered, and hence a finite number of assumptions or decisions will be made. Thus experts could be asked to formulate their output in such a way that individual decisions made are made explicit and their confidence in each measured: that is, they will produce a **list of decisions** made in formulating their conceptual model including some measures of the associated uncertainties.

For this some uncertainty indicators, or preference indicators, might be defined. However it must be recognised that while one particular expert group may assess its own decisions on its own interpreted (relative) scale, an absolute scale is not available unless a highly formal procedure is adopted. Such procedures might impose a range of constraints upon the experts; from simple voting (open or secret?), upwards.

There seems little point in overburdening the experts in asking them to produce, say, distributed probabilities, according to some highly structured practices. Instead we assert that any information gained cannot be sanitised by some highly mathematical procedure, and focus upon obtaining robust measures to be used for comparative purposes.

Is there any way to measure the divergence of concepts, prior to selecting some and performing any calculations? Such a measure would be useful in identifying those issues where a wide range of concepts are available, and those with a limited choice. However it should be stressed that divergent concepts may not yield divergent results. Once applied to a particular site, the alternatives may become irrelevant. Thus the task of estimating the sensitivity of assessment results to conceptual alternatives necessitates some prediction for intuition of, and experience of, their likely performance within the proposed assessment.

A central problem in dealing with uncertainty is defining indicators or scales, and asking experts questions they are competent and confident to answer. The definition of uncertainty indicators therefore needs to be robust, unambiguous and straightforward to enact. We need indicators which enable the identification of those conceptual modelling decisions which have the **least degree of support** from the available data, expertise, and/or wider peer review community, and which **may give rise to biases** in the system assessment results which cannot be accounted for by the incorporation of parameter variability (and should therefore be quantified).

For example, for each decision listed the experts should consider all available data and alternatives and assert the uncertainty that the process(es) under consideration are correctly represented by the adoption of the decision. The method by which this uncertainty is arrived at should also be logged (*e.g.* guess, vote, calculation for available site data, experimental basis, *etc.*).

This question could be addressed by asking groups of experts to classify the correctness of the following proposition by selecting a response from a predefined list of alternatives (absolutely certain, highly certain, ..., completely uncertain):

> The adoption of the decision correctly represents the range of behaviour of the (sub) system to be encountered within the assessment.

The use of predefined categories prevents experts spending too much effort in resolving their beliefs, and gives a clear indication of what is required.

Similarly, we might have experts estimate the potential of their listed decisions to induce bias, if they are subsequently found to be incorrect, as follows. A bias or biased result may be defined to be a calculated estimate for an output parameter, of the sub-system, that differs significantly (more than 50 %) from the actual ('true', though unknown) value. The previous question represents the uncertainty surrounding the adoption of a decision. Suppose now that new data or knowledge becomes available which shows that the decision should not have been adopted (*i.e.* that one or more alternatives, previously thought unlikely or unknown, would have been adopted in preference). How will the results of the sub-system change?

The expert group should be asked to represent the possibility of the decision inducing bias within the assessment by selecting a response from a predefined list (no reasonable possibility of bias, bias unlikely, ...). The information (list of decisions) from the expert groups could be ranked in a number of ways, by defining preference maps over the two dimensional range of responses to the above two questions. There is no unique way to do this, so a number of different proceedures could be used and compared. If a single method is adopted then this would impose an interpretation concerning the nature and quality of the information, that is unintended by the experts. The whole point of ranking such lists of decisions is that it is to be expected that those decisions (representing modelling assumptions) at the bottom of the list will be completely reasonable (acceptable, well founded, representative) ones; whereas those at the top will be uncertain and likely to have biased the assessment, if found to be incorrect. These decisions require further investigation, with alternative possibilities being made the subject of scoping calculations, or else they should be called out explicitly within the assessment documentation with a suitable caveat made.

10.1.10 Verification

Uncertainties arise because the computational implementation of a model might be flawed. These are generally dealt with by verification exercises – the aim being to reduce residual uncertainties to negligible levels. Two points should be made immediately. Firstly, verification of a code is only half of the effort needed to ensure that correct results are obtained – the data input in a particular case must be checked too. Second, verification is an on-going process – although a minimum level of verification must be achieved before any code results are trusted or used in an actual assessment.

Formally, verification of a code is confirmation that it performs according to its specification. In highly critical software very strict QA procedures ensure code performance – and a tight specification is a prerequisite. In scientific computing specifications are often more open-ended. For example, a code that claims to solve 1-D advection-diffusion equations for radionuclide decay chains will have a domain of applicability but this will rarely be stated. It is implicitly assumed that tiny half-lives or massive Peclet numbers will not be used. Also, codes which require user decisions on grids and/or timesteps leave open scope for the user to misuse them – even if they are fully correct when used properly.

Thus, verification is not absolute, but is really a continuing confidence building exercise. When difficulties are encountered these might be caused by coding or data errors. In the former case the appropriate response is correction of the error. In the latter case it may still be appropriate to modify the code so that it rejects extreme data. Of course, no code can read the user's mind and small data errors are most likely to go undetected.

The most important tool for assuring the good working of scientific codes is the scepticism of users. Anything that seems wrong or counter-intuitive or surprising should be followed up until it is understood. It is all too easy to simply disregard a surprising result especially if it is not crucial to the outcome of the calculations – this temptation should be resisted – the next time the bug (if there is one) shows up might be on a crucial run.

This applies at various levels. If a check is being made against an analytic solution then one should be aware of the accuracy that is to be expected. If 5 significant figures are expected and only 3 are obtained then there is something to be checked – next time it might be that a completely inaccurate result is obtained.

More often than we might care to admit, data errors cause the trouble. Data should be double checked and one should have no faith in one's ability to type without making mistakes.

The worst time to find a problem with a code is when it is being used on an important calculation. The ideas behind many international verification exercises is to provide an opportunity to test codes when they can be fixed. Also, the fact that multiple attempts on the same problem are made gives more opportunity to spot difficulties than in a single run. An analytic result is most useful – otherwise it can be hard to see who has the true answer. The success or otherwise of such exercises is hard to quantify, but overall they provide a useful verification opportunity. Too often, however, results differ not because of coding or even data errors but because of different interpretations of a poorly defined test case. Such problems are a waste of valuable resources but seem to be an inevitable part of the process, if only because participation is often not given the priority it needs within national programmes.

In summary, verification is not a substitute for the testing and QA which occurs during code development. It is an on-going process which is driven by formal exercises and, more importantly by suspicion in all code results. In finding coding errors, it pays to test the obvious too – such as conservation, monotonicity, and continuity.

10.2 Information from natural systems

10.2.1 Introduction

The issue of validating the models used in performance assessment was discussed in Section 10.1.9. Apart from a purely technical requirement to validate that models adequately represent the system they describe, there is a less easily defined requirement to demonstrate the credibility of both the models and their predictions to a wide variety of audiences. Given that the whole concept of geological disposal is based largely on our intuitive understanding that something buried in stable rock formations should be safe, it is perhaps not surprising that we turn to the natural geological environment when we seek to provide examples that enhance confidence in assessment results. Clearly, much geological information is derived from field observations, laboratory and underground experiments. These data, in particular site-specific data, are generally used as input for the models in which we are seeking to build confidence. This type of information is thus rather different from that required to increase the credibility of assessment results. Consequently, this section deals with two further aspects of the use of information from natural systems in support of performance assessment:

- the use of 'natural analogues' in building and providing constraints on the results of performance assessment models;
- the use of natural analogues, studies of natural geochemical fluxes, and palæohydrogeological studies to provide illustrations and contextual material to help us to digest the significance of performance assessment results.

The discussion ends by making some suggestions as to how information from this second aspect might be treated by decision-makers evaluating a safety assessment, and on how future developments in regulatory standards might regard semi-quantitative evidence of performance and safety. Much of the material in this Chapter is derived from current reviews and analyses, with reports and papers in press (Miller *et al.*, 1993; Chapman, 1992; Chapman and McEwen, 1992).

10.2.2 Natural analogues

Much of the information used in performance assessment models comes from field and laboratory observations or experiments which are of short duration and limited spatial scale. Extrapolation of such information to the time and space scales addressed by the assessments can be a source of considerable uncertainty. While field and laboratory experiments are the most direct way of obtaining information, recourse must be made to studies which can provide a broader framework. Studies of processes in the natural environment which are similar to those of interest in performance assessment, and which have been operative over commensurate time and space scales are known as 'natural analogues' in the jargon of radioactive waste disposal. Natural analogues can serve a number of functions (Miller *et al.*, 1993; Chapman and McKinley, 1990) but, broadly, they are studied:

- in order to ensure that models of processes, such as waste dissolution or radionuclide-rock interactions, are comprehensive and realistic, or
- to provide illustrative material which demonstrates, to a wide range of audiences, our degree of confidence in understanding long-term mechanisms and the scale of their effects.

Such studies have been carried out for more than a decade, although the application of information from them has only really become scientifically well-ordered in recent years. The principal forum for discussion during this period has been the Commission of the European Communities Natural Analogue Working Group (Côme and Chapman, 1986a, b, 1989, 1991).

Despite the fact that explaining the behaviour of the natural environment unambiguously is extremely difficult owing to the many potentially important parameters involved and the ill-constrained boundaries of the system under study, interest has grown steadily in the use of analogues in performance assessment. Although such poorly-controlled 'experiments' are somewhat distasteful to the theoretical or engineering instincts of many performance assessors, it has become increasingly clear that there are really no alternative tests of long-term predictions.

10.2.3 Defining natural analogues

An early definition (Côme & Chapman, 1986a), states that a natural analogue is

> '... an occurrence of materials or processes which resemble those expected in a proposed geological waste repository.'

More recently McKinley (1989) has suggested that

> 'The essence of a natural analogue is the aspect of testing of models – whether conceptual or mathematical – and not a particular attribute of the system itself.'

The IAEA (1989) states

> 'Natural analogues are defined more by the methodology used to study and assess them than by any intrinsic physico-chemical properties they may possess.'

This change of emphasis reflects a maturation in the understanding and appreciation of natural analogues.

The need for a definition has arisen owing to operational difficulties where it is important that natural analogue studies are not confused with repository site characterisation studies, even though a large-scale natural analogue study can provide a valuable 'dry run' of the site characterisation methodology. Site characterisation provides a different type of input to repository performance assessment defining the nature of the natural environment into which the repository is fitted. Natural analogue studies provide information on the subsequent behaviour and effect of the repository, and may use data from the site itself or, more likely, from other sites. These may be in diverse environments, but are linked back to the repository concept by some well-defined similarity of process or material. It follows that not just any geochemical study is a natural analogue; the vital aspect is the design of the study in terms of how it contributes to or tests models of repository effects.

10.2.4 Where natural analogues have been studied

A number of broadly definable geochemical environments have furnished natural analogue information. These are discussed by Miller *et al.*, (1993)

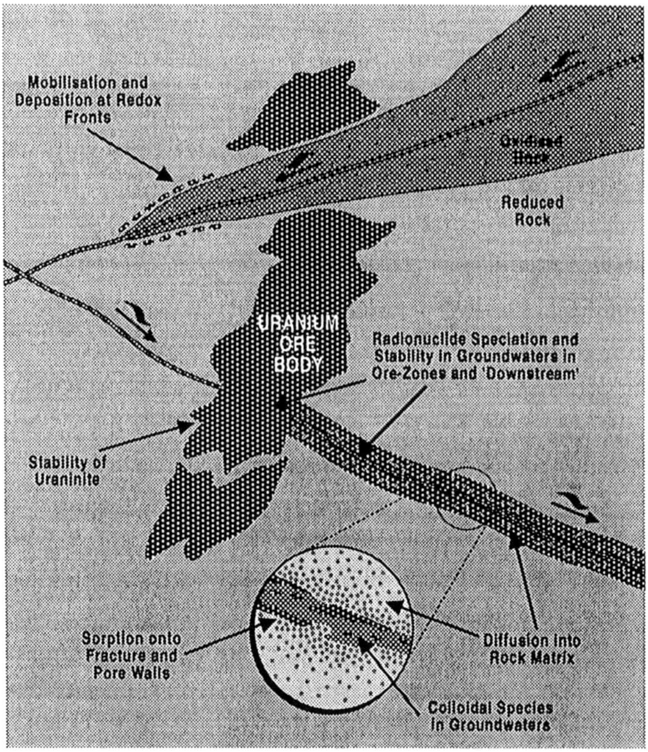

Figure 10.9 Principal processes of interest in natural analogue studies of uranium ore bodies.

and are outlined below.

Uranium ore bodies

Economic and subeconomic primary and secondary concentrations of uranium occur in various rock formations. Their principal interest lies in the mechanisms that have been responsible for their original deposition, and any subsequent remobilisation of natural series radionuclides (Figure 10.9). These processes are analogous to those that might be expected to occur principally in the near-field of spent-fuel or vitrified high-level waste repositories. In evaluating the transport processes responsible for their formation, secondary deposits are of more interest than primary deposits, as the processes generally occurred at much lower temperatures, more closely analogous to repository conditions. Remobilisation phenomena can be studied in both types of deposit. One limitation is that a number of the better studied sites are at relatively shallow depths, with high present-day and recent fluxes of oxygenated groundwaters which have dominated transport processes. For processes where it is interesting to extrapolate to low-flux, reducing conditions, this requires very thorough characterisation of the hydrochemical history of the site. The main features of potential value in uranium ore bodies are:

1. The composition, long-term stability and corrosion/dissolution behaviour of uraninite as an analogue to spent-fuel.
2. The role of redox processes in mobilising and trapping radionuclides (including redox fronts and other geochemical discontinuities).
3. Speciation and solubility controls of radionuclides in groundwaters, including colloid formation and behaviour,
4. Downstream retardation processes of remobilised radionuclides; sorption phenomena on various surfaces, diffusion into the rock matrix porosity, *etc.*
5. The ability to use natural decay series disequilibria to estimate the longevity of various mobilisation and deposition processes.

Hyperalkaline environments

Natural waters with abnormally high alkalinities may be analogous to the pore waters found in cements and concretes. They offer the opportunity to study the hydrochemical behaviour of radionuclides and trace elements in conditions similar to those of some repository near-fields. Hyperalkaline waters are uncommon and may have their origin in, for example, complex rock/water interactions involving the alteration of ultramafic rocks or thermally metamorphosed limestones and marls. To be reasonably representative of cementitious pore waters the pH must be greater than 10. The principal features which have been studied are:

1. The solubilities and speciation of radionuclides and other relevant trace elements and chemical analogues in high pH conditions where solubility calculations suggest that trace element concentrations will be very low.
2. The interaction of high pH fluids with surrounding rocks analogous to near-field porewaters migrating from a repository, and the consequent impacts on rock physico-chemical properties affecting radionuclide transport.
3. The nature and viability of microbially-mediated geochemical processes at high pH, with respect to their potential significance in waste-form breakdown and near-field radionuclide mobilisation.
4. The nature of colloidal species formed in and stable in high pH waters.

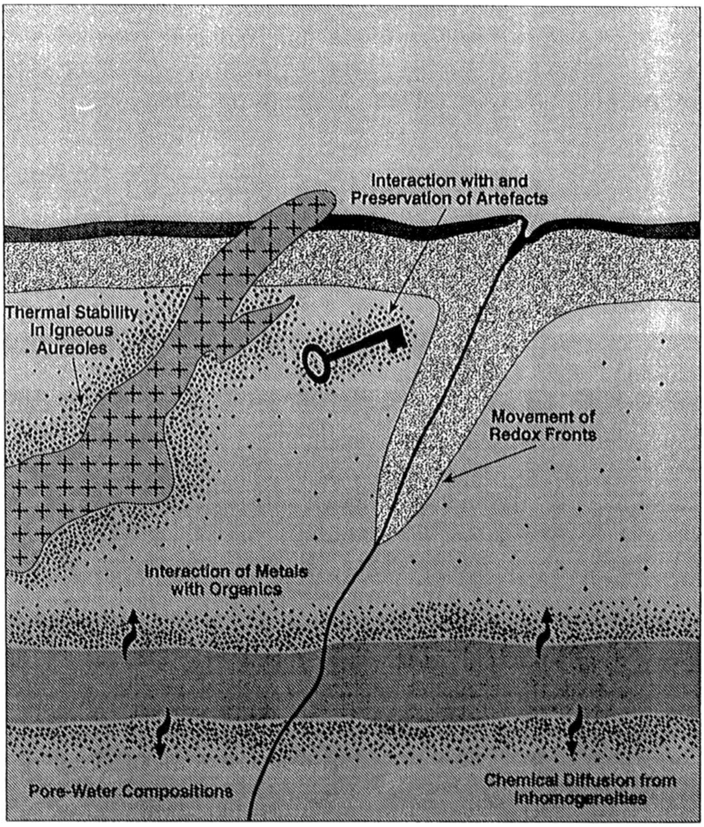

Figure 10.10 Some features of interest in natural analogue studies of clays.

Geochemical discontinuities in clays

Clays may be used as either a backfill/buffer in a repository, or may form the host-rock itself. Geochemical transport processes in clays are not well understood, owing largely to the complexity of the multiple coupled processes of solute/clay interactions which not only drive chemical migration but also control the movement of water and the development of the hydraulic properties of the medium.

Where geochemical discontinuities occur as a result of heterogeneity of a clay-rich sedimentary sequence, or of the intrusion of material into a clay body, they offer the opportunity to study small scale migration and other processes (Figure 10.10), principally:

1. The relative roles of diffusion and advection and small scale physical heterogeneities.
2. The estimation of elemental diffusion coefficients,
3. Movement of redox fronts (including movement along fractures in clays).
4. The thermal stability of clays, and thermal effects on transport properties (where igneous bodies have intruded into clays).
5. Chemical complexing behaviour of radionuclides and other trace elements in pore fluids of different compositions, and in contact with organic material in clays.

Hydrothermal systems

Natural hydrothermal systems have frequently been advanced as being analogous to the near-field of high-level waste repositories during the early period after disposal when temperatures are high. It is suggested that they provide the opportunity to study geochemical transport processes in warm, near-field fluids. Unfortunately, the analogy is generally weak, as repository host-rock temperatures in most concepts never reach more than about 80 to 100 °C, which is lower than the temperature of most hydrothermal systems. Further, most disposal concepts aim at complete containment of the wastes in the canisters until such high temperatures have declined. In addition, most natural hydrothermal systems have been flushed with very much larger volumes of hot fluids over time periods which may be several orders of magnitude greater than the thermally active epoch of a repository. Combined with the generally higher temperatures in natural systems this means that the supposed near-field rock-water interactions and mass transport analogy may be grossly exaggerated both in terms of the scales and amounts of materials involved, and the nature and kinetics of the reactions observed. Attempts to extract pertinent information for performance assessments from hydrothermally altered granites are usually made difficult by the superimposition of several hydrothermal events at various temperatures. A review of recent field studies is provided by Ménager *et al.* (1992). Thus, it may be possible, although very difficult to obtain usable information on elevated temperature near-field processes from some active or fossil hydrothermal systems, but the limitations must be borne in mind, and the earlier exhortation in this chapter to 'define the end-use' must be remembered:

1. Evaluation of the alteration of rock physical properties (*e.g.* pore and fracture enlargement or blocking) and chemical properties (*e.g.* fracture and pore surface mineralogy) of near-field rocks which could occur during a thermal transient and may affect transport and sorption of radionuclides released at a later time. This is only appropriate for high temperature

repository concepts.

2. If it is assumed that package failure occurs during the peak of a high-temperature thermal transient, exposing the waste to hydrothermal fluids then it may be useful to know something of the solubility and speciation of radionuclides and relevant trace elements in such fluids (in active geothermal systems). In this case, information on elemental mobilisation and transport or deposition down a *steep* temperature gradient, for example, where hydrothermal processes have affected an existing ore body or region of rock with elevated concentrations of relevant elements, may be useful.

However, both of these applications appear highly tenuous, which, perhaps, explains the lack of utility of almost all the observations made in past studies of these phenomena.

Archæological analogues

The progressive decay of man-made artefacts can be a very direct analogy to the behaviour of repository materials, and can be studied at carefully chosen archæological sites. The types of artefacts can range from small objects, such as jewellery, to larger features, such as buildings. The time period of study is, naturally, constrained to a few thousands of years at most for the majority of relevant materials, with the advantage, however, that the timescales may be very tightly constrained. Although the term archæological is used, some man-made systems that are only a few decades old also may yield useful information on the rates of processes that may be of significance in the early life of repositories.

The greatest difficulty with using archæological materials is that of determining the relevance of the environment in which they have been located to that of a repository. Most artefacts have been situated either on the surface or at shallow depths, often in conditions which, over comparable timespans, have been more chemically active than those expected in a repository; for example, in terms of chemical fluxes and redox environment. Clearly, if the environment has been highly chemically active, most artefacts will not have survived, so the general approach to archæological materials is one of trying to estimate rates and mechanisms of degradation as a function of the chemistry of the environment of preservation, and to extrapolate or interpolate the data from one or more sites or objects to the chemical conditions expected in a repository. Care must be taken in such evaluations, but the range of materials for study is potentially wide:

1. Corrosion of metallic objects analogous to waste containers or waste matrices.
2. Corrosion of glasses and cementitious or bituminous materials analogous

to waste forms.

3. The long-term evolution of physico-chemical properties of cements and other building materials analogous to repository structures.
4. The decay and breakdown products of organic materials and complexation with trace elements, analogous to waste degradation.
5. Chemical interaction of buried objects with surrounding rocks and soils which may be analogous to near-field processes.

Natural occurrences of repository materials

This category includes a wide range of materials analogous to wastes or other repository components, which are generally found as isolated and often uncommon occurrences in nature. Typical examples are glassy igneous intrusions, tektites, natural bitumens, iron meteorites and native metals. Superficially these materials are appealing analogues, but frequently they are compositionally inappropriate when compared to repository materials, to an extent that may make their properties and behaviour fundamentally different. Some good analogues can, however, be found, although careful evaluation is essential. As with archæological analogues, a critical aspect is the physical and chemical nature of the environment in which they are found, and its relevance to that of a repository. The range of studies that can be envisaged is similar to that presented above for archæological materials.

10.2.5 Applying natural analogue information

The two applications of natural analogues mentioned at the outset are the technical support of the performance assessment process, and the illustrative use in presenting and setting a context in which to evaluate the results of these assessments. Looking back two or three decades, deep burial of reactor wastes seemed fundamentally sound as it could be regarded as simply putting the material, albeit considerably modified, back in the environment from which it had been extracted. The concept of the radioactivity in a spent-fuel repository decaying after a period of time to the level of the original uranium ore was used frequently in presenting disposal options. It is now recognised that such a simplistic approach is as meaningless in safety assessment terms as comparing apples with oranges, but it does, nevertheless, indicate that natural analogues were being used almost unconsciously to propose a solution to the waste problem.

It is now appreciated that safety must be demonstrated on the basis of predicting what will actually happen in a repository, although perhaps the scientific community is less sanguine about its ability to do this than it was a few years ago. However, the whole conceptual basis of geological disposal continues to rely on the natural evidence for the long-term stability and sluggish evolution of deep geological environments. It is common practice, where possible, to utilise natural materials rather than man-made substances in repository engineered barriers, because we believe that they are more likely to be stable and have predictable behaviour when surrounded by other natural materials in a geological environment.

More fundamentally, decisions based on observations of the natural environment are deeply hidden and frequently unstated in the lower levels of performance assessment, in that they provide the whole conceptual basis for what shall be modelled; in particular, transport phenomena. Natural analogue studies generally are performed to fulfil performance assessment requirements, either in a qualitative manner simply to help understand the processes and couplings between processes as they operate in natural systems or, in a more quantitative manner by supplying data for direct input to models or, alternatively, as a tool for validating the models in blind predictions. As noted earlier, we will first examine their use in the three areas of development of performance assessment models. Their illustrative role is examined in section 10.2.6.

Natural analogues in performance assessment

Within performance assessment, analogues can be useful in:

- model construction,
- data acquisition, and
- model testing or validation.

The first step is essentially qualitative, the second two are quantitative. The level of importance of analogues in assisting in each step has been rather different.

Model construction. The development of a model is frequently natural analogue led, as it is only after close observation of natural systems that it is possible to make the necessary decisions regarding which processes to model. Thus, natural analogues tell the model builder:

- which processes to include;
- which processes are likely to be dominant and which of secondary importance;
- which process interactions to model;
- the scales (both spatial and temporal) with which the model should cope;
- whether the basic premises of a model are appropriate (*e.g.* linear extrapolation to long-time frames).

Data acquisition. Natural analogue studies would appear to have the potential to provide quantitative information in the form of various critical parameter values under repository relevant conditions. This is the area where, historically, modellers have had high expectations of natural analogues, to provide 'missing' data they require for input to their models. Unfortunately, the inherent complexity of natural systems sometimes makes it difficult, if not impossible, to obtain such precise results 'on demand' for modelling purposes.

Although it has become clear, from analogue studies, that certain laboratory derived data may be inappropriate, it has not often been possible for analogues to fill the consequent gaps. Extraction of well-controlled parameter values for assessments is fundamentally very difficult when compared with laboratory data derivation. Laboratory studies are thus likely to remain the principal means of data acquisition for performance assessments. However, analogues can identify when laboratory derived data are appropriate and when not and, as such, are an adjunct to laboratory tests. They are also invaluable in providing bounding values (maxima and minima) for processes which are too slow to observe in the laboratory. For example, they can allow statements to be made on the maximum rates of processes.

Model validation. This is an area which has recently grown considerably in importance, as models and databases used in performance assessment increasingly require evidence of their validation. Thermodynamic solubility and speciation codes and databases have been extensively tested in a number of natural analogue studies. The issue at stake with solubility and speciation models is whether the solubility controlling mineral species and complexes in solution, which are generally specified from generic hydrochemical databases or from theoretical or laboratory information, are appropriate to the site-specific conditions being modelled. At the present level of development, the modeller must make use of the available generic conceptual models and databases. These will need to be validated in a site specific sense at any potential repository site, in terms of the unique mineralogical and hydrochemical conditions found. That this will be necessary, and that we cannot use 'off-the-shelf' thermodynamic codes and databases with total confidence to make important predictions, has been well illustrated by analogue projects such as those in Oman (Bath *et al.*, 1987), Jordan (Khoury *et al.*, 1992) and Poços de Caldas (Chapman *et al.*, 1992).

In all of these studies a rigorous 'blind prediction' approach was taken to validating the suitability of thermodynamic codes and data-bases for defining elemental behaviour in solution. The approach adopted in each project was to pose a problem in essentially the same way that it will be put to a performance assessor of a repository site (see Figure 10.11). The basic geochemical properties of the rocks and waters are provided, as they would be derived during any type of thorough site characterisation exercise,

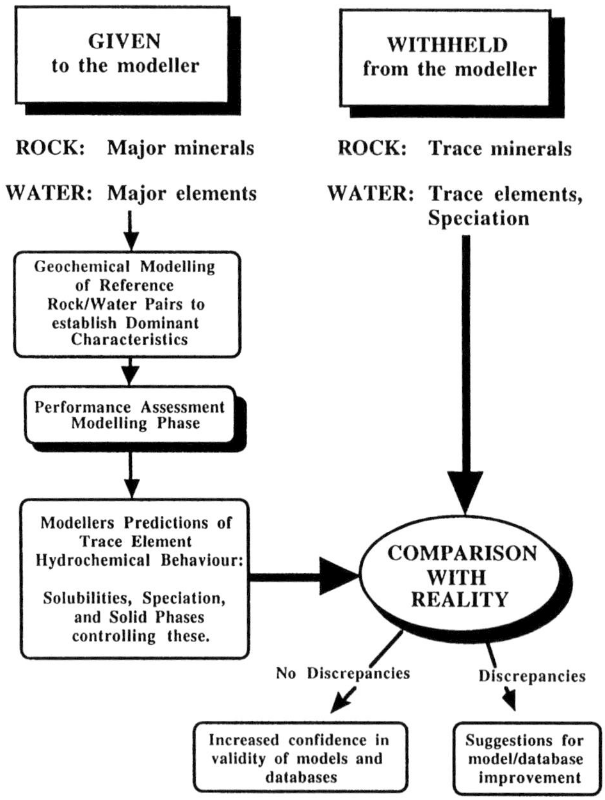

Figure 10.11 Procedure adopted for the blind modelling tests of solubility and speciation in the Poços de Caldas study. Reprinted from *Journal of Geochemical Exploration*, N.A. Chapman *et al.*, 'The Poços de Caldas Project: Summary and Implications for Radioactive Waste Management', Volume 45, pp 1-24, Copyright 1992, with kind permission from Elsevier Science Ltd., The Boulevard, Langford Lane, Kidlington OX5 1GB, U.K.

and the modeller is asked to use these data to predict how specific trace elements will behave. In the analogue case, these trace elements are naturally present in the waters, and their real speciations and concentrations can be measured separately and compared with the modellers predictions. In the case of a repository performance assessment, the trace elements will be introduced into the system from the waste, and the modeller needs to be able to predict how they will behave under the measured conditions at the site, and also over long periods into the future during which hydrochemical conditions may change. The lesson from each of these projects has been that, while the basic codes used appear adequate (provided the conceptual model of rock-water interactions in the system is appropriate), the databases used, and the assumptions made by the modeller can cause errors in prediction. There appears to be scope for extending this approach to the testing and validation of other types of model, particularly of system dynamics, rather than just static chemical equilibria. A limited attempt has been made on models of redox front movement, for example. Perhaps the most obvious target would be solute transport models where validation is currently based on short-term artificial tracer migration tests. The design of a predictive model validation exercise on this issue in an analogue study would require a simple and very well characterised system, and extensive planning, but is clearly worth the effort.

Thus, of the two quantitative applications of analogues in the model building process, model testing is seen as being far and away the most significant.

Actual application in published assessments

To date, the use of natural analogue data in published performance assessments has been rather limited. Analysis of several integrated assessments carried out in the last ten years suggests that:

- the role of natural analogues in setting a conceptual basis is apparent but largely unacknowledged,
- there are probably only two examples of parameter values being provided by natural analogues for direct input to an assessment model (pitting ratios for copper containers for spent-fuel and matrix diffusion penetration depths),
- a 'soft quantitative' use of natural analogues has been in providing somewhat diffuse limits to the ranges of parameter values used or results calculated, in the sense of checking their likelihood of being correct,
- natural analogues are only just starting to be used in a 'hard quantitative' role as test-beds against which to validate models, and only in the limited field of static geochemical modelling. These developments are so recent that they have not yet been fully absorbed in a published assessment.

An extra benefit: analogue 'spin-off'

Quite apart from the performance assessment related information that analogues can supply, some of the larger recent studies have demonstrated an additional value in that they can simulate, to some extent, the philosophy and methodology which must be applied to a repository site investigation.

A well-founded understanding of rock-fluid interactions at a repository site forms the basis on which predictive transport models are built, and the principal evidence supporting models of very long term hydrogeological fluxes and evolution of the site. Consequently, obtaining high-quality, well-chosen and representative rock-groundwater pairs for analysis is a very important aspect of site investigation. This is also what forms the basis of most natural analogue studies, and considerable effort needs to go into defining representative sample points and adequate temporal monitoring programmes.

One aspect found at any site, and which presents some problems in interpreting geochemical processes, yet has also been the cause for many of the driving forces behind these processes, is the amount of heterogeneity in the physical and chemical properties of the rocks and waters. In this respect analogue sites are little different to any site that will need to be characterised for disposal purposes. Heterogeneity occurs at all scales, and it has not always been possible to characterise it adequately during limited site investigations. At Poços de Caldas (Chapman *et al.*, 1992), for example, in order to interpret sparse information, quite robust models had to be constructed, in a similar fashion to what might be needed when characterising a repository site.

A further problem facing any investigation is to distill out a manageable, high-quality and comprehensive data set from the mass of available material, and to ensure that no bias is reflected in this selection. Bias can be introduced through preconceptions of the operative geochemical processes, and can lead to missed opportunities to collect data relevant to real, as opposed to perceived, processes. Examples are the positioning of boreholes along assumed hydraulic gradients/flow streamlines that subsequently prove more complex, or inadequate sample spacing across geochemical discontinuities.

Something which is probably almost universally true of any analogue site is that, although there are clear and obvious processes of interest to study, the site is inevitably more complex, ill-bounded, and of poorer quality in terms of multiple superimposed mechanisms and perturbations to the system than was expected. This is also worth pondering by modellers looking for simple, robust models of the repository far-field. It should be borne in mind that no matter how obvious and interesting a process may appear, it is not always possible to constrain it adequately enough for fully quantitative interpretation; a site cannot be forced to yield all the data that one might want.

A final point concerns the question of how 'site-specific' the results of an analogue study are. 'Transferability' of findings to other sites is obviously important. There is no obvious answer to this question. At Poços de Caldas, for example, many of the results concerning the hydrochemical trace element modelling, together with mechanisms at redox fronts, are non site-specific or, at least, could justifiably be transferred elsewhere with judicious, but non-controversial, modifications. Other results, for example those concerning colloid studies, reflect the specific physical nature of the rock mass in terms of weathering, clay content, porosity, *etc.* They tend to add to a growing database and understanding, rather than giving unequivocal answer, and more observations of the kind made at Morro do Ferro, as part of the Poços de Caldas project, are required, from diverse environments, before the 'colloid problem' is finally solved.

Concluding remarks on performance assessment applications

- Analogue studies will doubtless continue to feature in most national programmes. Given their importance, it is essential not to dilute their value by calling *any* geological or geotechnical study a natural analogue.
- Three distinct applications to performance assessment modelling have been suggested: conceptual model building, provision and verification of data, and use as test-beds against which to validate models. The first, qualitative, function is apparent, but has rarely been acknowledged. The second function is quantitative, but it now appears that analogues rarely have anything to offer in the way of directly assimilable numbers to fit into models. Their only role within this function is semi-quantitative, in providing somewhat diffuse limits for parameters or calculational results, or in verifying their correctness, or 'fitness for purpose'. The third function is extremely powerful, but, as yet, experience is restricted to static geochemical model testing. It could usefully be extended to other areas of performance assessment modelling.
- To be of most use it is important that the end-user of analogue information is involved at the outset in designing the study to ensure that the data flow and timing allow the maximum element of testing and validation. It is not enough simply to send someone out to characterise an interesting looking locality, and then hope to use the data for rigorously controlled testing.
- Certain issues, notably our still hazy appreciation of colloid behaviour, would bear frequent overlapping studies to build up progressively a convincing database that would allow colloids to be treated with confidence in an assessment.

10.2.6 Building confidence in assessments

Regulatory authorities are the key agencies that need to be convinced of the adequacy of a safety assessment. Such assessments may be presented from time to time as part of a rolling programme of research and development, or with the actual licensing of a repository in mind. Most agencies have, by now, set some blend of quantitative and qualitative standards against which the results of an assessment can be judged, although others are proceeding cautiously towards developing firm criteria. Quantitative standards are usually set in terms of acceptable levels of release of specific radionuclides at set points and times, or radiation doses or health risks to individuals or groups, again at various times in the future. Most waste disposers are now striving to demonstrate that specific disposal options can meet the quantitative targets set. However, a purely mechanistic engineering model that produces dose versus time projections always causes problems of belief in any audience. The lay-public are not alone in requiring some help in order to grasp the meaning of such information. Thus, acceptance is not necessarily as simple as complying with numerical targets, and a number of questions are raised:

- how confident is the assessor in the results of their assessment?
- how has such confidence been demonstrated and are the demonstrations credible?
- besides the regulators, what are the views of other concerned groups, including the scientific peers of the assessors?
- what other evidence, including qualitative information, supports the numerical conclusions of an assessment?

The evaluator of an assessment, and this could be anyone reading it, has to understand that the quantitative predictions presented are really only illustrations of what could happen, be they scenarios or time-dependent models. The biggest uncertainty comes not from future environmental evolution, but from the development of society, over a very much shorter timescale of decades rather than millenia. The most useful way of addressing these illustrations and answering the questions above is by putting them into context with parallel processes in the evolving natural environment. The three avenues currently open to facilitate this are discussed below.

The illustrative role of natural analogues

Apart from their role in building and testing performance asssessment models, discussed earlier, natural analogues can provide purely qualitative illustrations that the end result of complex, coupled processes over very long times does give the sorts of qualitative result suggested by a performance assessment model; such things are observed in nature, the assessor is not completely misguided in assuming that they might occur in a repository. They also have a much wider appeal as providers of illustrative and, in some cases, non-technical information to a broad range of audiences. From outside the confines of the performance assessment community, this aspect is probably perceived as the only reason natural analogues are studied. Natural analogues, or comparisons with natural systems, are frequently mentioned as important components of the process of evaluating and accepting disposal proposals. Among all levels of reviewer, from technical peer review panels, to non-technical audiences, there is a clear belief that predictive safety assessments are only credible if shown to have strong natural parallels. For many audiences, the nature of predictive safety assessments themselves are difficult to understand and accept. Take the following quotation from Blowers *et al.* (1991):

> 'Enormous scientific effort has been expended in Europe and North America researching and demonstrating the proposition that such repositories will be safe, for all practical purposes, for ever. Yet clearly the assertion is preposterous. The safety of an untried method cannot be proven until repositories have been constructed and monitored over many generations and the radionuclides have decayed to safe levels. Running such an empirical experiment is inconceivable. Sophisticated geological analysis, risk assessment or modelling of repository behaviour must rest upon heroic assumptions and are no substitute for empirical knowledge. Scientific predictions for periods of 10 000 years or more lie in the realm of fantasy, not rationality. In conditions of such uncertainty it must be concluded that there is no technical solution to the problem of radioactive waste.'

Although the authors come from a non-technical background, the strength of disbelief in long-term predictions voiced by all kinds of audiences cannot be dismissed by scientists. In the field of prediction, everyone has a more-or-less valid view, based on generations of widely publicised experience. No amount of scientific argument or proof is going to convince many people of the shining elegance and truth of a safety case that is inherently very complex and extends far into the future. Doubts are not the prerogative of the non-technical audience, and are pervasive throughout the concerned community.

Given this difficult position, those who have to present safety cases, or to evaluate them and base decisions on them, have found some very important uses of natural analogues. Broadly, these fall into two areas:

- As straightforward *factual examples, with a time context,* that the processes being modelled in performance assessments are not unusual or novel in the natural world, and that there is some sound basis for the time-scales involved. Such illustrations need not be quantitative, nor need they even be conclusive of any particular behaviour. They reassure that, for example, when we go to a lot of effort to model the rate of corrosion of metal canisters, then there is some basis for indicating that metals can survive in the natural environment for commensurate periods. In other words, we are not being totally misled by theory and chasing red-herrings.

- As *illustrations* that the overall result of waste disposal has natural parallels in terms of there being similar concentrations of natural radioactivity moving about in the environment, again on similar time-scales.

These two types of application are at their most valuable when used in a matter-of-fact and disinterested way to show that at least there is some sound basis in the real world for believing that geological disposal is not simply another ambitious technological enterprise, doomed to choke on its own complexity. A number of useful brochures and films have been using this approach (*e.g.* the U.K. Nirex Ltd. film 'Tunnel in Time', the Nagra booklet 'Observations of Nature' and the forthcoming internationally sponsored natural analogues film). There is, however, an acute danger in either being condescending to the reader or viewer, or overstating the case; usually both can arise because material is misrepresented or simply abused. The nuclear industry itself has not always represented natural analogues honestly, sometimes tending to oversimplify and overstate the facts. No doubt it may be argued that some simplification is necessary to make the scientific data readily digestible, but there is a fine line between necessary simplification and misrepresentation of the facts. The natural analogue that is most frequently oversimplified is Oklo, often to the point of being misleading. To the scientist, Oklo does *not* provide unequivocal proof that a repository will be safe; the situation at Oklo is simply too complex for that conclusion to be reached. Yet, Oklo is often presented in nuclear industry literature as 'proof'. An example of 'going too far' is provided by the following quotation from 'industry' literature:

> 'The Oklo reactors ran gently at the kilowatt-power level for millions of years. They never blew up. The radiation and waste from them did not deter surrounding life forms. Over immense timescales, the waste has barely moved away from the reactor site. As a result, scientists today are confident that waste in man-made stores and repositories is likely to move even less ... '

Such statements challenge belief at all levels, apart from leaving themselves open to scientific ridicule if ever they have to be defended. Caution is essential.

Natural geochemical fluxes

The core of a performance assessment looks at geochemical fluxes over long times. It is very useful for someone evaluating predicted radionuclide fluxes to have an idea of what other geochemical fluxes are occurring naturally at the same time. Over periods of 10^4-10^5 years or more, sediment diagenesis, deep and shallow rock-water interactions, erosion, weathering, sedimentation, mineralisation and other processes are busily moving elements around within the same system that is being modelled by the assessor, independently of the

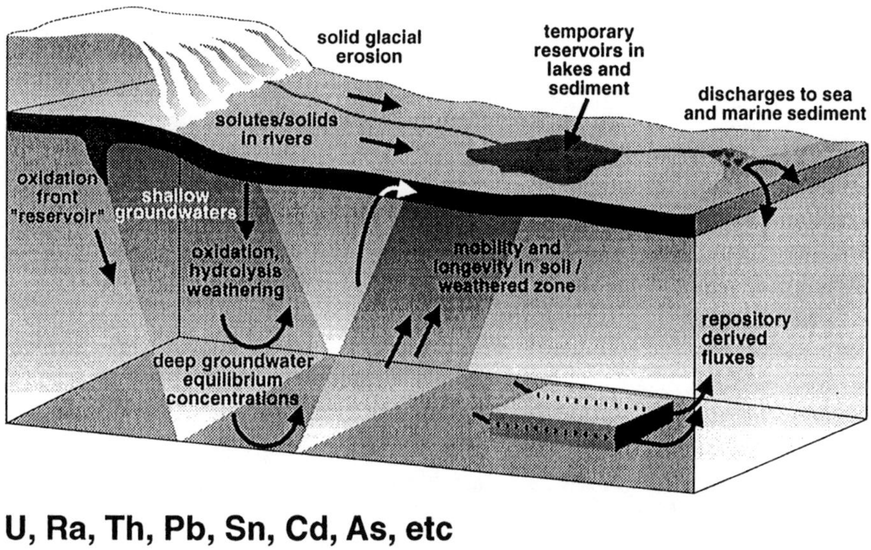

Figure 10.12 Natural geochemical fluxes occurring on the same timescale over which an assessment is performed may provide useful contextual material for evaluating assessment results.

presence of the repository (see Figure 10.12). Comparatively, vast amounts of material are on the move, including several of the elements central to the assessment. At the limit, where the best geochemical and epidemiological data are available, it is possible to compare the health effects of these parallel processes. But, even where this is impractical or illogical, it is extremely useful to be aware that these processes are active, simply to be able to put fluxes from the repository into context.

As an example it is worth looking briefly at the global flux of natural uranium in the geological environment. Typically crustal rocks such as granites, gneisses and most sedimentary rocks contain only small (1 to 10 ppm) concentrations of uranium. However, when large volumes of rock are considered, the total mass of uranium present can be considerable. To put this into perspective, some 15 tonnes of uranium would be present in the rock excavated to build the conceptual U.K. Nirex repository for L/ILW, if it were built in a typical granite.

The global average concentration of uranium in surface and groundwaters is 0.4 μg/kg but is highly variable locally. This variation is due to the variation in the uranium content of the local source rocks and also the physico-chemical environment which controls the solubility of uranium. In a typical oxygen-rich, shallow groundwater uranium would, of course, be soluble and have the potential to be highly mobile. In contrast, at depth in the environments considered for a repository, where groundwaters would be reducing, uranium would be poorly soluble and considerably less mobile. This change in uranium solubility is clearly demonstrated in the Stripa site characterisation where shallow groundwaters contain up to 100 μg/kg uranium whilst the deeper waters contain less than 1 μg/kg uranium. The mobility of uranium is also strongly dependent on the presence and nature of colloidal species.

The ultimate sink for all the uranium leached from continental rocks is the pelagic oceanic sediments. The present-day water flux off the continents is 3.8×10^{16} kg/a and, with an average uranium content of 0.4 μg/kg, this means that 15 000 tonnes of uranium are transported to the oceans each year. The uranium in solution in the ocean's waters has a residence time of approximately 300 000 years before being added to the sea-floor sediments. There is currently an equivalence between the mass of uranium added to the oceans by continental erosion and the mass of uranium lost by the oceans by precipitation to the sea-floor. The point of this example is to demonstrate that, once uranium and other radionuclides escape from the near-field of a repository and begin to migrate through the far-field, the concentrations of these repository-derived radionuclides in the rock will be little different from the concentrations of their natural equivalents. Clearly, actual radiological impacts from all radionuclides need to be evaluated, but putting the overall fluxes of each element (for which it is possible to make such calculations) into a natural context, such as illustrated briefly above for uranium, can considerably aid any reviewer to grasp the significance of the processes being modelled in a repository. This point is important to communicate, for almost all audiences will have no 'feel' for the numbers (concentrations, doses and timescales) generated by performance assessment calculations.

Concentrations of natural radioactivity

The concentrations of natural radioactivity found in everyday materials could, perhaps, be associated with the estimation of natural geochemical fluxes in that it provides additional useful information to put waste-derived radioactivity into context. Baxter (1993) provides a useful review of the principal non-nuclear industries which contribute to our exposure to radiation as a result of the natural materials they process. These include power generation from burning fossil fuels (coal, oil and gas), inorganic by-products of hydrocarbon extraction at oil and gas wells, phosphate processing for fertilisers and detergents, metal ore smelting, titanium oxide production, mineral water abstraction, and paint, ceramic and colouring material production.

As an example, it has been estimated that the generation of 1 GWe of 'coal-fired' electricity results in the annual release of 10^9 to 10^{11} Bq of both ^{220}Rn and ^{222}Rn and 10^8 to 10^{10} Bq each of ^{210}Pb, ^{210}Po, ^{226}Ra, ^{228}Ra, ^{232}Th, and ^{238}U. These releases are less than or comparable to releases from a nuclear power station under normal operating conditions. Globally, coal-fired power generation adds 5000 tonnes of U and 8000 tonnes of Th, plus their daughter products, including about 600 TBq of alpha-emitters, to the environment. A typical annual critical group dose rate is up to 50 μSv. It is also estimated that the critical group near a gas-fired plant can receive an effective dose equivalent of up to 20 μSv, largely from ^{222}Rn progeny.

The fly-ash produced by coal-fired power stations contains relatively high concentrations of both alpha and beta/gamma emitters (see Table 10.1). Some is released to the atmosphere in particulate form, whilst the bulk material may be re-utilised in landfill operations or in the manufacture of building materials.

Oversaturated brines from deep geological formations containing hydrocarbons cause problems when they precipitate as scale on oil-field pump equipment. The minerals precipitated scavenge and concentrate ^{226}Ra and ^{228}Ra from the brines up to concentrations of 10^6 Bq/kg, and it is not uncommon that residues from scale removal are disposed of in the marine environment.

Discharges of phosphogypsum effluents in the region around Rotterdam gives rise to individual dose rates in critical groups of around 0.1 to 0.3 mSv, and the use of contaminated sediments as landfill can give rise to dose up to 1 mSv. In the UK, a phosphate processing plant was, until recently, discharging wastes to the sea with a significant inventory of ^{226}Ra and Th isotopes. This resulted in critical group doses of 0.3 to 3 mSv, in excess of any value resulting at that time from nuclear discharges.

According to Baxter, few, if any, of these enhancements in environmental radioactivity are really significant in a health context either on a global or a local scale. Table 10.1 provides some typical concentrations of natural radioactivity in common, everyday materials.

Palæohydrogeology

The validity of predictions made using groundwater flow models constructed using observations of present hydrogeological properties is difficult to demonstrate. These models again need to be put into context, as they look very uncomfortable in isolation. Palæohydrogeological reconstructions of past flow systems at a repository site are perhaps the best means of supporting such models (Pearson *et al.*, 1988; Chapman and McEwen, 1992). Palæohydrogeology is an increasingly well developed science in sedimentary formations, where it is closely linked to diagenetic studies and hydrocarbon reservoir modelling. It is less well-developed in fractured basement rock

Table 10.1 Typical levels of natural radioactivity in common materials. Concentrations in Bq/kg (after Baxter, 1993). Reproduced by permission of IAEA.

Material	Alpha	Beta/gamma
Igneous rocks	140	800
Sandstone	64	330
Shale	95	800
Limestone	36	80
Phosphate ore	1500	260
Soil	300-1000	440
Bricks	110	600
Gypsum & Cement	30-800	40–160
Sand & Gravel	200-2000	30
Books	30	100
Coal	80	170
Fly ash	1400	1100
Concentrated superphosphate fertiliser	4600	–
Hay	–	600
For comparison, low-level wastes typically contain of the order of 10^4 to 10^5 Bq/kg of alpha or beta/gamma activity		

environments, where it has had no commercial application.

Palæohydrogeology is a combination of observations on hydrochemical and isotopic differences in various groundwater zones or bodies, mineralogical data on the rock formations, and the hydraulic properties of the same formations, which are then compiled to allow interpretation of the evolution of the rock-water system over long time periods in the past (see Figure 10.13 and Table 10.2). Such interpretation will require assumptions about the 'initial' conditions of the system (*e.g.* in sediments, about the original porewater composition at the time of deposition) together with information on the rates of various processes (*e.g.* diagenetic reactions), some of which may be difficult to obtain.

There are inevitable ambiguities and difficulties in interpretation of so much intricately coupled information, and there are some well-known problems and complications that need to be resolved, and need to be borne in mind in any investigation. These include:

- the disentangling of isotopic data to give apparent 'ages' for groundwaters;
- the overlapping nature of many rock-fluid interaction processes that can

Table 10.2 Typical input data for a palæohydrogeological analysis. From Chapman and McEwen (1992). Reproduced by permission of NEA/OECD.

Area of Application	Information required
Palæoclimate (*i.e.* recharge temperatures)	Equilibration temperatures for Ne, Ar, K and Xe. Stable isotope compositions: $^{2}H/^{1}H$, $^{18}O/^{16}O$, *etc.*
Water-rock interactions in shallow crust: long-term	Determination of ^{40}Ar in groundwaters Determination of N production by thermogenic N and nitrate reduction. U concentrations and isotope ratios Comparison of $^{226}Ra/^{228}Ra$ activity ratio with U/Th
Groundwater migration, age and mixing, *etc.*	Radiogenic He, ^{14}C, ^{36}, ^{81}Kr, ^{1}H, excess ^{234}U decay in closed system, *etc.* $^{234}U/^{238}U$ ratio with U
Deeper crustal process	$^{3}He/^{4}He$ ratios to evaluate crustal migration of ^{3}He
Estimation of natural radionuclide migration rates	^{222}Rn, U content
Hydrogeological parameters *e.g.* fracture widths, porosities	^{222}Rn, He loss
Mineralogy	Detailed fracture fill history – fracture coatings – fluid inclusions – pore structures, sizes, *etc.*
Hydraulic testing	Hydraulic conductivities, heads, porosities Geometry of flow system Salinity effects
Geological history	Climate change, tectonics, basin evolution, mineralisation, *etc.*

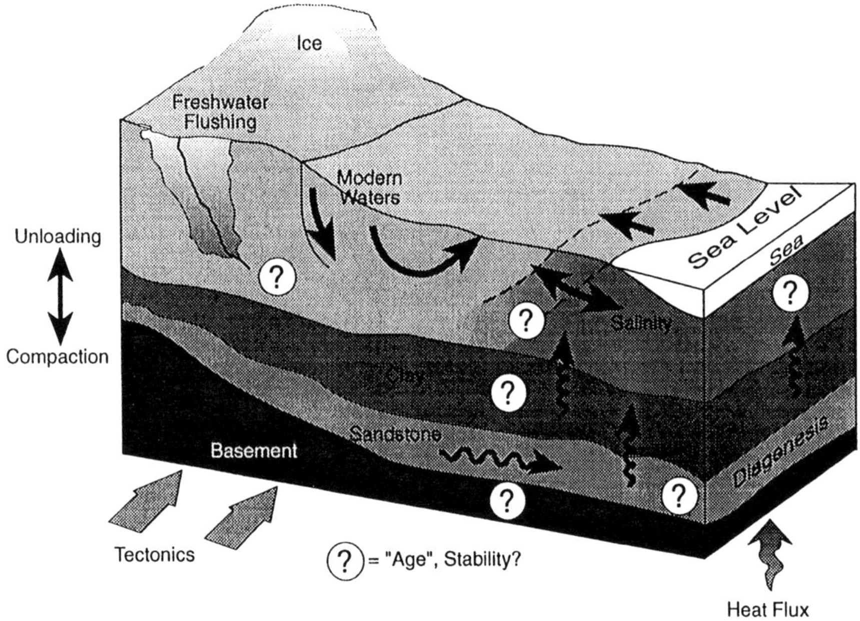

Figure 10.13 Schematic illustration of some of the processes and concepts to be accounted for or derived from a palæohydrogeological study. The question marks indicate zones of the groundwater system where it would be important to establish the 'age' or degree of mixing of groundwaters. From Chapman and McEwen, (1992). Reproduced by permission of NEA/OECD.

provide several mechanisms to explain certain observations;
- the natural *in situ* production of some environmental isotopes used in the 'dating' process;
- difficulties in the application of kinetic data for rock-water interactions;
- the representivity and purity of samples obtained (often with difficulty) from depth.

However, the technique is potentially exceptionally powerful because a system can usually be evaluated with a high level of confidence at a broad enough level of detail, that is adequate to define the main mechanisms and trends that have dominated its evolution. This then forms a firm foundation for any predictions made about future behaviour. The types of information which could be derived ideally from interpretation of a well-characterised site include:

- major bodies of water present in different regions of the deep geological environment;

- the residence times of these bodies of water in that region, and hence their 'stability' to changes in external conditions;
- the degree to which mixing of different bodies of water has occurred, and the geochemical transport mechanisms operating at their interfaces;
- the principal pathways for water movement in the system (higher conductivity formations or zones) and the extent to which these have influenced the overall evolution of the groundwater regime;
- regions where dilution of deeper-sourced waters may take place in shallow waters (with respect to repository releases), and the extent of such dilution;
- evidence for the major forces that have perturbed the system in the past (*e.g.* glaciation, sea-level change, permafrost), the depth and extent of the perturbations, and the response time of the system;
- the contribution of what are generally regarded as secondary mechanisms or sources to the evolution of the groundwater system (*e.g.* tectonic effects on flow; very deep sourced fluids affecting hydrochemistry, *etc.*);
- evidence for groundwater transit times from shallow to deep parts of the system, and vice-versa.

Application of palæohydrogeological information. At the very simplest level, palæohydrogeology can provide clear indicators of the suitability or otherwise of a site for a repository. The presence at depth of clearly 'young' waters, with high bomb tritium or anthropogenic surface water contaminants for example, and of rapid and widescale mixing of such waters with large parts of the deep flow system, would tend to exclude the site from further consideration. At this level, the information probably would not even be used in a detailed performance assessment, but would simply aid preliminary decisions on site suitability.

What the waste disposer would clearly hope for would be indications of the extremely stable nature of the system over periods of tens to hundreds of millenia, with a high buffering capacity to changes wrought by external climatic forcing factors, and an overall sluggish evolution of the chemical system, and of any advective transport processes in particular. If this type of picture emerges from a site investigation, how best can it be used in performance assessment? There may be two direct applications. The first is a potentially exciting and powerful approach that has not, to the authors' knowledge, been attempted. The second is the most widely accepted.

1. Testing predictive models. As discussed earlier, natural analogue studies have found that the most useful and credible information they can produce is the testing of PA models against a database from a natural system; where possible, being done 'blind'. It may be possible to devise a similar approach for using palæohydrogeological data. The idea would be to take the palæohydrogeological interpretation of the site, select some definable

state at a particular point in the past, and then use various component codes and models from the PA model chain to try to simulate the current, observed, state of the system, or some aspect of it (such as salinity profiles, or head gradients). The extent to which this could be done blind would depend very much on the way the site investigation was organised, and the teams involved in the interpretation, and it would clearly require some considerable amount of planning if it were to constitute any real quantitative test of the predictive capacity of a model.

2. A well-founded understanding of how the site got to be the way it is today provides what must be the best demonstration of confidence in its future behaviour. If past processes have been extremely slow at repository depth, then there is very good reason to believe that they will continue to behave in the same way. Critics of this approach may say that just because a site has been stable in the past does nor mean that it will remain so in the future. However, a large part of palæohydrogeology is the description of the effects of major perturbing processes. In northern Europe at least, we can be reasonably confident that any site will already have seen several cycles of all the natural processes that are likely to affect it in the next million years. This confidence cannot be extrapolated everywhere, unfortunately. Given this confidence in past stability, however, an assessment can be quite definite about the range of future conditions likely to be experienced by the site, its range and rates of response to those conditions, and, consequently, the rates of radionuclide transport and other important fluxes in the repository system. The predictive assessment transport model can then be nested comfortably in these broader descriptions of scales and rates of natural processes.

Finally, in this context, the effect of the repository itself in perturbing the natural hydrogeochemical system needs to be mentioned. Clearly this aspect has to be taken into account in projecting forward the results of a palæohydrogeological study. The two main issues which would need to be evaluated are:

- gas generation and migration;
- thermal effects on groundwater flow.

Other repository-produced effects, such as near-field/far-field geochemical interactions, localised stress relief around exacavations, and potential short-circuit paths caused by poor seals are less likely to cause more than localised perturbations in the overall flux field, although they may clearly lead to release scenarios that will require analysis.

10.2.7 Concluding remarks on confidence-building

To conclude, it is worth pondering the extent to which some of the 'indirect' semi-quantitative and qualitative information discussed above can really be used by a decision-maker. Most safety submissions must, inevitably, be based primarily on quantitative predictions of dose, risk, release, concentration at a certain point at a certain time, and so on. The recipient of these numbers might suspend reality for a while and take them at face value, checking how they compare with standards, or they may try to view them in some broader context. The context may be presented to them in the submission:

> 'these results must be taken as stylised indicators of the nominal results of a set of possible future scenarios, using present day model receptors to calculate likely reference doses ... *etc*'.

Alternatively a context may be provided which uses information on the rate of evolution of the geological environment, the direction, scale and effects of natural processes, or the natural geochemical cycling of radionuclides or other trace elements over comparable periods of time as yardsticks against which to gauge predictions of waste radionuclide behaviour.

Putting the results into the context of what is actually observed in the natural environment is vital. Workers who routinely model biological systems are familiar with the concept that a result must look intuitively correct. In order to say this, the reviewer clearly needs strong grounding in the appropriate natural sciences. The widely acknowledged non-linearity of behaviour of natural systems suggests that, whatever approach is used to produce quantitative assessments, then without this intuitive evaluation the numbers cannot be used to check compliance against a linear regulatory yardstick, no matter how much error and uncertainty treatment is included in their derivation. In other words, a safety case needs two legs to stand on.

All of this looks rather imprecise, of course, and one has to admit the difficulty of devising some formal process whereby the qualitative value of a performance assessment can be accounted for in parallel with the much easier task of checking numbers against a target. However, both, on their own, are insupportable as a means of judging safety and acceptability. Perhaps, at the end of the day, the answer is nothing more complicated than ensuring extensive peer review by people well-versed in long-term processes in both the geological and the surface environments.

An alternative way round this would seem to be the use by regulators and other decision-makers of a broader and more qualitative framework within which to place the numerical result of an assessment. Not only does this make it easier for everyone to understand, but it also clearly limits our expectations of what an assessment ought to be designed to achieve, indeed of what it is ever capable of demonstrating with confidence. There has been much effort recently to 'validate' our predictive radionuclide transport models. The fact that it is taking time to develop real consensus on how this can best be addressed, and that validation has not yet been widely acknowledged to have been achieved, should lead us to be more sanguine about both our quantitative predictive capabilities, and regulations based upon them.

10.3 Radioactive waste in perspective

10.3.1 Radioactive waste in perspective with other environmental hazards

There are three key reasons why it is useful to compare the hazards posed by radioactive waste with other environmental hazards:

1. **Compliance with criteria.** Chapter 1 has shown that various forms of comparison of repository derived radiation with natural radiation background are being considered or are already included in disposal criteria (The Radiation Protection and Nuclear Safety Authorities in Denmark, Finland, Iceland, Norway and Sweden, 1989). The advantage of such comparisons is that they avoid detailed consideration of some of the less readily quantifiable aspects of the biosphere and human behaviour assumptions required for many dose or risk assessments. However, for the comparisons to be useful they need to be relevant and, so far as possible, the comparisons should be made between truly comparable parameters. For example, one can argue that the repository flux to the biosphere ought to be compared with the natural flux in the same type of environment and not with some gross global average. There are many documents discussing natural radiation, but among the most comprehensive is the compilation by UNSCEAR (UNSCEAR, 1988), which also summarises the effects of other radiation sources.
2. **Presentation and explanation of results.** Some descriptions of risk calculations are necessarily complicated. Even the simple result, a risk of death of one in a million, is obscure in its implication. It can be useful to illustrate the significance of such a risk estimate to a range of audiences by showing what other risks are of a similar scale. However, it is hard to justify a proposal on the basis of risk comparison unless due care is given to the context of the risk estimate. It is important to consider:

    - the relative magnitudes of risks from all sources associated simply with staying alive;
    - voluntary as opposed to involuntary risks;
    - perception of individual industries and practices; and
    - worker as opposed to public risk.

Table 10.3 Average annual risk of death in the UK from some common causes (from NRPB, 1989a). Reproduced by permission of National Radiological Protection Board.

Cause	Risk of Death per year
Smoking 10 cigarettes a day	1 in 200
Natural causes, 40 years old	1 in 850
Accidents on the road	1 in 9,500
Accidents in the home	1 in 26,000
Accidents at work	1 in 43,500
Radiation exposure at a rate of 1 mSv per year	1 in 80,000

The tolerability of various types of risk has been examined by the U.K. Health and Safety Executive (Health and Safety Executive, 1987) and average annual risk of death in the U.K. from different causes are given in Table 10.3.

3. **Integrated pollution control.** With the development of integrated pollution control there is an increasing need to assess both the radiological and non-radiological environmental impacts of pollution. Thus, it is no longer sufficient to assume automatically that mixed radioactive and hazardous wastes can be treated as radioactive wastes, and that radioactive waste disposal regulations will automatically ensure that the wastes are disposed of in a safe manner. Comparison of such wastes' radioactive and other environmental hazards will allow better decisions to be taken concerning waste management and treatment options. It will, for example, facilitate decisions concerning two possible technologies; one which reduces a radiological impact but increases some other environmental hazard, and an alternative which does the opposite.

10.3.2 What other environmental hazards should be considered?

Given the need to place radioactive waste in perspective with other environmental hazards, a range of environmental hazards could be considered including risks from 'natural' sources (floods, hurricanes, lighting *etc.*), and anthropogenic sources (smoking, driving a car, factory explosions *etc.*). However, it is vital that any comparison which is used should be:

- relevant;
- comparable; and
- simple.

Given the above factors, it is particularly useful to consider the following two environmental hazards:

- other radiological hazards (for example those from natural background, the operation of nuclear facilities, and medical uses); and
- toxic waste disposal hazards.

10.3.3 Other radiological hazards

Comparing radioactive waste with other radiological hazards has the advantage that a common end point can be used; the dose and/or risk of death resulting from exposure to radionuclides. For example, Figure 10.14 shows the contribution of different sources to the average annual dose (2.5 mSv) to the U.K. population (NRPB, 1989b). It can be seen that 87 % of the dose is derived from 'natural' sources (65 % resulting from the background flux of radionuclides from the ground and building materials) and only 13 % from 'artificial' sources. Included in these 'artificial' sources are nuclear discharges which currently contribute 0.1 % to the 2.5 mSv total. Thus it might appear that doses from nuclear discharges (including those from the disposal of solid radioactive waste) are insignificant.

It should be remembered that these figures are averages and there is considerable variability in magnitude and composition of individual doses depending upon factors such as geographical location, occupation, life style, diet and health. For example, in 1983, a heavy consumer of seafood in the area around the Sellafield nuclear reprocessing plant would receive a total annual dose of 4.8 mSv, 47 % of which came from nuclear discharges. By 1987, the figures had fallen to 2.9 mSv and 12 % due to reductions in the discharge of radionuclides from Sellafield. At the other extreme, the average annual dose to a person living in Cornwall (a granite-rich area in the south west of England) is 7.8 mSv, 88 % resulting from the background flux of radionuclides from the ground and building materials (Figure 10.15).

The U.K. Authorising Departments (UKDOE, 1984) have set an annual risk target which is equivalent to a dose of 0.1 mSv for doses from a radioactive waste disposal facility at any time. This represents 4 % of the current U.K. average annual dose. Thus, if this target is adhered to, doses from land-based disposal facility should be minor in comparison to other radiation sources.

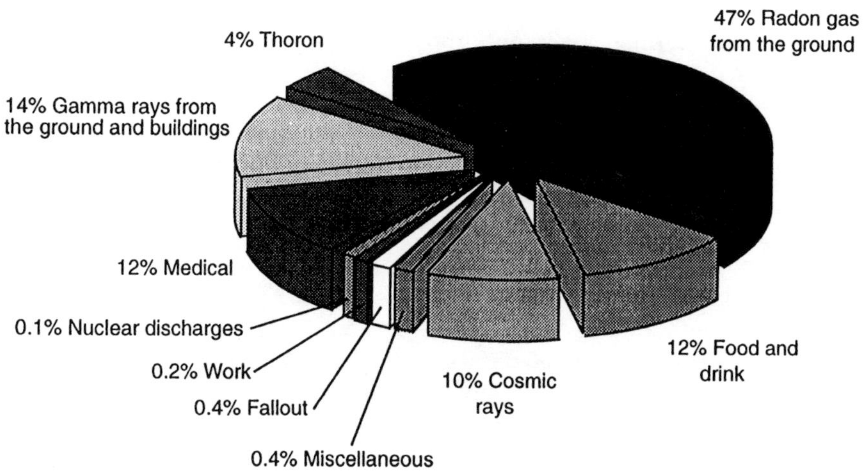

Figure 10.14 Average annual dose to the U.K. population.

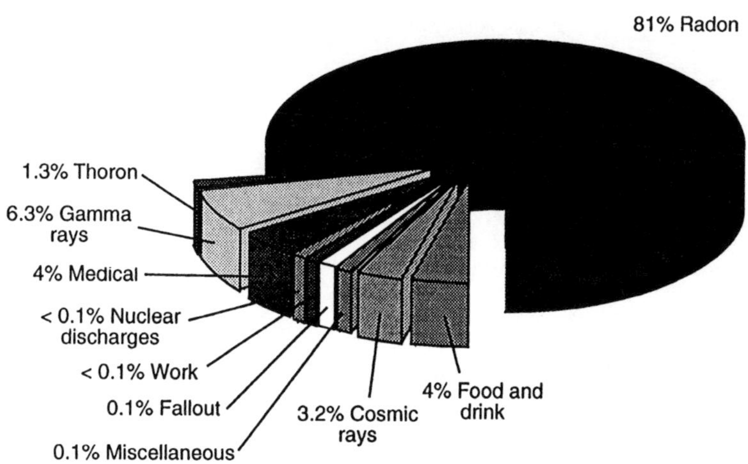

Figure 10.15 Average annual dose to a person living in Cornwall.

10.3.4 Toxic waste disposal hazards

Comparisons with hazards from toxic waste disposal are less straight forward than comparisons with other radiological hazards. This is mainly because of a poor understanding of the dose-response relationships for important non-radioactive pollutants. This prevents the health effects of pollutants from being assessed. This applies to both human and other types of dose-response. Nevertheless, some data do already exist, for example data collected by the United States Environmental Protection Agency (USEPA, 1992), and data are becoming more readily available due to research programmes such as the one described by Cardis *et al.* (1991). Thus for certain pollutants it is possible to use these data to assess health effects and express results in terms of a common end point – risk of death. This allows direct comparison with radioactive contaminants.

The use of risk as a common assessment end point for radioactive and non-radioactive contaminants has been illustrated by the work of the Uranium Mill Tailings Working Group in the BIOMOVS II (Biosphere Model Validation Study) exercise (BIOMOVS II, 1993). Lifetime average cancer incidence risks from a number of exposure pathways have been calculated for the U-238 chain and three heavy metals (As, Cr and Ni). Illustrative results from the use of the SACO (Safety Assessment Comparison) methodology (Torres *et al.*, 1993) are given in Figure 10.16 for this example. It is interesting to note that, in this illustration, the peak risk from ingestion of As is about two orders of magnitude higher than that from the U-238 chain.

Although a lack of data on health effects might, in some cases, limit the use of direct comparisons with toxic waste hazards, it is possible to use other quantitative end points such as:

- contaminant release rates from the disposal facility;
- contaminant concentrations in environmental media such as groundwater, soil, and river water; and
- contaminant intake by humans.

These can be compared against relevant regulatory limits. However, few methodologies have been developed which allow the assessment of these quantitative end points for non-radioactive waste (Little *et al.*, 1993). One exception is the SACO methodology which has been developed and used in a project funded by the Commission of the European Communities (CEC) and Empresa Nacional de Resíduos Radiactivos SA (ENRESA), Spain (Torres *et al.*, 1993). It is based on a generic approach developed from radioactive waste disposal assessment methods.

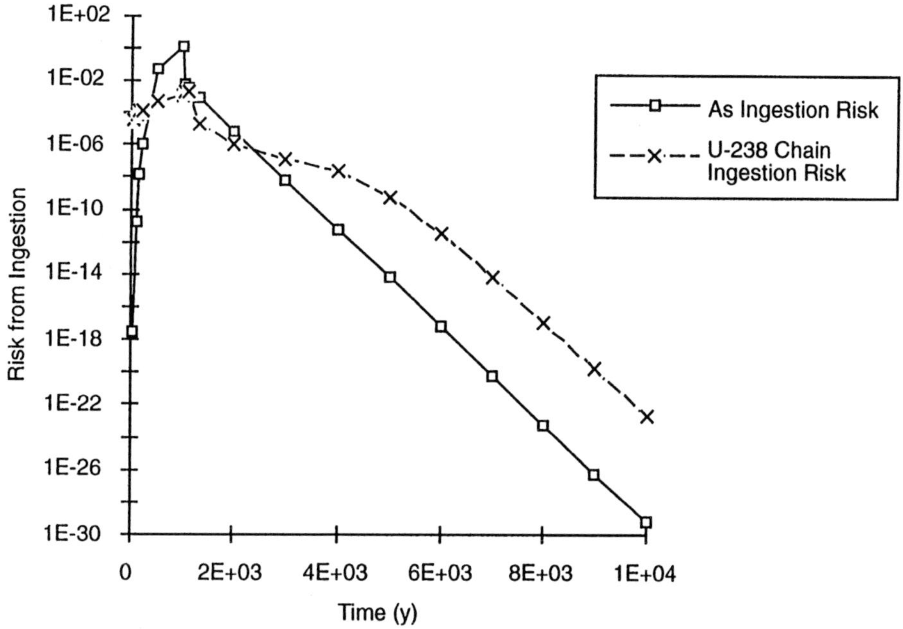

Figure 10.16 Lifetime average cancer incidence risk from ingestion of As and the U-238 chain.

10.4 References

Apostolakis (ed.) (1991) *Probabilistic Safety Assessment and Management.* Elsevier.

Bath, A.H., Berner, U., Cave, M., McKinley, I.G. and Neal, C. (1987) Testing geochemical models in a hyperalkaline environment. In *Natural analogues in radioactive waste disposal*, (eds. B. Côme and N.A. Chapman). *CEC Radioactive Waste Management Series*, EUR 11037, 167-178. Commission of the European Communities, Luxembourg.

Baxter, M. S. (1993) Environmental radioactivity: a perspective on industrial contributions. *IAEA Bulletin*, 2/1993, 33-38. International Atomic Energy Agency, Vienna, Austria.

BIOMOVS II (1993) Uranium Mill Tailings Working Group. Scenario V1.07, April 1993. Swedish Radiation Protection Institute, Stockholm, Sweden.

Blowers, A., Lowry, D. and Solomon, B.D. (1991) *The International Politics of Nuclear Waste.* Macmillan Academic and Professional Ltd.

Cardis E., Hours, M. and Fabry, J. (1991) A programme for exposure and epidemiologic surveillance of populations residing in the vicinity of industrial waste dumps in France. In *Proceedings of the Joint International Symposium on Environmental Consequences of Hazardous Waste Disposal*, Stockholm, May 1991. Swedish Radiation Protection Institute, Stockholm, Vol. 1, pp 269-278.

Chapman, N.A. (1992) Natural Analogues: The State of Play in 1992. In: *High Level Radioactive Waste Management; Proceedings of the 3rd International Conference,* **2:** 1695-1700. American Nuclear Society, Illinois, USA.

Chapman, N.A., McKinley, I.G., Shea, M.E. and Smellie, J.A.T. (1992) The Poços de Caldas Project: summary and implications for radioactive waste management. *Journal of Geochemical Exploration* **45:** 1-24.

Chapman, N. A. and McEwen, T.J. (1992, in press) The application of palæohydrogeological information to repository performance assessment. In: *Palæohydrogeological Methods and their Applications for Radioactive Waste Disposal.* OECD Nuclear Energy Agency, Paris, France.

Chapman, N.A. and McKinley, I.G. (1990) Radioactive wastes: back to the future? *New Scientist,* **1715:** 54-58.

Côme, B. and Chapman, N.A. (eds.) (1986a; 1986b, 1989, 1991) Natural analogue working group; First meeting, Brussels, November 1985. Second meeting, Interlaken, June 1986. Third meeting, Snowbird, June 1988. Fourth meeting and Poços de Caldas project final meeting, Pitlochry, June 1990. *CEC Nuclear Science and Technology Reports,* EUR 10315, 10671, 11725 and 13014. Commission of the European Communities, Luxembourg.

Health and Safety Executive (1987) The tolerability of risk from nuclear power stations. HMSO, London.

IAEA (1989) Natural analogues in performance assessments for the disposal of radioactive wastes. *IAEA Technical Report* 304. International Atomic Energy Agency, Vienna, Austria.

INTRACOIN (1984) International Nuclide Transport Code Intercomparison Study – Final Report Level 1 – Code Verification. *SKI Technical Report* 84-3. Swedish Nuclear Power Inspectorate, Stockholm, Sweden.

INTRACOIN (1986) International Nuclide Transport Code Intercomparison Study – Final Report Levels 2 and 3 – Model Validation and Uncertainty Analysis. *SKI Technical Report* 86-2. Swedish Nuclear Power Inspectorate, Stockholm, Sweden.

Khoury, H.N., Salameh, E., Clark, I.P., Fritz, P., Milodowski, A.E., Cave, M.R., Bajjali, W. and Alexander, W.R. (1992) A natural analogue of high pH cement pore waters from the Maqarin area of northern Jordan. I: introduction to the site. *Journal of Geochemical Exploration,* **46:** 117-132.

Little R.H., Torres C., Charles D., Grogan H.A., Simón I., Smith G.M., Sumerling, T.J. and Watkins, B.M. (1993) Post-disposal safety assessment of toxic and radioactive waste: waste types, disposal practices, disposal criteria, assessment methods, and post–disposal impacts. *CEC Report* EUR 14627 EN. Commission of the European Communities, Luxembourg.

von Maravic, H. and Smellie, J. A. T. (1993 in press) 5th CEC Natural Analogues Working Group Meeting, Toledo, Spain. October 1992.*CEC Nuclear Science and Technology Reports* EUR xxxxxx. Commission of the European Communities, Luxembourg.

McKinley, I.G. (1989) Applying natural analogues in predictive performance assessment. *Nagra Technical Report* NTB 89-02. Nagra, Baden, Switzerland.

Ménager, M-T., Petit, J-C. and Brocandel, M. (1992) The migration of radionuclides in granite: a review based on natural analogues. *Applied Geochemistry*

Supplementary Issue **1:** 217-238.

Miller, W.M., Alexander, W.R., Chapman, N.A., McKinley, I.G. and Smellie, J.A.T. (1993 in press) Natural Analogue Studies in the Geological Disposal of Nuclear Wastes. Elsevier, Amsterdam (publ).

NEA (1987a) *Uncertainty Analysis of Radioactive Waste Disposal Systems.* Seattle, February 1987. OECD/NEA. Paris, France.

NEA (1987b) PSAC User Group, PSACOIN level 0 intercomparison: an international exercise on a hypothetical case study for radioactive waste disposal systems. OECD/NEA, Paris, France.

NEA (1989) PSAC User Group, PSACOIN level E intercomparison: an international exercise on a hypothetical case study for radioactive waste disposal systems. OECD/NEA Paris, France.

NEA (1990a) The International Probabilistic System Assessment Group, background and results. OECD/NEA, Paris, France.

NEA (1990b) The International INTRAVAL Project, Background and Results. OECD/NEA, Paris, France.

NEA (1990c) GEOVAL 1990: Symposium on Validation of Geosphere Flow and Transport Models, Stockholm 1990. OECD/NEA, Paris, France.

NEA (1991) PSAC User Group, PSACOIN level 1a intercomparison: an international exercise on a hypothetical case study for radioactive waste disposal systems. OECD/NEA, Paris, France.

NEA (1992a) PSAC User Group, PSACOIN level1b intercomparison: an international exercise on a hypothetical case study for radioactive waste disposal systems. OECD/NEA, Paris, France.

NEA (1992b) The International HYDROCOIN Project, Summary Report.

NEA (1992c) The 4th International Symposium on the OECD/NEA Stripa Project held in Stockholm, October 1992. OECD/NEA, Paris, France.

NEA (1993) PSAC User Group, PSACOIN level S intercomparison: an international exercise on a hypothetical case study for radioactive waste disposal systems. OECD/NEA, Paris, France.

NRPB (1989a) *Living with Radiation.* National Radiological Protection Board, Chilton, Oxfordshire, U.K.

NRPB (1989b) *Radiation Doses – Maps and Magnitudes.* National Radiological Protection Board, Chilton, Oxfordshire, U.K.

Pearson, F.J., Balderer, W., Loosli, H.H., Lehmann, B.E., Matter, A., Peters, Tj., Schmassmann, H. and Gautschi, A. (1988) Applied Isotope Hydrogeology – A Case Study in Northern Switzerland. *Nagra Technical Report* NTB 88-01. Nagra, Wettingen, Switzerland.

Read, D. and Broyd T.W. (1989) CHEMVAL Project. Report on Stage 1: Verification of Speciation Models. *CEC Report* EUR 12237 EN. Commission of the European Communities, Luxembourg.

Read, D. (ed.) (1990) CHEMVAL Project. Report on Stage 2: Application of Speciation Models to Laboratory and Field Data Sets. *CEC Report* EUR 13124 EN. Commission of the European Communities, Luxembourg.

Read, D. (ed.) (1991) CHEMVAL Project. Report on Stages 3 and 4: Testing of Coupled Chemical Transport Models. *CEC Report* EUR 13675 EN. Commission of the European Communities, Luxembourg.

The Radiation Protection and Nuclear Safety Authorities in Denmark, Finland, Iceland, Norway and Sweden (1989) *Disposal of High Level Radioactive Waste, Consideration of Some Basic Criteria. A Consultative Document.*

Torres C., Simón I., Agüero, A., Little, R.H. and Smith, G.M. (1993) Safety Assessment Comparison Assessment Methodology for Toxic and Radioactive Wastes (SACO Version 1.0). In *Proceedings of the International Conference on Nuclear Waste Management and Environmental Remediation*, Prague, Czech Republic, September 1993.

U.K. Department of the Environment, Scottish Office, Welsh Office, Department of the Environment for Northern Ireland and Ministry of Agriculture, Fisheries and Food (1984) *Disposal Facilities on Land for Low and Intermediate Level Radioactive Wastes: Principles for the Protection of the Human Environment.* HMSO, London, U.K.

United Nations Scientific Committee on the Effects of Atomic Radiation (1988) *Sources, Effects and Risks of Ionizing Radiation.* United Nations, New York, U.S.A.

USEPA (1992) Health Effects Assessment Summary Tables. *NTIS Report* No PB92-921199. Office of Research and Development, Office of Emergency and Remedial Response, Washington DC, U.S.A.

11

Where next?

At present there are no operating deep geological repositories for high-level wastes or spent fuel in the OECD countries. A number of historic underground disposals of long-lived wastes do exist, largely resulting from activities during the 1940's to 60's. In general, these facilities ceased operation many years ago, do not follow the multibarrier conceptual and design approaches outlined in earlier Chapters and were not subjected to stringent performance and safety assessments of the kind described here. In some countries, notably the United States and states of the former Soviet Union, a legacy of problematic or contaminated sites now exists. Over the last fifteen to twenty years, considerable resources have gone into refining the deep geological disposal concept and to testing various aspects of repository design and performance assessment in the laboratory and in underground experimental facilities. This period has coincided with increasing concern about environmental matters which, combined with a lack of political will to solve the radioactive waste problem, has seen most national plans for deep disposal progressively postponed.

The result of this period of abeyance has been both negative and positive. In reality, the conceptual basis and design options for deep repositories have changed very little over the last ten years, and the widespread inability to get access to the underground for research purposes has hampered some programmes. However, a considerable amount has been learned about how to approach the non-technical aspects of disposal in this period. In countries which have grasped the political nettle and decided to resolve the waste issue, there has been a marked change from the early position of 'the state (or the industry) knows best', particularly with respect to finding sites for repositories. A much more open, consultative approach has developed in several countries, notably Canada, France, Sweden, Switzerland and the U.S.A., aimed at obtaining the involvement and consent of communities most affected by development plans. Although it has often been said that their are no votes in radioactive wastes, this welcome openness, combined with a clear policy to resolve the waste problem is actually having very beneficial effects. It is to be hoped that other countries will learn from the experience.

Another positive aspect to the delays of recent years has been significant progress in our ability to understand the evolution of repository systems into the distant future and our capabilities to assimilate data, and our many uncertainties, into performance assessments which describe this evolution. As a result, we are now in a very good position to begin to implement plans for deep disposal, in a cautious and stepwise fashion, filling in the gaps in our knowledge of potential sites as we begin the construction of repositories. Indeed, we have now reached a stage where such activity is essential; further delays are not likely to lead to any significant new benefits.

We believe that it is now fair to say that there really is a *scientific basis* for geological disposal, and that a solution to the waste problem actually exists. The next step is to implement this solution. In some countries this will be a slow struggle. As mentioned in the Introduction, it is not in the interests of the anti-nuclear lobby to see an operating solution to the waste problem, and in the ensuing confrontation, this issue is more likely to drive opposition to repositories than concerns about their long-term performance and safety.

During this period of uncertainty concerning deep repositories there has been significant progress in the design and construction of surface and near-surface repositories. In the latter case, we can now see two fully operational geological repositories in rock caverns excavated at depths of some tens of metres in hard fractured rocks at Forsmark in Sweden and Olkiluoto in Finland. It is interesting to note that many of the issues involved in designing and evaluating these 'less deep' geological repositories are little different from those concerning deeper disposal of other waste types. Consequently, while the high-level waste problem has been on the 'back-burner', there has actually been a quiet leap forward in solving the problem of the much greater volumes of lower activity wastes, simultaneously building confidence in the design and operation of underground facilities.

11.1 Future issues

Looking forward to the next decade, a number of new imperatives may emerge which will be superimposed on the complex social and political factors which already influence waste management practices.

The issue of long-term storage for certain types of waste is once again featuring in some national programmes. The aim is to place waste, in a conditioned form, in some kind of facility, either at surface or underground, where it can be retrieved, for some unspecified purpose, in the future. This is generally a counsel of inaction, appealing largely in circumstances where there is no proper policy on wastes, where procrastination takes precedence over decision-making, or where cash-flow dominates concerns over environmental impacts. What is clearly true is that long-term storage is simply passing on the problem of finding a final solution from one generation to another. It is certainly appealing to the anti-nuclear lobby since, as long as there is no disposal solution where the operator eventually intends to lock the door and walk away, then there is no demonstrated closure of the nuclear fuel cycle.

Many countries with small nuclear power programmes have spent considerable resources trying to demonstrate that a solution to the waste problem exists within their own boundaries, particularly a solution to high-level waste disposal. There is a growing feeling that international repositories for some categories of wastes would make sense in terms of economics, practicality and overall safety. A decade ago, it was felt that such a solution would be politically unachievable, and some nations were concerned that it would seem like passing on an inconvenient problem to their neighbours. Once it has been established that independent national solutions can be found, then this become less of an issue, and the concept of a series of 'local' international repositories, making use of the pooled experience of each country involved and managed under the auspices of a recognised international agency, begins to look more attractive. The issue still arises of finding a willing and acceptable host and obtaining the consent of the people of that area. However, there are possible economic and social attractions in having such an important and high-profile facility in any location, as has been found with several international high-technology developments. It is clear that the safety and practicalities of trans-boundary waste transport would feature large in any discussions over such a solution.

We now begin to see an additional source of radioactive wastes; those resulting from the large-scale destruction and decommissioning of nuclear weapons as a result of the end of the cold war. While many people may react negatively to solving the problems of the nuclear power industry, most would probably subscribe to the overall benefits of this latter activity. It is tempting to ask whether this positive feeling might encourage the search for an international disposal solution for these wastes, if not for others.

We must hope to see some spin-off from the huge investment in development of a methodology for radioactive waste management to the management of other environmentally hazardous substances. The underlying concepts of waste separation and conditioning, multibarrier disposal, and repositories tailored to the containment requirements of a specific type of waste clearly have applications elsewhere, as does the principle of carrying out proper safety assessments to evaluate realistic environmental impacts under various circumstances. We do not see this approach yet to other wastes, but it is to be hoped that this will only be a matter of time.

In looking at the resources that we have collectively invested in building a safe solution to nuclear wastes, it is very important to ask ourselves what we are really trying to achieve by such demanding solutions, and why we are going to such lengths. Should we spend similar resources on other problems with equivalent or very much worse real or potential environmental impacts, or should we now have learned enough from radioactive wastes to treat these problems more convincingly on a scale relative to their comparative hazards? Unfortunately, the reality is that there do not appear to be either the resources available, or the interest or will to handle these other problems as diligently as we are managing radioactive wastes. A critical problem is that the regulatory framework for other environmental hazards lags far behind that for radioactive wastes, and is comparatively undemanding in what it requires from a disposer. Individually and collectively we appear to be quite willing to make arbitrary compromises. Hence the question; what are we really intent on achieving?

11.2 A final thought ...

Throughout the book we have discussed the problems of making predictions of the long-term future behaviour of repositories. Essentially, there is no easy answer to this matter, and any quantitative assessments we do make must be regarded as purely illustrative of possible behaviour, rather than predictions of what will actually occur. Getting this idea across, not only to the reader of a safety report, but also to some of those involved in preparing it, still proves difficult and contentious. A particular difficulty is that none of us really understands the significance of future timescales at all the levels that are necessary when presenting a safety case. While the geologist may feel comfortable with millions of years, the historian with thousands and the sociologist with decades, few people have a grasp of the overall significance of an assessment which convolutes aspects of all these timescales. Figure 11.1 illustrates this problem by superimposing events in the past 100 Ma on typical timeframes of predicted radionuclide release curves. Apart from the obviously dynamic nature of the environment in which we place a repository, the brevity of human history compared to the times when we might predict first releases gives pause for thought. It is clear that our estimates of releases can be nothing more than illustrations to enable decision-makers to draw comparisons. They must understand quite clearly the time and uncertainty framework within which they working, and be able to relate these stylised future impacts to much shorter term imperatives and issues which affect the real safety of real people living today.

The biggest uncertainty in looking to the future thus concerns the development and activities of people, and the effects that this will have on the natural environment. It is probably true to say that people will be the deciding factor in the future of a repository, not natural processes, yet most of our safety analysis concerns protecting against natural changes in the disposal environment over very long timescales indeed; tens or hundreds of thousands of years or more. Going back to the original choice of deep geological disposal,

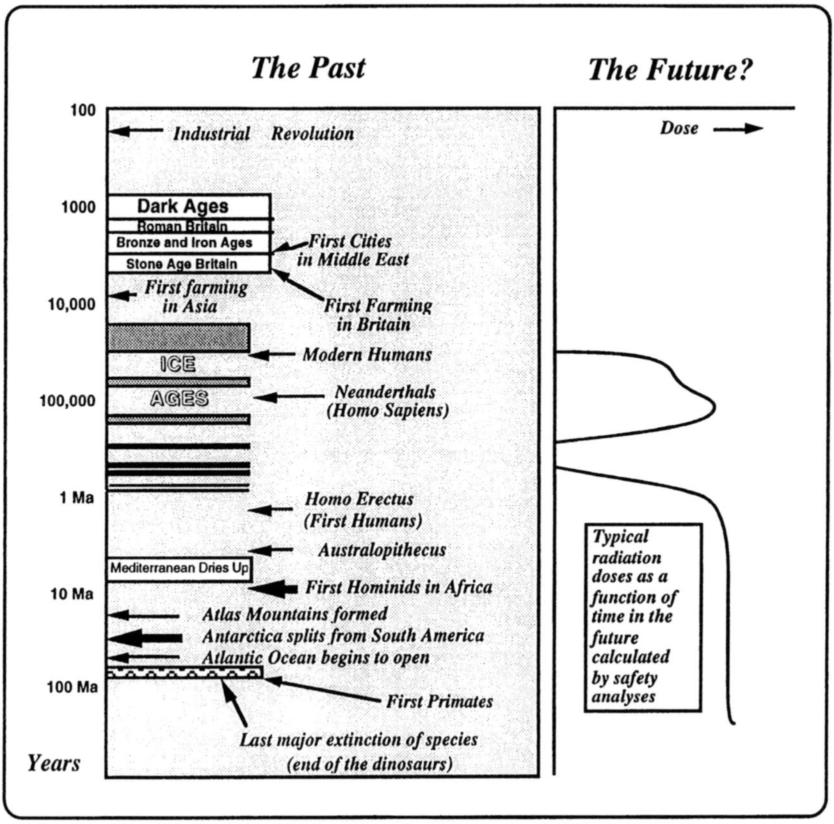

Figure 11.1 Typical results of calculations of radiation dose to future individuals during the next 100 million years, compared to events in the past 100 million years. Reprinted from *Journal of Volcanology and Geothermal Research*, N.A. Chapman, 'Natural Radioactivity and Radioactive Waste Disposal', Volume 50, pp 197-206, Copyright 1992, with kind permission from Elsevier Science Ltd., The Boulevard, Langford Lane, Kidlington OX5 1GB, U.K.

one of the main reasons for selecting this option is to isolate the wastes from the human environment. It must be a sobering thought to anyone who has been involved in the complexities of site characterisation and the mental gymnastics of safety assessment and regulatory criteria to consider that the carefully planned evolution of the repository is almost inevitably going to be prejudiced, for better or worse, by the inquisitive, venal or just downright destructive actions of our own species on a timescale of hundreds or thousands of years. Although we may intend that the wastes rest undisturbed for millions of years, the biggest uncertainty is ourselves.

11.3 References

Chapman, N.A. (1992) Natural radioactivity and radioactive waste disposal. *Journal of Volcanology and Geothermal Research* **50:** 197-206.

Index